Transnational Corporations, Technology and Economic Development

NEW HORIZONS IN INTERNATIONAL BUSINESS

Series Editor: Peter J. Buckley
Centre for International Business,
University of Leeds (CIBUL), UK

The New Horizons in International Business series has established itself as the world's leading forum for the presentation of new ideas in international business research. It offers pre-eminent contributions in the areas of multinational enterprise – including foreign direct investment, business strategy and corporate alliances, global competitive strategies, and entrepreneurship. In short, this series constitutes essential reading for academics, business strategists and policy makers alike.

Titles in the series include:

Japanese Subsidiaries in the New Global Economy
Edited by Paul W. Beamish, Andrew Delios and Shige Makino

Globalizing Europe
Deepening Integration, Alliance Capitalism and Structural Statecraft
Edited by Thomas L. Brewer, Paul A. Brenton and Gavin Boyd

China and its Regions
Economic Growth and Reform in Chinese Provinces
Edited by Mary-Françoise Renard

Emerging Issues in International Business Research
Edited by Masaaki Kotabe and Preet S. Aulakh

Network Knowledge in International Business
Edited by Sarianna M. Lundan

Business Strategy and National Culture
US and Asia Pacific Microcomputer Multinationals in Europe
Denise Tsang

Learning in the Internationalisation Process of Firms
Edited by Anders Blomstermo and D. Deo Sharma

Alliance Capitalism and Corporate Management
Entrepreneurial Cooperation in Knowledge Based Economies
Edited by John H. Dunning and Gavin Boyd

The New Economic Analysis of Multinationals
An Agenda for Management, Policy and Research
Edited by Thomas L. Brewer, Stephen Young and Stephen E. Guisinger

Transnational Corporations, Technology and Economic Development
Backward Linkages and Knowledge Transfer in South East Asia
Axèle Giroud

Transnational Corporations, Technology and Economic Development

Backward Linkages and Knowledge Transfer in South-east Asia

Axèle Giroud

Lecturer in International Business, Bradford University School of Management, UK

Edward Elgar

Cheltenham, UK • Northampton, MA, USA

Published by
Edward Elgar Publishing Limited
Glensanda House
Montpellier Parade
Cheltenham
Glos GL50 1UA
UK

Edward Elgar Publishing, Inc.
136 West Street
Suite 202
Northampton
Massachusetts 01060
USA

A catalogue record for this book
is available from the British Library

Library of Congress Cataloguing in Publication Data

Giroud, Axèle.
Transnational corporations, technology, and economic development : backward linkages and knowledge transfer in South East Asia / by Axèle Giroud.
p. cm. — (New horizons in international business series)
Includes bibliographical references and index.
1. International business enterprises—Asia, Southeastern. 2. Technology transfer—Asia, Southeastern. I. Title. II. New horizons in international business

HD2901 .G57 2003
338.959'06—dc21

2002034712

ISBN 1 84064 907 0

Printed and bound in Great Britain by MPG Books Ltd, Bodmin, Cornwall

For my parents,
Jean and Christine,
And for Antonio

Contents

Abbreviations

AALCC	Asian-African Legal Consultative Committee
ACTID	Action Committee on Tariff and Industrial Development
ADB	Asian Development Bank
AFTA	ASEAN Free Trade Area
APEC	Asia Pacific Economic Co-operation
ASEAN	Association of Southeast Asian Nations
AVL	Approved Vendor List
BOP	Balance of Payments
CCI	Coordination Centre for Investment
ECR	Export Credit Refinancing
EPZ	Export Processing Zone
FCBA	Flexible Circuit Board Assembling
FDI	Foreign Direct Investment
FMM	Federation of Malaysian Manufacturers
FTZ	Free Trade Zone
GDP	Gross Domestic Product
GFCF	Gross Fixed Capital Formation
GDCF	Gross Domestic Capital Formation
GSP	Generalized System of Preferences
HICOM	Heavy Industries Corporation of Malaysia
HRDF	Human Resource Development Fund
IIA	Investment Incentives Act
IC	Integrated Circuit
ILP	Industrial Linkage Programme
IPC	International Procurement Centres
IMF	International Monetary Fund
ISEAS	Institute of Southeast Asian Studies
ISIC	International Standard Industrial Classification
ISO	International Organization for Standardization
ITA	Investment Tax Allowance
JETRO	Japan External Trade Organisation
JIT	Just-in-Time
JV	Joint Ventures
KHTP	Kulim High-Tech Park
LDC	Less Developed Countries
LMW	Licensed Manufacturing Warehouses
MAI	Multilateral Agreement on Investment

MEE	Malaysian Electrical and Electronics sector
MIDA	Malaysian Industrial Development Authority
MITI	Ministry of International Trade and Industry
MNE	Multinational Enterprise
MRP	Material Requirement Planning
MSC	Multimedia Super Corridor
NEP	New Economic Policy
NIE	Newly Industrialising Economy
ODA	Official Development Assistance
OECD	Organisation for Economic Co-operation and Development
OEM	Original Equipment Manufacturing
OHQ	Operational Headquarters
PCB	Printed Circuit Board
PIA	Promotion of Investment Act
PIO	Pioneer Industries Ordinance
PSDC	Penang Skills Development Corporation
PVL	Preferred Vendor List
QCD	Quality Cost Delivery
RDA	Regional Development Authorities
RM	Malaysian Ringgitt
R&D	Research and Development
SEDC	State Economic Development Corporation
SITC	Standard Industrial Trade Classification
SME	Small- and Medium-sized Enterprise
SMI	Small and Medium Industries
SMIDEC	Small and Medium Industries Development Corporation
TNC	Transnational Corporation
TQM	Total Quality Management
TRIMs	Trade-Related Investment Measures
UN	United Nations
UNESCAP	United Nations Economic and Social Commission for Asia and Pacific
UNCTAD	United Nations Centre for Trade and Development
UNCTC	United Nations Centre on Transnational Corporations
UNIDO	United Nations Industrial Development Organisation

Figures

Tables

Acknowledgements

The most difficult part to write in a book is the acknowledgements. It would be almost impossible for any author not to mention people who have either inspired his or her thoughts, or indeed helped him or her at some stage. This book is drawn from the work carried out for my PhD thesis. The process of doctoral research was a personal challenge and achievement, a tremendous effort towards self-discipline and motivation, with a wonderful personal satisfaction upon completion. The work involved has been a true learning curve, a first step towards grasping the depth and complexity, and the usefulness of academic research. Times of frustration have been followed by enjoyable times. To me, doing this research has also been a unique human experience, shared with other doctoral students, discussion with academics, motivating comments by and discussions with the managers I interviewed, and not least trying to convince some of my friends that the whole experience was worth it. It would be impossible to mention every single person who helped me along this research path; I am grateful to all of them.

Some individuals deserve a mention for their trust, support and encouragement: Hafiz Mirza for his invaluable guidance, trust and patience; Peter J. Buckley who supervised my progress; Jean-Louis Mucchielli and Stephen Young, in their role of external examiners, for their challenging remarks and much appreciated encouragement. My two friends Simon Chapple and Séverine Chédor also contributed, by reading the first manuscript of this book and by giving me precious comments. My former colleague Anne-Wil Harzing inspired me greatly through her immense dedication to research. I also received precious support from many staff at the Bradford School of Management. There is a tendency, of course, to forget those who helped me in the earlier stages, therefore, I particularly wish to thank colleagues at the Bradford University School of Management, the French Ministry of Higher Education that financed me throughout this research, Nasser Noor and Mohammed Sulaiman who both helped me greatly during my field research in Malaysia, along with all the staff at Universiti Sains Malaysia. This book would not have been completed without the support, help, trust and cooperation shown by managers of transnational corporations and government officials in Malaysia. There has been much research carried out in Southeast Asia, although not all similar in nature. This does put pressure on managers who receive increasing requests for cooperation. Academic research in business and management relies heavily on the contribution of those managers, and this book relies on the quality of the information provided by them.

Finally, I wish to thank the whole of my family, my parents, Jean and Christine in particular, for being there in good and difficult times. My close friends in France, England, Malaysia and elsewhere in the world. To all of them, and especially to my son Antonio, I dedicate this book.

1. Introduction

BACKGROUND OF THIS RESEARCH

Globalisation has attracted increasing attention among scholars in international business. In an international business context, globalisation finds its roots in the increasing trade and investment flows between countries. Trade growth continuously exceeded the growth in both world commodity output and world gross domestic product (GDP). In 2000, the excess margins between the growth rates were high and above those observed during the early to mid-1990s. World merchandise trade grew by 12% compared to 4.5% for world merchandise output. The ratio of world trade in goods and services to world GDP reached 29% (http://www.wto.org[1]). Transnational corporations (TNCs) play a large role in the globalisation process, with an increasing control over world productive activities; and reverse, the globalisation trends impact on TNCs' strategies (Mucchielli, 1997; Dunning, 1998a, 2000; Levy-Livermore, 1998). Global foreign direct investment (FDI) flows surged by 18% in 2000 to US$1.3 trillion compared to US$331 billion in 1995 (UNCTAD, 2001) and this sharp increase resumed in 2001. UNCTAD has projected a 40% drop in world FDI inflows, mainly due to a slowdown of world economic growth and a decrease in cross-border mergers and acquisitions (http://www.unctad.org). During the 1990s, national governments acted as facilitators, mostly by trying to attract TNCs within their national boundaries, both in the developed and in the developing worlds, and by creating a favourable environment where created assets have had an increasing importance. Between 1991 and 2000, a total of 1,185 regulatory changes were introduced in national FDI regimes, of which 1,121 were in the direction of creating a more favourable environment for FDI (UNCTAD, 2001: 6). These included liberalising changes or changes aimed at strengthening market functioning, as well as increased incentives.

Increasingly, scholars are concerned with the impact of this world-wide rise in FDI on both the home countries and the host countries. The topic is even more crucial for developing countries, which in some cases rely heavily on foreign inflow of capital as a national investment generator. We can wonder if the current trend to globalisation of economic activity corresponds to further emphasis on inter-firm relationship. Globalisation comprises the spreading of international operations of firms and the increase in the amount of intra-firm relation, and the increase in inter-firm cooperation, notably in the high-tech-intensive industries. Thus, Buckley and Casson (1998a) mention the central place of cooperation through joint ventures (JVs) and business networks in the new dynamic agenda for research on the multinational enterprise (MNE).

East and Southeast Asian countries have shown high growth rates over the past decades, and have recovered promptly from the Asian crisis in 1997. They attract a large share of the world flows of FDI (ISEAS, 1995; Dobson and Chia, 1997; Mirza et al., 1997, 1999; UNCTAD, 2001). FDI flows to South, East and Southeast Asia increased from US$96.2 billion in 1999 to US$137.3 billion in 2000 and the sharpest increase took place for Hong Kong that attracted US$64.4 billion worth of investment. Outflows from the region decreased as a result of the Asian crisis, from US$49.6 billion in 1996, to US$34.4 billion in 1999, before picking up again in 2000 at US$83.6 billion. The region is clearly the first developing region in terms of FDI outflows, with 7.2% of world FDI outflows in 2000. Malaysia is a major host for FDI, inward FDI flows to the country increased in 1999 and 2000, after a decline in 1997 and 1998 following the Asian crisis. FDI inward stock in 2000 reached US$54.3 billion, compared to just US$5.1 billion in 1980, and US$10.3 billion in 1990. Malaysia is the third most transnational country among developing economies. Malaysia's transnationality index[2] increased from 35.4 in 1997 to 44.4 in 1998, coming close to Hong Kong and Trinidad and Tobago and above Singapore (UNCTAD, 2001). There has been an increasing concentration of activities in the electrical and electronics sector in Asia, which started in the 1970s, with operations from Western and Asian firms highly concentrated in a few Asian countries (Chalmers, 1991; Borrus, 1993; Hobday, 1995). The Malaysian electrical and electronics (MEE) sector is thriving, thanks to a large extent to foreign firms located in that sector (Linden, 2000; MIDA, 2002). In 1999, 25% of approved manufacturing investment projects by the Malaysian government were in the electrical and electronics sector (http://www.mida.gov.my).

The predominance of TNCs in the world economy, both in terms of trade and capital flows, has led to research into understanding this process and its wider implications. Many TNCs have created international networks, with intra- and inter-firms supply systems. Once implemented in a country, production operation functions not only on a national basis, but also on a regional and international basis. The manufacturing of different parts and components in one country and the assembly base in another is not unknown. The spreading of TNC activities has led to changes in the international division of labour, as well as in some cases the regional division of labour.

Currently, the vertical division of labour within an industry is very important; it derives from Adam Smith's idea of production, whereas previously conventional trade theory focused on the horizontal division of labour.

There is an extensive body of research on the impact of FDI on a host country's development (Dunning, 1994; Kumar, 1998; Wells, 1998; UNCTAD, 1999; Hood and Young, 2000; Takatoshi and Krueger, 2000). Whereas some studies seem to conclude that the FDI has played an important role in a host country's development, others do not. In the case of countries such as Taiwan and Singapore, FDI has promoted exports and growth. But each country's development experience is different. In Malaysia and Thailand, it appears that the export-led growth has prompted FDI, which in turn seems to have had a strong impact in terms of competitiveness and industrial relations. Despite all the uncertainties about the actual impact of foreign activity on the local economy, FDI is a substantial alternative for economic development for those countries that are less able to access and raise finance through bank borrowings, and for which official development assistance (ODA) has declined.

One aspect of a foreign operation's role in a host country's economy is its impact on the dynamic comparative advantage of the country and its international specialisation. On the one hand, widespread intra-firm operations may reduce inter-firm linkages; yet, considering the need created by TNCs for a particular product, a country can specialise in its production. Beyond the dynamics followed towards higher value-added types of comparative advantage in host economies, research pays more and more attention to created capabilities or competencies, as opposed to natural resources (Dunning, 1998b, 2000). TNCs interfere in the determination of countries' or firms' competitiveness, and contribute to the creation and improvement of such man-made comparative advantages, by injecting competition into the market and by developing factor conditions, demand conditions and the network of related and supplying industries. This topic is at the heart of this study, since we look at ways through which TNCs increase competitiveness of local suppliers after knowledge transfer has occurred. The relocation of a firm's operation in a host country is expected to increase the demand for local supply. Because of cross-border supplying, intra-firm trade and the lack of reliable indigenous production capabilities, local demand by TNCs in host-developing countries may be limited. In the case of Southeast Asian countries like Thailand, Malaysia and Indonesia, growth rates are substantial and local firms are more and more competitive, but these developments seem to be independent of the presence of foreign enterprises. The generation of competitive local supplier firms is not clearly linked to the presence of internationally competitive firms. Recent studies on backward linkages'[3] creation showed a beneficial impact of foreign firms on local suppliers' capabilities (Halbach, 1989; Fleury et al., 1994; and more specifically for Asia, Rasiah, 1989, 1995; Wong, 1991; O'Brien, 1993; IDE, 1994; Supapol, 1995; Awuah, 1997; Kelegama and Foley, 1999; Reinhardt, 2000; UNCTAD, 2001). Most of those studies are based on case study analysis, and do not allow for generalisation of results.

RESEARCH QUESTIONS

In view of the increasing presence of foreign firms in Asian Newly Industrialising Economies (NIEs), and the need for further study on inter-firm networks and relative impact on the host countries' development, we need to focus this study on backward linkage creation and impact on host countries. We acknowledge that the impact of the TNC in the host country depends on a number of factors, either internal or external to the firm.[4] This leads to our first research question:

> *Research question 1*: Do TNCs have an impact on host developing countries through backward linkages?

TNCs' impact on host countries has been the source of considerable discussion since the 1970s, when certain theorists started to lay out the specific nature of international business activities, as opposed to national business. With foreign firms operating in host economies, the need to analyse their behaviour and impact on host countries was accentuated by host government's response to inward FDI. To know whether foreign firms ought to be welcome and treated as local investors, host governments needed to understand better the process through which foreign firms impact on their economy. Within the vast existing literature on TNCs' impact on host economies, far less work has been conducted on the specific nature of buyer-supplier relations generated by foreign firms with local suppliers, that is backward linkages. To cover this gap, we intend to analyse theoretical concepts related to backward linkages borrowed from various existing theories related to the firm, and to international trade. We propose to draw up a framework of analysis to identify the impact foreign firms may have on host countries through backward linkages, with particular emphasis on developing countries.

Our attention is thus focused towards the nature of backward linkages, and whether they give rise to knowledge transfer. We intend to show whether or not the inter-firm relations through supply exchange lead to the creation of substantial links that may in turn be beneficial for the development of local indigenous supplier firms. Nowadays, the process of externalisation of the supply sourcing is more complex than a simple choice between 'make' or 'buy'. The external supply relations have become a key element in competitiveness and in the building of firms' competitive advantages. We therefore wonder if foreign firms adopt a strategic attitude towards local suppliers in developing countries, and consider sharing information with them through knowledge transfer. This is formulated in our second research question:

> *Research question 2*: To what extent do backward linkages' creation lead to knowledge transfer?

As mentioned before, very few studies have investigated transfer through backward linkages of foreign firms in host developing economies. From the results presented by these studies, it is extremely difficult to establish a clear pattern. Most studies use case studies in order to grasp the full complexity of transfer through backward linkages. As a result, no clear picture of transfer through backward linkages can be established. All studies focus on companies that have an existing network of local supply relationship, and analyse subsequent transfer occurring within this existing network. We would like, on the contrary, to grasp the full backward linkages' picture, by analysing firms that do not have local supply network, as well as firms' backward linkages, but not necessarily leading to close relationship with suppliers and subsequent knowledge transfer.

Because of the different groups of firms to be included in our analysis, we decided to shed light on the differences existing between firms with transfers through backward linkages and firms without. We expect that firms with no backward linkages in the host economy will not have any impact on the capability development of local suppliers. However, even when TNCs do establish backward linkages locally, such links may not bring substantial benefits to local suppliers. Thus, we thought it was important to study the factors that may influence firms to transfer knowledge through backward linkages. This leads us to our third research question:

> *Research question 3a*: What are the factors that may explain the existence of knowledge transfer through backward linkages?

Although it appears as extremely important to understand why a foreign firm may or may not transfer knowledge to local suppliers, there is no academic evidence provided on this topic, and indeed no other research has looked into this comparison. On the other hand, it should be strongly emphasised the necessity to understand fully why a firm operates transfer to local suppliers or not. We include in our analysis firms with and without transfer, even though no previous study has paid attention to firms without transfer.

What is often studied in the literature is the degree to which foreign firms operate transfer in the host economy. All studies on backward linkage' creation and knowledge transfer focus on firms with transfer and analyse the complex relationships established between buyer and supplier. This issue is closely related to the analysis of why transfer exists in the first place. Thus, our final research question can be phrased as follows:

Research question 3b: What are the factors that may explain the degree of knowledge transfer through backward linkages?

While the degree of backward linkages' knowledge transfer has been studied by various scholars, we wish to clarify the factors that significantly affect the degree of transfer. There is no general agreement on the factors that explain the degree of transfer, and we feel there is room to clarify which factors may or may not explain why transfer is substantial or not.

The advantage of our study is that we provide a complete picture of backward linkage creation, its impact on the host economy and subsequent knowledge transfer. With the knowledge acquired in previous studies, we can complement and clarify existing knowledge and pay particular attention to important variables explaining backward linkages. These include time, country of origin, size, product manufactured, market orientation, purchasing strategy and local environment.

RESEARCH PROCESS

This study analyses the nature of backward linkages, knowledge transfer to local suppliers and the factors that explain when and how much transfer through backward linkages takes place. In order to complete our objectives, we intend first to establish a theoretical framework of analysis, borrowing concepts from theories of the firm and theories of international trade. This will be completed by a review of the literature on TNCs' impact and backward linkages, with a specific focus on Southeast Asia.

Because we wished to include firms with and without backward linkages, and firms with and without transfer, we decided to focus on only one country and one sector of activity. Although we acknowledge that this decision is self-limiting in nature, we thought it was the only answer to obtain a full range of firms in terms of backward linkages. Thus, our emphasis and decision were focused on firms answering all aspects of backward linkages' creation. The targeted country is Malaysia. As mentioned above, this country is a major recipient of FDI, and the government is actively encouraging the development of endogenous production capabilities as well as attracting foreign firms to settle within its boundaries. The electrical and electronics sector is studied because it has shown a rapid growth in East Asia. It is one of the targeted strategic sectors by the Malaysian government, and it relies on a network of subcontracting work and on an international exchange of goods. For these reasons, we chose to study the MEE. We show whether or not inter-firm relations through supply exchange in the MEE lead to the creation of substantial links that may in turn be beneficial for the development of local indigenous supplier firms.[5]

Since we wish to analyse all types of backward linkages, we thought we ought to adopt a variety of research techniques. The triangulation method is adopted in this research, bringing together in-depth case study analysis and a series of logistic and linear regression models to provide both a detailed description of TNCs' backward linkage creation and a set of statistical results to define the context behind which buyer-supplier transfer takes place. Data are collected by means of semi-structured interviews and a mail survey. Our study therefore provides a complete analysis of TNCs, backward linkage impact on host-developing countries, through knowledge transfer. Because the study of the existence of backward linkages and the related influencing factors has not been performed before, we systematically discuss the issues considered and the choices made, and try to mention all the problems encountered in this research.

STRUCTURE OF BOOK

Figure 1.1 presents a visual representation for the structure of this book. Each chapter is linked to the other, and brings essential information for the study of TNCs, backward linkage, and related transfer.

Following this introductory chapter, Chapter 2 presents the theoretical framework. All the relevant theories existing in the field have been considered, in order to establish a theoretical framework to be applied as a basis for the research. One key assumption in the research is that we need to link various theories together to investigate backward linkages-related transfer. The theories of the firm, such as internationalisation theory, and network analysis, as well as concepts borrowed from the stages of growth approach, are all relevant in understanding backward linkages by TNCs in a host-developing country, and their subsequent knowledge transfer. We provide an integrative view of these theories as a theoretical background to the study of knowledge transfer through backward linkages.

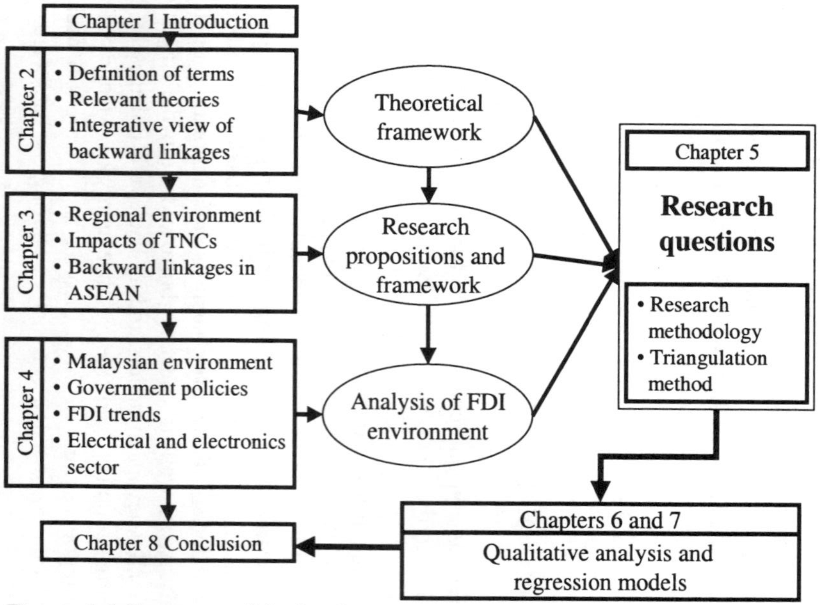

Figure 1.1 Structure of the book

Chapter 3 takes a regional approach looking at the existing literature on the impact of FDI on host countries. The overall situation in which foreign firms operate, taking into account the local environmental conditions and their necessary link with the possible impact, is presented. The analysis then draws on the specific impacts related to backward linkages and knowledge transfer under supplying contracts. Out of this process, the research question is refined, supported by existing empirical research carried out on the topic of backward linkages. We present a series of propositions linked to backward linkage creation and related transfer.

Chapter 4 specifically looks at Malaysia as a host for FDI, and at Malaysian policies towards foreign firms and their operations. The shift in government policy orientation, towards increased awareness of the useful interactions that could be generated between foreign firms and local indigenous firms, is discussed. The case for soft intervention, with subtle encouragement for backward linkages' effects, is looked at. Trends in FDI inflows and outflows are described. Finally, the electrical and electronics sector is considered and this provides key information about the context of this research. This chapter helps to narrow down the research propositions, by providing all the necessary information about the targeted country and sector.

In Chapter 5, with the in-depth knowledge about the FDI environment provided in Chapter 4, research propositions are refined into research questions to be answered for the particular case of backward linkages in the MEE. The methodological strategy applied for the empirical analysis is also presented. Several complementary strategies are adopted to examine all aspects of the topics. The triangulation method uses a qualitative approach obtained from interviews and quantitative models using data collected as part of this study from mail questionnaires, and finally combines findings from both methods. This method is the key innovative concept adopted in this research. Preliminary results from the interviews and the mail survey are also examined in this chapter.

Chapter 6 and 7 provide the description and analysis of the data gathered in Malaysia during the field research carried out in 1996 among foreign firms' subsidiaries operating in the electrical and electronics sector. Chapter 6 draws on the information provided by respondents through semi-structured interviews. Although some information is presented case by case, most of the analysis is cross-interviews, drawing out conclusions through a comparative framework. Chapter 7 employs logistic and linear regression models using data obtained during the mail questionnaire survey. The quantitative approach helps substantially in this research by providing a list of factors significantly affecting the existence and the degree of transfer created between foreign subsidiaries and their local direct suppliers, through the inclusion of a sufficiently high number of firms in the analysis.

Chapter 8 applies the triangulation method, brings together both sets of results, and reflects on earlier studies carried out on backward linkages. This concluding chapter links together all the chapters by providing a summary of our own results, and considers the contributions and limitations of the study. The chapter ends with suggestions for further research.

NOTES

1. Data published by the World Trade Organization (WTO) in the International Trade Statistics 2001are available at http://www.wto.org
2. Transnationality index is an average of four shares: FDI inflows as a percentage of gross fixed capital formation for the past three years (1995-97); FDI inward stocks as a percentage of GDP in 1997; value added of foreign affiliates as a percentage of GDP in 1997; and employment of foreign affiliates as a percentage of total employment in 1997.
3. Backward linkages refer to relations that foreign firms establish with local suppliers in the host country in which they operate.
4. We make a number of assumptions related to the TNCs' impacts on a host country. First, on a theoretical perspective, we need to draw upon various theories. Theories related to the MNE help to understand the profile, operations and strategy of international firms. The network analysis theory draws upon inter-firm linkages to understand the nature of buyer-supplier relations. If we want to grasp the impact on the host economy, we need to borrow concepts developed in the stages of growth theories. The second assumption refers to the role of FDI in the host country. We assume that FDI is beneficial, inasmuch as the host country is better off when the investment is made than if the foreign investment had not taken place. Third, the fact that TNCs operating in a developing country have ownership advantages in high-tech sectors that local firms may have not yet acquired. It is therefore assumed that TNC subsidiaries bring their bundle of assets with them, such as financial, production, marketing and organisational assets. Thus, the TNC brings its soft and hard technology, its learning experience and organisational competence to the host country, some of which we assume is passed on to the local suppliers.
5. This research does not address itself to local firms as it is difficult to evaluate the actual level of technological capabilities of local suppliers (human resource capabilities, technological orientation, productive capabilities or source of technological information). Instead, we focus on the view the subsidiary has of the local suppliers and the way it deals with its partners.

2. TNCs and Backward Linkages in Host Countries: A Theoretical Approach

INTRODUCTION

This theoretical chapter draws on the various existing theories useful to define the relationships existing between foreign subsidiaries and their local suppliers in a host country. The chapter begins with a definition of terms used in this book and the delimitation of our research strand. Theories related to the operations, strategy and activities of transnational corporations (TNCs) are then covered, and include the transaction cost theory, the internalisation theory and the industrial networks theories, as shown in Figure 2.1. The theories that arise from the economics of the firm are looked at in the light of the theories derived from the economics of international trade. These are the competitive advantage of nations, the asset accumulation and restructuring paradigm and the stages of growth paradigm. The logic behind this differentiation is that, although related phenomena, foreign direct investment (FDI) and TNCs consistently come under differentiated theoretical strands. Figure 2.1 is an attempt to show the relationships between TNCs, FDI, local subsidiaries and their local suppliers, from a theoretical perspective. It shows why each angle is a necessary link when tackling the concepts of foreign subsidiaries' backward linkages within a host economy. FDI is a necessary link between TNCs and their local subsidiaries. Whereas TNC-related theories explain the behaviour of the firms, theories related to FDI are a key component in the explanation of the impact of foreign firms on a host economy. Finally, the relations between subsidiaries and their suppliers are often conceptualised in the above-mentioned theories, but the links and the type of relationships are best described in organisation theories. The specific concepts of trust and bonds help in understanding the type of relationships that may be created, how and why, and the subsequent spillover effect on the host economy.

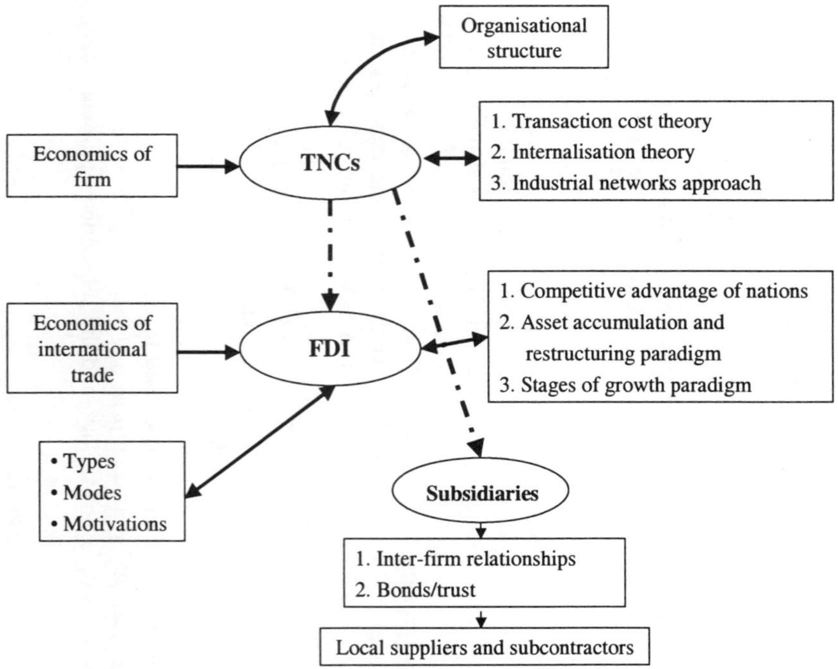

Figure 2.1 Integrative figure of existing theoretical framework in the context of the current study

This wide theoretical coverage enables us to build a comprehensive and integrative theoretical framework for the analysis of TNCs, their backward linkages and the subsequent impact on host countries' technological capabilities development.

TNC NETWORK STRATEGIES: BETWEEN MARKETS AND HIERARCHIES

In the attempt to analyse the role of TNCs in economic development through the creation of backward linkages and technology sharing, it soon becomes apparent that more than one theoretical stream is needed to develop a conceptual framework of analysis. The main challenge resides in the necessity to borrow concepts from a variety of TNC-related theories, some deriving from the economics of the firm, and some from the economics of international trade. Faced with such an eclectic background, one needs to first explain the key terms used in this book.

Concepts of TNCs, Knowledge and Backward Linkages

There often exists a plethora of words for the same meaning, or alternatively a plethora of meanings for the same word. Therefore, it is essential to define each key word and concept before embarking on research.

The transnational corporation and its various facets

Buckley and Casson (1976) provided the basic definition of a multinational enterprise. It is an enterprise that owns and controls activities in different countries. In the light of this study, the analysis of TNC activity needs to take into account the dynamic interaction between the comparative advantage of countries, or even, following Porter's diamond, the competitive advantage of countries and the activities of the TNC (Porter, 1990). Although Porter does not place the emphasis on TNCs, he claims that countries, like firms, have competitive advantages.

A more complete definition is to look at the TNC as 'a [...] major vehicle for the transfer of mobile resources (technology, capital, management skills) to areas with immobile (or fixed) complementary inputs (markets, raw materials, labour)' (Buckley and Casson, 1985). This definition is useful to this study, since the Ricardian endowments are taken into account, thus following the analysis of their role in attracting FDI, and vice versa: the role of FDI in promoting the host country endowments. Market imperfections are also major factors explaining the existence of multinational firms. Hence, the unique characteristic of the multinational firm is that it is engaged in multi-activities and in the internal transfer of intermediate products outside its national boundaries. There exist differences between domestic and international market failures, and these differences distinguish the multinational from the uninational multi-activity firm (Casson, 1987).

Buckley and Casson (1998a) look at multinational enterprises as flexible entities. To cope with the inherent volatility of the international business environment, corporations need to adopt flexible strategies. These strategies are associated with efficient information processing to cope with increasingly complex decision-making processes. This being said, some factors make some multinational firms more 'transnational' than others by virtue of the nature of their organisation, their centralisation of decision-making and authority, their global strategy or their ability to act as one cohesive unit under changing circumstances. For Lall and Streeten (1977), a 'truly' TNC thus acts as an organisation maximising one overall objective for all its units; it treats the whole world as its operational area, and it is able to coordinate all its functions in any way necessary for achieving maximisation and operationality.

As more complex organisational structures and new strategies have occurred, companies that go international have to opt for a strategy. The multinational strategy corresponds to a strong local presence through responding to different national market requirements. The global strategy refers to a competitive search through cost advantage from centralised global-scale operations. Finally, the international strategy is associated with the ability of the subsidiary to exploit the parent company knowledge (Bartlett and Ghoshal, 1989). Harzing (1999) provides a typology of existing studies on multinational companies and proposes five terms to characterise the latter. She uses the term 'multinational' as a general term for a company that operates internationally. 'Multidomestic', 'international', 'global' or 'transnational' would then describe different configurations of multinational companies (Harzing, 1999: 34-6).

Faced by a changing world environment, companies need to adapt their strategies to remain competitive and to develop simultaneously efficiency, responsiveness and learning as strengths for success. The TNC seeks efficiency to achieve global competitiveness. In this case, no specific strategy, such as one of those mentioned above, will be carried out. Instead, the most effective strategy for each different situation may be adopted. The company needs to be responsive to the environment. This results in an internationally integrated network of the firm's operations. 'The attributes of the transnational are internally consistent and mutually reinforcing. The integrated work configuration; the differentiation of subsidiary roles and responsibilities; and the simultaneous management of multiple innovation processes collectively constitute an integrated and viable organisation system' (Bartlett and Ghoshal, 1998).

As firms start operations abroad, the definition of a 'foreign affiliate' in the host country needs precision. Some authors consider a firm as a foreign subsidiary when foreign control is a minimum of 25% shares held by the parent firm (Kumar, 1990). For others, a majority shareholding by the foreign firm is not necessary for it to exercise control. Indeed, in appropriate circumstances, and even without an explicit management contract, a particular investor can exercise control with an equity share as low as 10% (Lall and Streeten, 1977; Moosa, 2002). Many authors consider foreign ownership to be an element of market structure, like concentration, product differentiation and entry barriers affecting other aspects of market structures, conduct and performance. It can be expected that the behaviour of a foreign firm in a host country will differ according to its ownership structure (Kumar, 1990). A wholly-owned subsidiary may have less contact with local businesses, although after a long period of time, this discrepancy is expected to decrease, as the foreign firms learn about the host country's environment.

Having acknowledged the various types of international organisational structures, it is not the purpose of this study to push forward and refine the definition of a company that has internationalised. Throughout this research, we do not differentiate firms according to their various organisational forms. For our definition and for the purpose of the research, the Ricardian approach to firms in foreign markets is emphasised. The term 'TNC' is adopted and refers to the United Nations' definition.

> Transnational Corporations are incorporated or unincorporated enterprises comprising parent enterprises and their foreign affiliates. A parent enterprise is defined as an enterprise that controls assets of other entities in countries other than its home country, usually by owning a certain equity capital stake. [...] subsidiary enterprises, associate enterprises and branches are all referred to as foreign affiliates or affiliates (UNCTAD, 2001: 275).

Subsidiary enterprises and affiliates are all referred to as foreign firms and their subsidiaries, in the context of the host country. The essential link between the parent company and the subsidiary in the host country is that of FDI flows, and these are discussed in the following sections.

FDI

In defining FDI and what it encompasses, the first step is to know what type of FDI the study is dealing with. Initially, one identifies whether the firm conducts direct production and/or marketing activities in the host country. Because of the nature of this research, this book addresses the first type, and we define FDI as the process whereby residents of one country acquire ownership of assets for the purpose of controlling the production (Moosa, 2002). Additionally, from the differentiation between market transactions and internalisation referred to until the 1980s, new forms of economic arrangements have sprung up, such as non-equity-based corporate alliances, which link up through various types of arrangements (such as joint ventures, subcontracts or marketing arrangements), TNCs and national firms (UNCTC, 1988). Hence, other linked aspects include the mode of entry (greenfield venture or other), and the organisation form (joint venture (JV), wholly owned affiliates). Additionally, when carrying out a foreign activity, the firm can choose between horizontal or vertical foreign investment. In the former, the TNC carries out the same stage of production abroad as it does in the home country. In the latter case, the firm then keeps control of downstream raw material, other factor inputs or a forward stage of production.

The major motivating factors behind FDI are the avoidance of uncertainty and the creation of barriers to entry. More specifically, motives for investing abroad vary according to the type of investment the firm wishes to carry out. Resource-based or rationalised investments are motivated by the availability of indigenous resources and capabilities, as well as by the avoidance of cross-border transportation costs. Market-seeking investment is an answer to the possible loss of an existing market or an endeavour to expand to growing and profitable markets. In the case of the initial foreign production, a defensive motive is for example an endeavour to maintain a competitive position, whereas the aggressive motive leads the firm to be present in a market in which growth is anticipated. The defensive position is also found in the strategic asset-seeking investment (Dunning, 1993b; UNCTAD, 1998). FDI is expected to be prominent in those sectors/branches of industries in which intangible assets with high internalisation and locational incentives are important; this differs from licensing, which is expected in sectors with intangible assets (Kumar, 1990). This research targets FDI in the manufacturing sector, and subsequent relationships established between the foreign subsidiary and its suppliers in the host economy, thus forming backward links.

TNCs, backward linkages and spillover effects

When a TNC proceeds to FDI and establishes a subsidiary in a host country, the latter creates inter-firm linkages. There are two major directions to these linkages. Backward linkages refer to all the relations established with supplier firms, whereas forward linkages describe the relations established with the customers in the host country (UNCTC, 1981; Dunning, 1993b; Blomström et al., 2000; UNCTAD, 2001). The spillovers occur when local firms benefit from the TNC affiliate's superior knowledge of product or process technologies or markets, without incurring a cost that exhausts the whole gain from the improvement (Blomström et al., 2000). In this research, the main focus is on TNCs' backward linkages. These are of particular importance for host-developing countries because first they provide opportunities for production and employment by domestic suppliers, but most importantly because they constitute a direct channel for knowledge diffusion. Linkages can become substantial channels for transferring knowledge and skills. With time, transfers assist in upgrading domestic suppliers, through increased technological and other capabilities, with spillover effects on the rest of the economy (Giroud, 2000; UNCTAD, 2001). Hence, backward linkages lead to the transfer of knowledge and technology to local suppliers and subcontractors.

Knowledge transfer

The most common form of transfer is known as 'technology transfer' and has been widely studied in the literature (Lall, 1993a; Chen, 1994; Buckley et al., 1997; Kumar, 1998; among others). Technology transfer involves the transfer of physical goods, such as capital goods in the form of machinery and equipment, and the transfer of tacit knowledge, such as managerial skills, technical skills and know-how. The latter is becoming more important and involves acquiring new skills and technical and organisational capabilities (UNCTAD, 1999). In fact, TNCs transfer their own business culture to the host countries (Mirza, 1998b). Knowledge and technology transfer involves transfer from one firm to another, and possible benefits through long-term relations and the exchange of information. Direct channels of knowledge transfer include FDI, turnkey operation, purchase of machinery and equipment, licensing and management agreements, technological assistance contracts, direct employment of foreign experts and training of local staff. Indirect channels include transfers to suppliers or customers. One of the crucial contributions by foreign subsidiaries to host countries lies in this flow of knowledge from TNC systems to local firms (UNCTAD, 1995, 1999, 2001). Through the specific case of subcontracting, backward linkages include knowledge transfer, technical, financial, procurement, locational, managerial, pricing flows and flows related to sales arrangements or export help (UNCTC, 1981). For the purpose of this research, we use knowledge transfer as a generic term to cover hard and soft technology that may be shared among business partners, that is technological knowledge, know-how, specific skills and business culture. Knowledge transfer through backward linkages, between TNCs' subsidiaries and their local suppliers in the host country, constitute the core of this research, and is investigated throughout this book.

TNC-related Theories: Plea for Integrated Approach

In the 1970s, Hymer (1976) contributed widely to the understanding of the multinational operation phenomenon by making a distinction between portfolio investment and direct investment. This distinction is important, inasmuch as FDI leads to foreign production and other foreign activities and to the maintenance of the control of the invested assets. Hymer viewed FDI as an extension of the industrial organisation; it aimed to make profits and could no longer be explained by theories on capital flows and differential interest rates. Various theories on TNC activities have emerged, each bringing further understanding of these activities, yet authors have come to the conclusion that until now there is no unifying theory able to cover all aspects of FDI. Most of these theories are only a partial explanation of FDI, because their explanatory powers are limited to specific aspects of international investment (Casson, 1990; Dunning, 1993b).

Focusing on the activity itself abroad rather than on its investment, the theory on TNCs appeared because of the inability of traditional theories to explain:

- the behaviour of firms in an imperfect environment, within which the markets of goods and factors were imperfect;
- the reasons why firms wish to establish and control foreign activities outside their national boundaries.

Studies on TNCs result from two theoretical approaches. First, the concept of the TNC takes after the economics of the firm. The economics of the domestic firm provides the base for the explanation of the activity of the TNC. Coase's (1937) transaction cost theory focuses on networks of trading relations and on economics as being intertwined by contracts between and within firms. On the other hand, the organisation theory originated by Williamson (1971) emphasises the notion of institutional factors within the firms. Following this theoretical base, in order to study the relations between the foreign subsidiaries and their local suppliers, the transaction cost theory, the internalisation theory, the industrial network theory and the inter-organisation analysis of the activity of TNCs are examined first. Second, the economics of international trade is the root to the debate on substitutability versus complementarity between international investment and international trade. On the one hand, obstacles to free trade lead companies to resort to foreign investment as a substitute to international trade. These obstacles result from imperfect competition on the goods' and services' market as well as on the factors of production market, or from competition created by government policies. On the other hand, Linder (1961) showed the rising heterogeneity of traded goods and services by the existence of higher income rates. Numerous interpretations on the nature and complementarity of international investment with international trade take their roots in this theory, leading to the re-actualisation of Ricardo's concept of comparative advantage. Such analyses have paved the way to the stages of growth analysis of the activities of TNCs and to the geographical agglomeration analysis.

The present study is based on the adoption of the dynamic elements of the theories on the rise of the firm, with specific emphasis on the growing international production networks and economic linkages. Therefore, it is useful to combine concepts such as the internalisation of activities, market structure and competitive advantages. Various topics are addressed; they are mainly the particularities of the Asian NIEs' multinational firms, the analysis in terms of networks, the role of strategic management and the dynamics of international production. The two bodies of research described provide an overall view for the study of the effects of TNC on indigenous capabilities of host countries, notably through the existence of backward linkages. While the theories of the firm highlighted the relationships between TNC subsidiaries and their suppliers, the theories on international trade help understanding of TNC-led spillover and the development process of the host countries.

Approaches related to the theory of the firm

Four theories are to be considered in this section. The transaction cost theory is concerned with institutional arrangements, while the internalisation theory describes the reasons why a firm may wish to internalise its activities across national boundaries. The industrial networks' approach centres around the definition of relationships among firms. Finally, the discussion draws upon the move towards a network theory of the TNC.

Transaction cost theory

The transaction cost theory finds its roots in the work of Coase (1937) who tried to explain the factors responsible for the boundaries of the firm, and the reasons why all the production is not carried out in only one major firm. Hierarchies are substitutes for the market when transaction costs are high. The economy is described as a network of trading relations and is intertwined by contracts between and within firms. Transaction cost economics is predominantly concerned with institutional arrangements, which are often referred to as 'governance structures'. An 'institutional arrangement' is an arrangement between economic units that governs the ways in which these units can cooperate and/or compete. These institutional factors within the firm are often included in the organisation theory (Williamson, 1981, 1985, Williamson and Winter, 1991). Bounded rationality and opportunism are the two critical behavioural assumptions under study.

Given bounded rationality, all complex contracts are unavoidably incomplete. Given opportunism, contract-as-promise unsupported by credible commitments is hopelessly naive. Taken together, the following organisational imperative obtains: organise transactions so as to economise on bounded rationality while simultaneously safeguarding transactions against the hazards of opportunism (Aoki et al., 1990: 12).

Given these assumptions, the firm is a nexus of implicit and incomplete contracts that govern relationships with suppliers of inputs and customers of output. If a production process describes the technical relationships that map inputs into feasible outputs, and two production processes are 'successive' if the output from one is an input into the other. A firm is then the global entity that owns assets, and enters into commercial and financial contracts. A firm organises a production process if it purchases or owns the inputs used in production (Riordan, 1990: 94). This leads back to the notion of the 'boundaries' of the firm, and to the 'make' or 'buy' decision. Transactions spread from discrete market exchange at one extreme to centralised hierarchical organisation at the other, with a myriad of mixed or intermediate modes filling the range in between. This shows the changing character of economic organisation over time within and between markets and hierarchies (Williamson, 1985). The 'efficient boundary' not only varies by the properties of the transaction, but the boundaries also have to be adjusted over time as the actors move along the learning curve. Actors gain more experience, asset specificity is reduced and more market contracting is expected. This is found within many mature industries. Examples include increased use of subcontractors for various types of components (as high as 75% in the case of automobiles and electrical appliances in Japan), and more reliance on joint venture and cooperative agreements (Reve, 1990).

When studying the supply activity of the firm, one has to refer to the question of 'vertical integration'. Vertical integration could protect valuable knowledge developed in a 'learning-by-using' process. The main problem is the interdependence of the incentives for the users to transmit their knowledge to the producer and the institutional relations between the agents (Lundgren, 1990). Vertical integration is often related to the necessity to protect knowledge as a result of 'information impactedness' (asymmetrical distribution of information among exchanging parties). A clearer distinction between vertical integration versus 'buy' decision is introduced by the 'transaction characteristics', that is 'asset specificity', 'uncertainty' and 'frequency' (Williamson, 1985; Williamson and Winter, 1991). Asset specificity refers to the investment required by a transaction specifically for a particular exchange relationship. In some cases, specific asset investments can be carried out by the supplier as well (Joshi and Stump, 1999). The uncertainty is related to the ambiguity in defining the transaction and its performance, often because of the inability to predict partner behaviour and changes in the external environment. There exist several types of uncertainty, and not all are positively related to vertical integration (Krickx, 2000; Slater, 2000). The frequency refers to the number of transactions being undertaken.

The transaction cost theory has clear implications for 'make' or 'buy' decisions, and for the contractual relationships between buyer and seller. Market transactions are by definition external to the firm, and do not require any particular skills or incentives to complete, as the price mechanism conveys all the necessary governance information. The relationships being established, the manufacturer needs to consider supplier opportunism. One way to avoid the latter is to adopt bilateral partnerships, long-term relationships, strategic alliances, hybrids and relational exchange (Joshi and Stump, 1999). Figure 2.2 represents schematised transaction cost characteristics that are asset specificity, uncertainty and frequency of transaction and the resulting type of relationships between firms.

However insightful the transaction cost theory is in the analysis of the buyer-seller relationships, recent research on the governance of capitalist economies has shown that the analysis must extend beyond markets and hierarchies to consider other governance structures, such as various forms of networks and the trade associations. These forms are neither controlled by markets nor are they hierarchies, and thus have not yet been adequately examined in transaction cost terms. Serious questions persist regarding the scope and applicability of transaction cost economics (Schneiberg and Hollingsworth, 1990).

Internalisation theory

The internalisation theory results from theoretical attempts since the 1970s to explain the international transactions of intermediate products by multinational hierarchies, as opposed to market forces. The internalisation theory was first described by Buckley and Casson (1976), before being sharpened in the 1980s (Buckley and Casson, 1985; Rugman, 1986), and in the 1990s (Buckley and Casson, 2000). The theory follows earlier attempts to explain the growth of domestic firms by Penrose (1958), but predominantly arises from Coase's ideas (1937), and is influenced by Williamson (1971, 1985, Williamson and Winter, 1991), as seen earlier in this chapter. Thus, the internalisation theory uses relative costs and benefits of internal organisation compared to market transactions to explain the international horizontal and vertical integration of value-added activities.

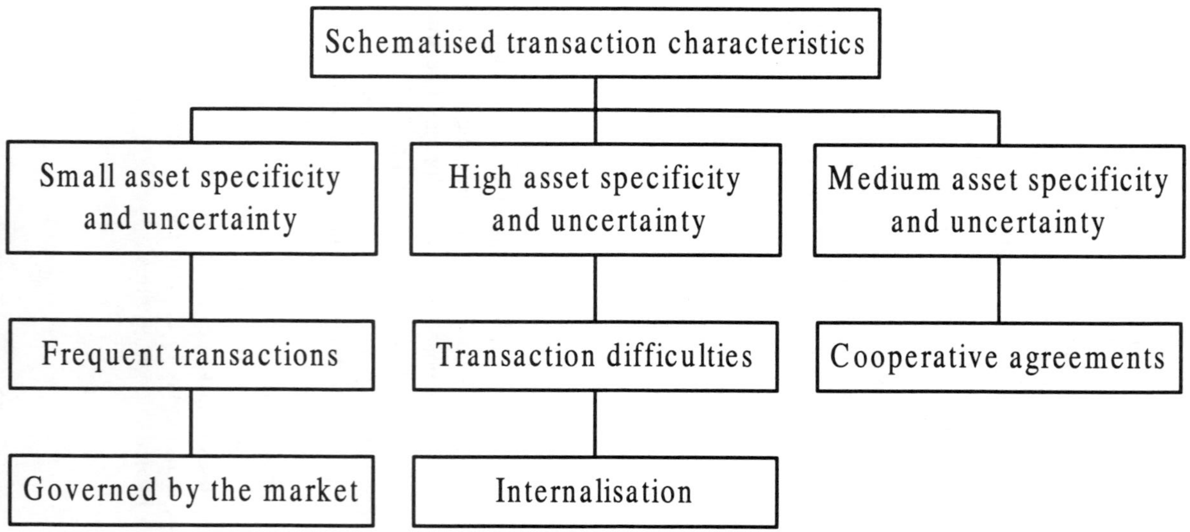

Figure 2.2 Schematised transaction cost characteristics

The central postulate is that TNCs create and possess proprietary assets that they try to exploit internally by extending their own organisation and control across national boundaries, preferably by means of wholly owned subsidiaries. It is primarily concerned with market failure and the costs of failures. Several types of uncertainties may lead the firm to internalise its activity; on the one hand, it can be the need to ensure the quality of the final product; on the other hand, the need to ensure regular supplies without wide price variations (Buckley and Casson, 1976). As a result of the above, these three postulates can be applied:

1. firms maximise profits in a world of imperfect markets.
2. imperfections lead to creating internal markets.
3. internalisation of market across national boundaries generates TNCs.

A market is likely to be internalised when the benefits of internalisation surpass the costs. Recent advances of the internalisation theory stand in the integration of aspects from the international management of international business, such as learning processes, trust and hierarchies within the firm. Indeed, the two approaches have much in common, including joint elements of intellectual heritage (Buckley, 1994: 102). Hence, the internalisation theory is rapidly evolving, widening its span of topics, and is evolving towards the integration of different disciplines, which is the wish of various scholars (Dunning, 1994).

Although the internalisation theory turns out to be the core theory of the study of multinational activity, it has been admitted by some of its main protagonists and other scholars that it does not give a full record of multinational operations of the firm on its own. It needs to be coupled with the locational theory (Buckley, 1988, 1990); in other words, with the theory of international trade. Although the theory is based in economics, it adds a behavioural element to economic explanations of TNC conduct (Graham, 1991), by enlarging its scope to other concepts (trust, collusion and market power and mainly network behaviour), which are described in the systems' view (Casson, 1990, 2000). There are four sub-topics that require further attention: (1) the type of integration chosen by the firm; (2) the psychic distance and level of trust faced by the firm in a foreign environment; (3) the entrepôt potential of the host country and (4) the systems' view approach.

Vertical versus horizontal integration. Different types of intermediate products will give rise to different types of internalisation; the intra-firm links between activities will vary accordingly.

> The internalisation of tangible intermediate product flows between upstream and downstream production can explain vertical integration. That of intangible flows of know-how leads to a combination of horizontal and vertical integration. Internalisation also applies to marketing and distribution. It therefore generates both a vertical link between production and distribution and horizontal links between the various distribution facilities that use a common source of supply (Casson, 1991: 5).

The amount of goods purchased by a firm depends on the extent to which the affiliates internalise the markets for the inputs required for their value added activities (how vertically integrated they are). As mentioned earlier, several types of uncertainties may lead the firm to internalise its activity; this can be the need to ensure the quality of the final product or the need to ensure regular supplies without wide price variations. The same types of uncertainties that are faced by the firms are likely to intervene in the relationship existing between a firm and its suppliers.

Psychic distance, level of trust. When choosing to go international, the firm is faced with an unknown market, which involves costs of adaptation, or in other words, the cost of doing business abroad. It is part of the managerial learning process. An integrated TNC's experience of one geographic area can be transferred to another. This adaptation process is also included in the Swedish approach to the internationalisation of the firm (Johanson and Mattson, 1988; Petersen and Pedersen, 1997). Related to the mode of internationalisation is the level of trust in the economic environment in which the foreign firm operates, and in the culture of the firm itself. Trust is important in the internalisation decision; it has various levels according to countries and cultures, and widely influences international business. Operation difficulties do exist in a foreign country that can arise due to the lack of knowledge about the new environment and the disparities between the home and the foreign cultures. The level of trust is likely to determine the type of economic coordination adopted by firms. Thus an inter-firm network is favoured when economic coordination takes place through the market; these are external (or arm's-length relationships) in a situation of high trust. Each member has confidence in the other members to conform to the arranged agreement (Casson and Cox, 1992/3). This level of experience gained by a TNC in a host country, as part of the adaptation process, and the knowledge related to the host market, hence of trust, will therefore be important determinants in the external links created by a TNC in the host economy. Thus, as a TNC has been operating in a host economy for some time, it may have increased its level of trust with the local firms, and higher levels of local supply may be achieved as a result of this, So, the time spent in a host country is a key to explain the relationship between the foreign firms and local partners.

Entrepôt potential. The process of development of a country includes traditional concepts of resource endowments such as land, labour and capital, as well as the level of education and training available. Buckley and Casson (1991) emphasise the need for the host country to have a certain level of entrepôt potential and entrepreneurial culture. The entrepôt potential covers geographic determinants such as good transportation, internal division of labour, ability to participate in a wider division of labour beyond its boundaries, interface between the internal and external division of labour, the condition most favourable to industrialisation (natural harbour close to shipping route, good internal communication between port and hinterland, livestock farming in the hinterland, abundant endowment of both mineral and primary energy resources, coastline suitable for the disposal of pollutants), scientific outlook of systems thinking (not possible where mysticism and superstition prevail), competitive individualism or voluntary association (in less developed countries (LDCs), lack of individualism thinking) and geographical and cultural aspects of the global trading system. The entrepreneurial culture provides an economy with flexibility, in particular the cultural flexibility, to cope with changes in the division of labour. Entrepreneurial culture is technical or moral.

While it is important for the TNC that the entrepôt potential of the host economy is sufficient to facilitate its operations, this potential is also crucial to the possibility of maximised backward linkages and their subsequent beneficial impact on the managerial and technical endownment of the host economy. Therefore, the level of economic development of the host country is another key determinant when looking at the impact of FDI on host economies. This concept is further discussed in the last part of this chapter.

Systems' view of firms' activities and creation of cross-effects. The systems' view of firms' activities synthesises the economic and cultural factors within a system of international production (Casson, 1990, 2000). The basic idea of the systems' view arises from the distinction between activities and markets on the one hand, and between facilities and linkages on the other hand. An activity can occur across boundaries, sometimes leading the firm to establish facilities in more than one country. Linkages concern relations between facilities, with facilities belonging to one only or to several firms. This reveals the need for the internalisation theory to be linked to the location theory to explain fully international business behaviour. The location theory has received renewed attention, with the emphasis on the changing nature of location attractive factors (Dunning, 1998b).

The systems' view approach pays attention to partial internalisation and its possible implications, insofar as a linkage between two activities is rarely independent and it leads to cross-effects, therefore 'when each linkage is embedded in a system of other linkages, cross-effects are created' (Casson, 1991: 7). Two firms are considered, each operating business with suppliers. If one decides to internalise the other firm, it implies that it now has to deal with the supplier. This is a simple cross-effect.

On an international basis, the cost of dealing with this externality may be higher, as the firm does not yet have the knowledge of the foreign environment. A greater market power may occur for the firm that has internalised against the downstream plants, through collusive agreements. As a result, it is considered that decisions concerning the acquisition and divestment of facilities typically involve the simultaneous internalisation or externalisation of several linkages (Casson, 1990).

The internalisation theory provides the basis to explain why a firm may wish to establish operations across national boundaries. Although it has recently expanded, a disadvantage of this theory remains to be mostly concerned with the sources of the TNC's internal or firm-specific advantages (Bartlett and Ghoshal, 1990, 1998). While all these transaction cost concepts are utilised to explain the internalisation process of the firm, the theory does not explain the types of external relationship that both TNCs and their subsidiaries create.

Industrial networks' approach

> Internalisation involves the absorption of other organisations or their tasks. At one end of the spectrum is what we may call the open market. At the other we find the firm which is relatively self-sufficient in terms of vertical or functional integration. In some ways these distinctions are analogous to Williamson's (1975) markets and hierarchies although he would include as part of 'markets' a number of in-between forms where we would rather apply the generic term NETWORKS (Thorelli, 1986: 37).

The industrial networks' approach gathers the work of many scholars interested in explaining the 'in-between' forms of relationships existing between market players. The analysis presented below is a record of the different themes tackled by industrial networks' theorists.

Formal/informal type of relationship. In the line of Thorelli's description of in-between forms of relationships beween firms, this research focuses on the concept of inter-organisational networks as a set of formal or informal contractual relationships among firms (Thorelli, 1986; Jarillo, 1988; Alter and Hage, 1993), and more specifically, in terms of obligational linkages such as subcontracting or referrals (Schneiberg and Hollingsworth, 1990). Networking refers to the activity of creating and maintaining organisational exchanges. Organisations engage in informational collection and exchange in order to reduce risk or uncertainty, or to share resources or expertise needed to develop innovative solutions, and networks provide the necessary vector for informational exchange (Robertson et al., 1996). This complex set of inter-linkages represents the network, and if the network goes beyond national borders, one may speak of an international production system (Morsink, 1998: 21).

International business comprises a formal type of relationship, with the establishment of JVs, licensing, co-production agreements and management contracts, and an informal type of relationship between a supplier and a customer who engage in a lasting, mutually advantageous relationship (Håkansson and Johanson, 1988). For the latter, Achrol (1999) follows the definition of a network organisation as distinguished from a simple network (or informal type of relationship) by the multiplexity and reciprocity of ties among its members, a sense of membership, mutuality and a system of social control. Miles et al. (1992) highlight the distinction between the various types of network organisation as being internal, stable and dynamic.

The 'internal network' captures entrepreneurial and market benefits, but the company does not resort much to outsourcing. In other words, most of the activities in the value added chain are carried out internally, and the company minimises its external relations. In a 'stable network', partial outsourcing is used and allows flexibility into the overall value chain. Several firms own various assets, all dedicated to a particular business. A set of vendors is nestled around a large 'core' firm, either providing inputs to the firm or distributing its inputs. The 'dynamic network' is a network in which a 'lead firm' identifies and assembles assets owned largely (or entirely) by other companies. Lead firms typically rely on a core skill, such as manufacturing, research and development (R&D) and design, design and assembly, or sometimes, pure brokering. The distinction fits best with the present study, to analyse the external network created by the TNC in a developing host country.

The TNC has a bundle of competitive advantages often unavailable to local indigenous firms. This dynamic network is also 'strategic', since the 'hub firm' (Jarillo, 1988) sets up the network, and takes a proactive attitude in the care of it. Such strategical importance is put forward in the case of supplier-customer relationships in the industrial market (Johanson and Mattson, 1988), where the central focus is not on products or on markets any more, but rather on buyer-seller relationship.

Whatever the type of network, the central idea stands in the existence of 'nodes', 'positions' and 'links' created by the interaction of firms. The network is not solely governed by abstract economic rationality, the key underlying notion is that of embeddedness (Granovetter, 1985; Yeung, 1998), that is ongoing social relations. At the individual level, social or personal networks usually refer to a collection of individuals where exchanges occur based on shared norms of trustworthy behaviour or simply as a support system that provides access to information or assistance (Zhara et al., 1998: 71)

Power and trust as central concepts of network analysis. There are two main points about firms being part of a network. First, the positioning of the firms depends on the power of the partner through its economic base, its technology, its expertise, trust and legitimacy. Trust may be viewed as the confidence in the continuation of a mutually satisfying relationship. It is based on reputation and on past performance, and it is built on personal friendship and social bonds. It is manifested by mutual feelings of belonging and interdependence.

While position is a structural concept, strategy is more closely associated with links or inter-positional relationships. Links may involve economic performance, technology transfer, diffusion of know-how and expertise, and forging or exploitation of trust and the flow of legitimacy (Thorelli, 1986). Each link therefore generates a bundle of flows, which are embodied in multiple 'bonds' (Boissevain, 1974) that relate the leading firm to its suppliers (bonds are fully described below).

Relations with external suppliers: discrete or close. A customer firm has two types of relations with external suppliers: discrete relations with money on one side and an easily measured commodity on the other, or closer relations between the partners built on network linkages (Dwyer et al., 1987; D'Cruz and Rugman, 1992, Dyer, 1996). The latter relational exchange includes key dimensions such as future collaboration being based on implicit and explicit assumptions, trust and planning.

Evolving relationships. A network being dynamic, these relationships are subject to evolution. In parallel to the four dynamic processes in the network membership that are the entry, positioning, repositioning and exit (Thorelli, 1986), relationships evolve through the awareness of the partner's existence, the exploration phase, the expansion of the relationship phase, the commitment phase with the durability and consistency of the relationship and finally the dissolution phase (Dwyer et al., 1987). The recent trend in purchasing is to reduce the number of key suppliers and to treat purchasing as a competitive element in the firm (Spekman et al., 1994; Lewis, 1995), suppliers become long-term business partners and the dissolution phase is less frequent. The 'evolving relationship' is important to understand why foreign firms would endeavour to treat local suppliers as business partners, which would involve closer relations between both partners.

Existing bonds between buyers and suppliers. When operating in a host country, the TNC has an impact on local firms. One way of evaluating the effect of the TNC on its local supplier is to concentrate on the information flows. As mentioned above, the notion of bonds linking customers to their suppliers corresponds to information flows in various areas, such as economic performance, transfer of technology, the diffusion of know-how and expertise and finally the creation of trust. Interactions between buyers and suppliers are based on business, social and information exchange. Bonds and investment are developed as a consequence of commitments and the use of resources in a specific purpose in order to achieve stability, closeness, security and long-term relationships with other firms in the network. These bonds are indicated in Table 2.1. They have often been referred to in the literature (Håkanson and Johanson, 1988; Holmlund and Kock, 1995), and help in the analysis of the existing backward linkages of firms.

While economic informational flows include the type of contractual relationships established between the two partners and the technological flows are easier to evaluate, functional informational flows are more difficult to measure quantitatively. These bonds depend on the establishment of trust between the partners.

Concepts of nodes, positions, power, trust and bonds/information flows lead us to alternative models of TNCs as open interaction networks willing to selectively share control, technology and markets when it is optimal to do so (Contractor and Lorange, 1988; Dyer and Singh, 1998; D'Cruz and Rugman 1992, 2000).

Towards a network theory of TNC

Keeping in mind the various particularities of TNCs and the inter-firm relationships we have already seen, the inter-organisational network paradigms draw upon the network theory, but are applied to the activity of the TNC and its FDI. In this recent literature, the multi-domestic firm is distinguished from the global firm.

Table 2.1 Existing bonds between buyers and suppliers

Information	Bonds	Concepts
Economic	1. Legal bonds	1. Contracts, legal ownership
	2. Financial bonds	2. Payment patterns, facilities
	3. Administrative bonds	3. Administrative routines and procedures of the firm
Functional	4. Planning bonds	4. Implementing management agreement, planning deliveries
	5. Knowledge bonds	5. Knowing about respective firms involved in relationship
	6. Social bonds	6. Getting to know employees of partner firm
Technical	7. Technical bonds	7. Technical adaptation of product, materials, equipment ...etc.
	8. Training bonds	8. Exchange of technological knowledge

Source: compiled by the author from Håkanson and Johanson, (1988), and from Holmlund and Kock (1995).

The location of activities is crucial, insofar as a country needs to be a favourable location for the development of a competitive advantage; and the number and type (upstream and downstream) of the plants is of particular importance in the overall strategy of the multinational firm. Those emphases come from the fact that for most TNCs, limited organisational capability represents the most critical constraint in responding to new strategic demands that are the increasing complexity of the international environment.

A TNC will have to integrate efficiency, responsiveness and ability to exploit learning if it wants to become successful (Bartlett and Ghoshal, 1987). In other words, the primary advantage of the TNC (as differentiated from a national corporation) lies in its flexibility to transfer resources across borders through a globally maximising network (Kogut, 1983). As large sophisticated TNCs have stimulated more effective local competition in national markets around the world, many of them have been forced to transfer more resources, technology and personnel abroad. They established more integrated local operations to protect export sales that had often become of vital strategic importance (Bartlett et al., 1990). These new trends have led to the new body of theory. A TNC becomes a complex entity comprising an inter-organisational network. The firm is embedded in an external network consisting of all other organisations such as customers, suppliers and regulators with which the different units of the multinational interact. (Bartlett and Ghoshal, 1990, 1998; Casson, 1990, 2000; Rugman and D'Cruz, 2000).

Bartlett and Ghoshal (1990) develop a model of a TNC as an internally differentiated inter-organisational network. A parent company has relationships and linkages with its subsidiaries, each of which has relationships with suppliers, buyers, regulatory agencies, and each of which competes for resources with an identifiable set of competitors. The overall relationships of the subsidiary are called its 'organisation set'.

Different subsidiaries may be interconnected, as a supplier to one subsidiary may be a supplier to another subsidiary of the same TNC. Because of such linkages among the different local organisation sets, all members of the organisation sets of the different units of the TNC collectively constitute the external network within which the TNC network is embedded.

There exist 'within densities', that is, the density of ties within each local organisation sets, and 'across densities', that is, density of ties within the total external network. 'Within density' can have some effects on the national organisation sets. Indeed, the density of linkages among key players in a national industrial context greatly influences industry performance and company strategy (Bartlett and Ghoshal, 1990: 264). Such linkages among the different actors may involve different kinds of exchange, such as those involving funds, people or information.

When 'within densities' are high, the company has to fragment its activities and locate more of the different kinds of resources in each market so as to provide the variety necessary to match the structures of the local organisation sets. In this case, dispersal increases while specialisation decreases, and vice versa.

'Across density' will also have an effect on the external network. If the 'within density' is low and the 'across density' is high, the TNC will have a low dispersal and a high specialisation, and its resources will be located according to resource niches in different countries. This means that overall, the resource-seeking concentration will decrease. On the other hand, if the 'within density' is high, together with a high 'across density', resource configuration will be influenced by the nodal characteristics of the complex external networks.

One of the strengths of this approach is to grasp the possibility that the country-specific advantages may actually contribute to the long-run development of new firm-specific advantages, through their leverage use in the corporation.

D'Cruz and Rugman (1992, 2000) have developed a similar model through which the TNC is seen as a 'flagship firm'. The analysis is based on the existence of strategic clusters, that is, groups of firms within a small geographical region, all of which participate in the same industry or a closely related group of industries. Cluster development can take place at a regional level, and its development can be intrinsically linked to the presence and activities of TNCs (Carrie, 2000; Enright, 2000). 'The business network' is defined as a web of strategic relationships that tie the members of a cluster together and are achieved through the harmonisation of the strategies of the firms within the cluster. The 'effective network' will be embodied by an overall structure, including suppliers, customers, competitors and the non-business infrastructure.

The authors differentiate network linkages from commercial relations between the flagship firm and its various partners. Network linkages involve the development of closer relations, due to two trends in the case of suppliers: first, many firms have decreased their overall number of suppliers; second, there is a general increase in the supplier's share of value added when the supplier becomes a network partner.

Despite such attempts to view the TNC as a network bundle, there is very little said about the specific relationships and the exact linkages existing between the various parties of the network. There is no comprehension of the dynamic between the links, although it is part of the overall strategy of the TNC to ensure long-term relations with partners, and optimal profitability. This represents the gap this research endeavours to fill. We now develop further the concepts studied in this section in our theoretical framework.

Approaches related to international trade theory

In the specific study of inter-firm relations between the multinational firms' subsidiaries in host countries with their local suppliers, the theories presented are useful to grasp a full image of the study. However, these do not provide us with a uniquely international theory of the firm. The interaction between the activities of the TNC in a host country with the development process of the latter are specifically tackled by approaches based on the international trade theories. In this section, we cover the competitive advantage of nations, the asset accumulation and restructuring paradigm and the stages of growth approach, but first we emphasise the theoretical background of these paradigms.

The product cycle theory (Vernon, 1966, 1971) first pointed out the importance of the life cycle of the product to give an explanation for the international production. The initial production location is decided when the new product is launched on the market. As the product reaches maturity, the firm has to choose the way to supply the foreign market: it can be by export or direct investment in the targeted country. Following these steps, when the product is standardised, foreign and domestic production compete against each other (Vernon, 1966). This theory was later refined to include oligopolistic behaviour of firms (Vernon, 1971). The main virtue of this theory is to include factors of supply and demand, thus integrating the interaction between the firm and the market. However, the investments that are not linked to substitution of exports remain unexplained in this theoretical setting. Still, for some LDCs, the theory can provide a useful insight to explain the geographical distribution of the production processes.

Some paradigms give a general frame to explain the interactions between direct investment and economic development. They are useful in providing a global view for the present study. Various analytical frameworks can be given to identify the impacts of multinational firms on the host country's economy. Overall, they all follow the principles already enunciated by Ricardo (1772-1823) on the need for a structural change of an economy during the development from the primary sector to the manufacturing sector.

Competitive advantage of nations

The Porter (1990, 1998) model of the diamond of competitive advantages is based on four country-specific determinants and two external variables that are chance and government. The four determinants are natural resources, the type of production the consumer is asking for, the presence of local suppliers and the rivalry between firms, the domestic firms' strategy and structure and the nature of domestic rivalry. The four determinants and the two outside forces interact in the 'diamond' of competitive advantage, with the nature of a country's international competitiveness depending upon the type and quality of these interactions. The two outside forces provide the missing elements, the government is critical as an influence on a direct entry barrier penalising foreign firms or it can use subsidies as an indirect vehicle to penalise foreign-based firms.

In earlier work, Porter differentiated the multi-domestic industry from the global industry. The multi-domestic industry is one that is present in many countries, but in which competition occurs on a country-by-country basis. In this case, the competitive advantages of the firm are largely specific to each country. The global industry is an industry in which a firm's competitive position in one country is significantly influenced by its position in other countries. The international industry is a series of linked domestic industries in which the rivals compete against each other on a truly world-wide basis; such industries include electronics products (Porter, 1986). These descriptions relate to the idea developed in the inter-organisational network paradigms; indeed, a global competitor must view its international activities as an overall system, but still has to maintain some country perspective.

In the value chain, the procurement activity of firms is part of the support activities, together with firms' infrastructure, human resources and technology development. Procurement is the obtaining of purchased inputs, such as raw materials, purchased services, machinery, and so on. Procurement stretches across the entire value chain because it supports various activities. There are typically many different discrete procurement activities within a firm, often performed by different people (Porter, 1986: 20).

The purpose of this research is to see how support activities intertwine. A firm has a stock of know-how about how to perform each activity. Technology development involves a variety of different discrete activities. Some of those are performed outside the R&D department, some for the human resource management and some for the firm's infrastructure (that is, accounting, legal, finance, strategic planning and all the other activities decoupled from specific primary or support activities, but essential to enable the entire chain's operation). The diamond is useful when one is using locational advantages to explain the international production when these advantages are part of the TNC success in the host country, and when subsidiaries' activities help to promote the technological and organisational efficiency of local firms. It has been used and extended by numerous scholars (Davies and Ellis, 2000). Inward and outward investments may affect the configuration of the diamond (Dunning, 1994, 1997). However, the scale of the impact of FDI on the competitive advantage of nations or on the evolution of this advantage is not yet proved, despite recent research focusing on the topic (Narula, 1996).

Asset accumulation and restructuring paradigm

This paradigm is an extension of the model of cumulative technological change. The nature of technological accumulation comes from the firm itself; it is cumulative and differentiated (Pavitt, 1988). This concept can be used to explain the sectoral evolution of comparative advantages of the innovative activities of domestic firms (Cantwell, 1989) with organisational resources and assets restructuring. Taken after this paradigm, the TNC expansion from LDCs or NIEs is schematised by three sequential stages of the geographical diversification of the domestic firms' activities. In stage one, the new TNC concentrates on simple activities (raw material extraction, less capitalistic manufacturing activities, and so on) to start with. The investment is carried out in neighbouring and/or ethnically close countries. This stage declines as the firm acquires experience in international production, and as the home country reaches a higher level of industrial development. In stage two, the activities of the TNC spread towards other developing countries that are geographically or ethnically close. The forms of investment remain basic. The specific advantages of those TNC lie in products or adapted innovation processes applicable to other developing countries that are on a lower level of development. In stage three, the sectoral distribution of the international production processes improves in favour of more sophisticated manufacturing investments. The investment takes place in countries with higher geographical, economic development and psychic distances. Overall, the rise in technological knowledge and in the number of TNCs from developing countries is linked to the domestic industrial development, and to the accumulation through which firms create their own technological experience (Tolentino, 1993, 2000).

Stages of growth paradigms

Several paradigms on the interaction between FDI and host country economic development have been established. The eclectic paradigm (Dunning, 1988a, 1993a, 1997, 2000) suggests that as a firm invests in a country, this capital inflow has an influence on the existing advantages of the host country (that is, the ownership advantage, the location advantage and the internalisation advantage). Recently, Dunning (1993b) describes a more dynamic process than the stages of development, but he only includes the rise of domestic foreign investment and per capita GDP. Although strategic considerations are taken into account, the means through which FDI is actually linked to the development of local foreign investors remain unexplained.

The industrial restructuring paradigm, on the other hand, describes the interaction between TNC activity and economic development and/or the restructuring of countries (Balassa, 1989) among the Asian countries, which is said to be following the 'flying geese' formation pattern. This pattern first originated in the work of Akamatsu (1962) who describes economic development in terms of domestic and international stages with countries following each other's pattern in the shape of a reversed 'V', the 'flying geese' pattern. In this model, it is the capital accumulation, together with forward and backward integration of firms, which allows for the change in comparative advantage of countries, and in addition, allows for the upgrading in industrial structure (Fukushima and Kwan, 1995; Dowling and Cheang, 2000; Kojima, 2000; Sikorski and Menkhoff, 2000). This emphasis on forward and backward linkages was given little attention in most models. Ozawa (1991, 1992) uses the stages of the restructuring model to explain the changing role of inward or outward investments in the accomplishment of the economic development. The first stage of development corresponds to the multiplication of industries of the Heckscher-Ohlin type; they are labour-intensive and lead to the elementary stage of foreign relocation. The second stage of development concerns the rise of non-differentiated Smithian industries, which allow for large economies of scale. A new type of investment arises, because of the lack of natural resources and the need for raw material. This multinationalisation stage corresponds to the Ricardo-

Hicks gap. The third stage of development is characterised by differentiated Smithian industries with mass production of durable goods. This is the export-substituting-cum-surplus-recycling stage of multinationalisation. It corresponds to the rise in assembly-based industries, which are less dependent on resources, and essentially cover electronic and car goods. The fourth stage of development shows the rapid spread of technological innovation and the Schumpeterian industries. While describing the development of Japan, Ozawa (1992) identifies five major sources of competitive advantages for the Japanese success in the automobiles and electronics goods. These were the effective use of multi-layered subcontracting, the practice of the just-in-time (JIT) inventory system and 'total quality control', the trade of new technology from Japan R&D, the good feedback of information from domestic consumers and the vigorous inter-firm competition among Japanese manufacturers. The experience of each country's development is different. The object of this study is to further the first source of success to other developing countries.

The theories described above have provided a framework of analysis for the impact of foreign firms' operations on host countries through FDI. In the following section, we concentrate upon both the economics of the firm, and the economics of trade theories to investigate the process by which TNCs have an impact on the host countries' production capabilities through backward linkages.

CRITICAL ANALYSIS OF THEORETICAL CONCEPTS RELATED TO BACKWARD LINKAGES

In this section, we summarise and further elaborate on the theories and concepts presented earlier in the chapter. The heterogeneous nature of this research means that it is difficult to follow the theoretical stream alone. This is the reason why, following the analysis of the major theories in the previous section, we now concentrate on establishing the key concepts to be considered in this research using, as appropriate, aspects of the theories discussed, as well as additional ideas. We first examine the notion of backward linkages as networks created by TNCs in host countries, followed subsequently by an examination of transfers to local suppliers. Finally, we focus on the impact of TNCs on host countries' level of economic development.

Backward Linkages as TNCs' External Networks in Host Countries

In order to understand fully the process through which TNCs initiate and conduct backward linkages in host countries, relevant concepts are borrowed from approaches related to the theory of the firm. Table 2.2 summarises, for each theory described earlier in the chapter, the central idea, key concepts for the analysis of backward linkages and influential factors identified.

In this research, backward linkages are defined as external networks created by TNCs in host countries, leading to inter-firm transfers. To grasp fully the complexity of this process, the TNC is first defined as a lead firm, its subsidiaries' role in the network are then commented upon, before emphasising the inter-firm network strategy and the influential factors affecting these networks.

The TNC is identified as a lead firm in a particular business environment because it possesses proprietary assets that it tries to exploit internally by extending its own organisation and control across national boundaries (Buckley and Casson, 1976). To overcome market failure and the subsequent cost of failure, the firm favours internalisation of its activities across borders. Internalisation eliminates uncertainties, inasmuch as the firm can ensure quality of its final product and regular suppliers with variations in prices. Internalisation of activities is also explained in terms of the transaction cost theory. Transactions can take place through a centralised hierarchical organisation or through discrete market exchanges. The firm needs to decide on its efficient boundary; this can be adjusted over time, as the firm gains experience, as asset specificity decreases and marketing contracting increases.

Whereas the internalisation theory best explains why firms internalise their activities across borders, it does not help in understanding the process through which the subsidiary's local links with suppliers are determined. There is the need to clarify what surrounds the firm, once it has internalised its activities across borders. The lead firm[1] is one around which organisational sets (Bartlett and Ghoshal, 1990, 1998), strategic clusters (D'Cruz and Rugman, 1992, 2000) or a system of activities (Casson, 1990, 2000) can be defined. All these terms are referred to as external networks created by the subsidiary, once activities have been internalised by the parent firm. The lead firm might be in a position of power in a developing country perspective. The positioning of the firm depends on its power, embodied by its economic base, its technology, its expertise, trust and legitimacy (Thorelli, 1986).

Table 2.2a Theoretical concepts and backward linkages: approaches related to theory of the firm

	Transaction cost theory	Internalisation theory
Central idea	Economy as network of trading relations and intertwined by contracts between and within firms	Relative costs and benefits of internal organisation compared to market transactions to explain international integration of value-added activities
Key concepts for analysis of backward linkages	1. *Institutional arrangements:* governance structures linked to bounded rationality and opportunism 2. *Efficient boundaries*: (make/buy decision) between discrete market exchange and centralised hierarchical organisation 3. *Vertical integration*: information impactedness 4. *Transaction characteristics*: asset specificity, uncertainty, frequency	1. *Proprietary assets*: firms concerned with market failure and linked cost, i.e. uncertainty 2. *Locational theory*: necessary complement 3. *Vertical or horizontal integration*: tangible versus intangible flows. Position in value added chain 4. *Psychic distance/level of trust*: influence inter-firm network 5. *Entrepôt potential*: essential for host country to maximise backward linkages 6. *Systems' view/cross-effect*: activities and markets versus facilities and linkages
Influential factors	1. *Time*: linked to learning curve and adjustments 2. *Industry*: whether mature or not, influences the network of relations	1. *Experience*: managerial learning process on international activities 2. *Nature of flows*: tangible products or intangibles

Table 2.2b Theoretical concepts and backward linkages: approaches related to theory of the firm

	Industrial network approach	Approaches linking networks and TNCs	Towards an integrative approach
Central idea	Emphasis on 'in-between' forms of relationships existing between market players. Networking as activity of creating and maintaining organisational exchanges	Focus on firms with international and regional networks and subsequent organisation	Backward linkages as network created by TNCs in host countries, and subsequent transfers
Key concepts for analysis of backward linkages	1. *Formal/informal relations*: lead to obligational linkage and information exchange 2. *Network*: internal, stable or dynamic within nodes, positions and links 3. *Power and trust*: determine position and strategy 4. *Discrete or close relations*: exchange or network linkages 5. *Evolving relations*: entry, positioning, repositioning, exist through awareness, exploration, expansion, commitment, dissolution 6. *Existing bonds*: commitment for stability, closeness, security, long-term relation	1. *Primary advantage of TNCs*: flexibility to transfer resources (can be called flagship firm, or core firm [...] within strategic clusters) 2. *Organisational set* (effective network): subsidiary (or parent), customers, suppliers, regulators (or non-business infrastructure), etc. 3. *Within and across densities*: links within a set or across all organisation sets of one TNC	1. *TNC as lead firm*: possesses proprietary assets and creates internal and external networks across borders → gives it a lead position in a developing country environment 2. *Subsidiaries*: create external network locally and across borders 3. *Inter-firm network strategy*: types of contracts, existence of bonds
Influential factors	1. *Time*: linked to type and position of network 2. *Bonds*: for technology transfer, technical and training bonds	1. *Location*: of organisation set	1. *Time*: international and local experience 2. *Industry*: mature or not 3. *Location*: entrepôt potential

External linkage creation depends on psychic distance and the level of trust in the host environment (Johanson and Mattson, 1988; Casson and Cox, 1992/93) and on the entrepôt potential of the country (Buckley and Casson, 1991). Internalisation theory provides useful insights into the intra-firm relationships across national borders.

The nature of the network is best defined by borrowing concepts from the transaction cost theory and the industrial network approach. If there is small asset specificity and little uncertainty in the market, transactions are frequent and are governed by the market (Williamson, 1985, Williamson and Winter, 1991). The price mechanism conveys the necessary governance information. What the transaction cost theory does not describe is why some suppliers enter as part of a strategic network, where close partnership with the buyer is established. The industrial network approach helps to understand the mechanism through which the subsidiary creates an external network in the host country. The external network is composed of discrete market exchanges in the transaction cost perspective, and of in-between forms of relationship. These in-between forms of relationships, as defined by Thorelli (1986), are the focus of this research, since they often lead to information exchange between parties. The network is the necessary vector for informational exchange between firms (Robertson et al., 1996), yet not all suppliers benefit from close relationships with the subsidiary.

The external network is, following the classification defined by Miles et al. (1992), stable or dynamic. In the stable network, partial outsourcing is used and allows flexibility in the overall value chain. In the dynamic network, the lead firm identifies and assembles assets owned largely (or entirely) by other companies. The subsidiary may also have an internal network, with goods internally exchanged within various entities of the firm. The balance between the internal and the external networks appears clearly in the Miles et al. (1992) categories.

The external network is strategic, as the lead firm will take a proactive attitude in the network (Jarillo, 1988; Johanson and Mattson, 1988). The external network strategy adopted by the subsidiary is closely associated with inter-firm links' creation, or network linkages (Dwyer et al., 1987; D'Cruz and Rugman, 1992), involving exchange of information. The external network evolves over time, so do the relationships within it between the entry/awareness phase and the exit/dissolution phase (Thorelli, 1986; Dwyer et al., 1987). As suppliers have become more strategic within the firm, the dissolution phase is less frequent (Spekman et al., 1994; Lewis, 1995). Yet, one question is not answered by the industrial network approach. When would the bonds' creation be most likely in the evolution of the relationships? By focusing on technology and knowledge transfer, we know that technical bonds are important and comprise technical bonds (see Table

2.1) and training bonds. What remains unanswered is how these bonds emerge in the external networks' creation, and how much information is included within these bonds or transfers. Further, in the Miles et al. (1992) concept of internal versus external networks, the concept of intra-firm exchange of goods is important to define the level of external linkages created by the subsidiary. While the intra-firm exchanges are taken into account, with the distinction between the internal and the external networks, the clear mechanism that may link internal and external networks is not defined. To what extent does the internal network of the firm (that is, the company does not resort at all, or does not resort much to outsourcing, versus the company that does rely on outsourcing and on its external network) influence the external network of the firm and subsequent transfers?

Backward Linkages and TNCs' Impact on Host Country

The inter-firm relationships are now defined as the external network of the firm. The next question addressed in this research is that of its impact on the host economy. Approaches related to the international trade theory take a macroeconomic perspective on the impact of firms in host countries. The unit of analysis is FDI, rather than individual firms' behaviour. Table 2.3 summarises the key concepts for the study of backward linkages borrowed from the competitive advantage of nations, the asset accumulation and restructuring paradigm and the stages of growth approach.

Porter (1990) mentions four country-specific determinants that are the factor conditions, demand conditions, firm strategy/structure and rivalry and related and supporting industries. In this context, the TNC can influence the competitiveness of the host country by influencing the competitiveness of local suppliers. The specific nature of the TNC-supplier relationship is not clearly analysed by Porter, as he emphasises contractual relations within the economy as a whole, with no focus on TNC *per se*. Dunning (1994) mentions that inward and outward investments may affect the configuration of the diamond, but the specific links are not described, nor is the scale of the impact of FDI on the competitive advantage of nations (Narula, 1996). Porter, however, provides a structure for the country's competitive advantage, and within this structure, the related and supporting industries and the suppliers appear as a key element on which TNCs can have an impact.

The level of technology of the host country is key to its capacity to develop. The asset accumulation and restructuring paradigm puts the emphasis on country and company technological accumulation (Pavitt, 1988; Cantwell, 1989). This can be compared with innovation as a source of competitiveness in Porter's work. Taking the example of developing countries' firms, Tolentino (1993) shows that the rise in technological knowledge and in the number of TNCs from developing countries is linked to the domestic industrial development, and to the accumulation through which firms create their own technological experience. The level of technology is important at the level of the country as well as at the level of the firm. In the stages of growth approach, Ozawa (1991, 1992) links the stages of economic growth with types of industries. The fourth stage of development is linked with technological innovation and Schumpeterian industries, hence the local technological capabilities.

There are levels of analysis to be borrowed from these approaches. First, FDI may impact on the types of industry in the host country, but also the level of technological knowledge in the host economy determines the degree to which technology may be assimilated. Second, depending on which industry is concerned, TNCs show various levels of technological knowledge, and the interaction with suppliers may be affected by this level of technology.

The exact process through which TNCs impact on the technological capabilities of local suppliers is not described by the approaches related to the international trade theory. However, the concept of technology at the national and firm levels, and the concepts of stages of growth and links with FDI are to provide the basis of the analysis of the impact of TNCs on host economies through backward linkages.

Influential Factors

Tables 2.2 and 2.3 summarised the key concepts and ideas raised by various theoretical streams. Additional factors are worth mentioning for their particular impact on the concepts defined in the theories.

Time is very important in the analysis of backward linkages. First, as the TNC internalises its activities across borders, time influences decision inasmuch as it is part of the experience process of the TNC. Experience might refer to the international experience, which is linked to the number of activities the firm has overseas or to how long the firm has had international operations. Experience might also be related to the accumulated knowledge on the host country itself. Second, contractual relationships vary over time, just as the type of networks created by the TNC in the host economy. Finally, time also intervenes when talking about economic development, and stages of economic development. Although each country may not follow the same development path, and some may develop faster than others, the concept of time is present in the analysis.

Table 2.3 Theoretical concepts and backward linkages: approaches related to international trade theory

	Competitive advantage of nations	Asset accumulation and restructuring paradigm	Stages of growth approach	Towards an integrative approach
Central idea	Countries, governments and even private industry can influence conditions within a country to aid competitiveness of firms	Emphasis on country and company technological accumulation	Interaction between FDI and host country economic development	TNCs' backward linkages and impact on host country's level of economic development
Key concepts for the analysis of backward linkages	1. *Country-specific determinants*: factor conditions, demand conditions, firm strategy/structure and rivalry, related and supporting industries 2. *External variables*: chance and government 3. *Suppliers*: importance of close working relationships, proximity to suppliers, timeliness of product and information flows 4. *International industry*: series of linked domestic industries 5. *Innovation*: source of competitiveness	1. *Technological accumulation*: from the firm, cumulative and differentiated. Linked to comparative advantages and innovative activities 2. *Stages of international expansion for NIE companies*: Stage 1: simple activities in neighbour countries Stage 2: spread to other developing countries Stage 3: sophisticated manufacturing investment 3. *Firms' competitiveness* linked to domestic industrial development and technological accumulation within the firm	1. *Countries within international framework*: 1.Domestic and foreign investment/inward and outward investment 2.Domestic and international development stages 2. *Stages of economic development and FDI*: Stage 1: Heckscher-Ohlin type industries Stage 2: non-differentiated Smithian industries Stage 3: differentiated Smithian industries Stage 4: technological innovation and Schumpeterian industries	1. *Supporting industries*: TNC can influence competitiveness of local suppliers 2. *Level of technology*: of host country, of firms in host country and of TNC 3. *Stages of growth*: of host country and interaction between TNC and local suppliers
Influential factors	*1. Industry and product level*	1. *Absorption capacity* of firms and countries 2. *Time*: levels of technological development	1. *Industry*: specificities and impact on local endogenous production capabilities 2. *Location*: development level of neighbour countries	1. *Industry* and type of *product* manufactured 2. *Time* 3. *Industry* 4. *Location*

The type of industry is looked at to explain the extent of external contractual relations created by the firm. Following the transaction cost theory, in mature industries such as electrical appliances, there is an increased use of subcontractors, JVs or cooperative agreements. The level of technology embodied in the industry may determine the relationship between FDI and level of economic development, in the stages of growth approach. Thus, the industry, but more specifically the level of technology, requires particular attention.

The location factor is key to the explanation of where the firm may wish to internalise its activities. Hence, the entrepôt potential and technological capabilities also constitute a level of analysis for the study of backward linkages and their impact on host country. The location includes the neighbouring environment, and the level of development of countries close to the host country.

All these concepts and factors are now integrated in a theoretical framework as a basis for the research of TNCs, backward linkages and their impact in host countries.

TOWARDS INTEGRATIVE VIEW OF BACKWARD LINKAGES: HOW TNCS PARTICIPATE IN UPGRADING PRODUCTION CAPABILITIES OF LOCAL SUPPLIERS

Impacts of TNCs on host economies are wide ranging, they can affect the host country's balance of payments, the local investment rate and stock of capital, the employment level and local firms' competitiveness. TNCs may cover for market imperfections, for instance by bringing resources difficult to find in specific markets, such as technology, management know-how, skilled labour, access to international production networks, access to major markets and established brand names. The foreign firm can impact directly on the economy through its own activities, or indirectly through activities that might be generated in the rest of the economy as a result of its operation. Hence, in terms of technology transfer, the use of new technology by the subsidiary is only one benefit. Other benefits can result from the diffusion of technology and skills to the rest of the host economy. Diffusion of technology cannot be priced and paid for. The spillover effect would affect a variety of actors in the host economy, such that technology diffusion would occur among suppliers and customers, non-related firms and/or competing firms in the host economy. While we acknowledge the importance of spillovers generated by TNCs in host countries, this research focus is on transfers through backward linkages.

In previous sections, we elaborated on the theories relevant to the study of the TNC and backward linkages in host countries. Concepts and approaches are combined into an integrative view, to provide a theoretical framework for our study. This section first presents the comprehensive view of backward linkages. After relating backward linkages with the TNC and with the upgrading of a host country's production capabilities, the framework for analysis is designed.

Comprehensive View of Backward Linkages

Backward linkages by TNCs in host countries depend on the existing international, regional and national succession of production processes of the firm. This includes internal and external linkage creation by TNCs at each level of analysis.

Succession of production processes

Figure 2.3 is a schematised representation of the global view of backward linkages drawn from all the concepts tackled in the relevant theories. The creation, substitution and/or succession of production processes within and outside the TNC show the analytical steps to be followed in the exploration of the TNC impact on host country suppliers' capabilities. The established succession of production processes depends upon a number of factors, including the time (such as the time spent in the host country, and the economic environment at a particular time) and the industry concerned.

From internal linkage creation…

The parent firm exercises power and control on its subsidiary firms, as embodied in hierarchical relations established within the TNC network. Direct control of the subsidiary is possible through the ownership structure and the organisational structure of the TNC. The dotted arrow linking the parent firm to its local subsidiary, once FDI has taken place, includes the various aspects of this parent-subsidiary relationship. Preliminary aspects to the new internal linkage refer to the international vertical or horizontal integration of the firm in order to avoid market failures and uncertainties. The parent firm possesses proprietary assets, which give it specific ownership advantages. Such assets are key advantages for the TNC, particularly in developing host countries, where technological, managerial and organisation assets tend to be less refined. In the early stage of the investment process, the proprietary assets of the parent firms enable it to exercise substantial control over its subsidiary, often implying a very limited decision-making power.

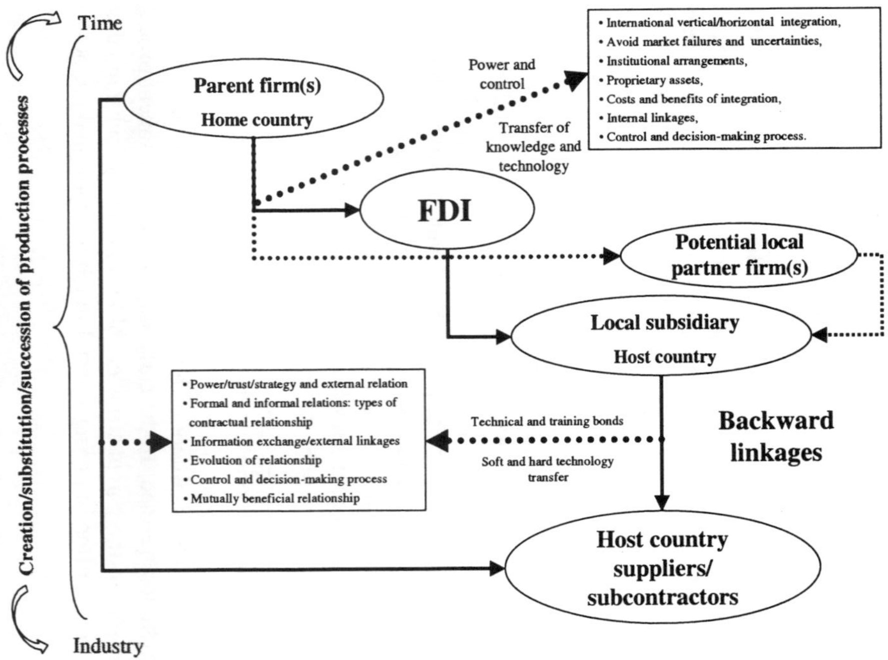

Figure 2.3 Comprehensive view of backward linkages

Such a limited decision-making power includes supply-related decisions, and a high level of internal supply is common when a subsidiary is initially established in a host developing country. Supposedly, the rational behaviour of the parent firm leads to maximisation of the benefits of integration. Internal linkages are highest at an early stage of integration. Issues of market knowledge, psychic distance and limited trust in an unknown environment influence the parent firm to exercise this close control on its subsidiary's activities.

...to external linkage creation

Once established in the host economy, the subsidiary firms will start generating links with local businesses. The extent of external links' creation will depend to a large extent on the power and control exercised by the parent firm, as knowledge creation in the local environment increases gradually, and local linkage creation is dependent on the level of independence of the subsidiary in the decision-making process. The level and extent of backward linkage creation is dependent on the existing succession of production processes. The production process is a dynamic structure, which changes according to various factors such as time and· industry, and the resulting linkages created will therefore be dynamic. The transfer of knowledge and technology schematised by the dotted arrow takes effect when local subsidiaries or the parent firm start generating contractual relations with the host country's suppliers and subcontractors. The level and relevance of these transfers will be dependent on a number of factors, but these are likely to be substantial when they operate between the locally established subsidiary rather than directly through the parent firm.

The external linkage creation follows various steps. The first step is dependent upon the level of trust and knowledge accumulated by the subsidiary, as it operates in the host country. The first step also comprises the specific advantages of the subsidiary as compared to local firms, and the resulting power it can exercise on partner firms. From the level of trust and power derives the purchasing strategy adopted by the subsidiary and/or the parent firm, if substantial independence in decision-making is not granted.

The second step in external linkage creation is the formation of formal or informal relations with local firms. External linkages' creation does not necessarily imply informational exchange between partner firms. When the linkage creation is not binding for either of the firms, a simple exchange of goods can occur without further implicit or explicit obligations. In this case, informal relations will not lead to extended linkage creation. The net benefit for the host country is of a monetary nature alone; the first step can then be repeated. On the other hand, the linkage creation can be binding for both parties if long-term relations are intended/created. The subsidiary may wish to maintain stable relationships with its local partners, or the local partners may posses sufficient proprietary advantages to generate interest in the formation of a formal linkage. In this case, the resulting external linkage creation is linked to the exchange of information.

Step three of external linkage creation is a dynamic process between partners. The relationship follows an evolutionary pattern. In step two, some links with suppliers may be solely monetary in nature, and may not involve substantial interaction between the business partners. In step three, although information exchange may occur, a long-term relationship is not always applied. The relationship may evolve and both partners may end their relations, which would lead to a stop in the information exchange. On the other hand, when stable long-term relationships are created, increasing information flows may occur, sometimes leading the subsidiary to exchange power over its suppliers, hence, the possible overtaking of decision-making with the subsidiary exercising control over key organisational decisions of the suppliers.

All along the external linkage creation process, it is assumed that links and information flows are the result of a mutually beneficial relationship. Exchange of information and practices occur as both parties are satisfied in the positive outcome of the relationship.

Backward Linkages and the TNC

Suppliers, customers, competitors and the non-business infrastructure all intervene in the TNC portfolio of linkages' creation. Supply of goods can originate either from the intra-organisational linkages of the TNC as schematised in Figure 2.3 or from the inter-organisational linkages. A closer relationship implies information flows between the customer and its supplier. The various types of relationships will vary according to the succession of production processes and the steps reached by the partners. The formal relationship established by the subsidiary can take the form of subcontracting agreements or partnership agreements; these are schematised as the key subcontractors and the key suppliers in the inter-organisation linkages portfolio.

Exchange of goods is supplemented by exchange of information with these key partners. Figure 2.4 shows how the TNC may be the key generating force behind a complex succession of linkages, often associated with information exchange.

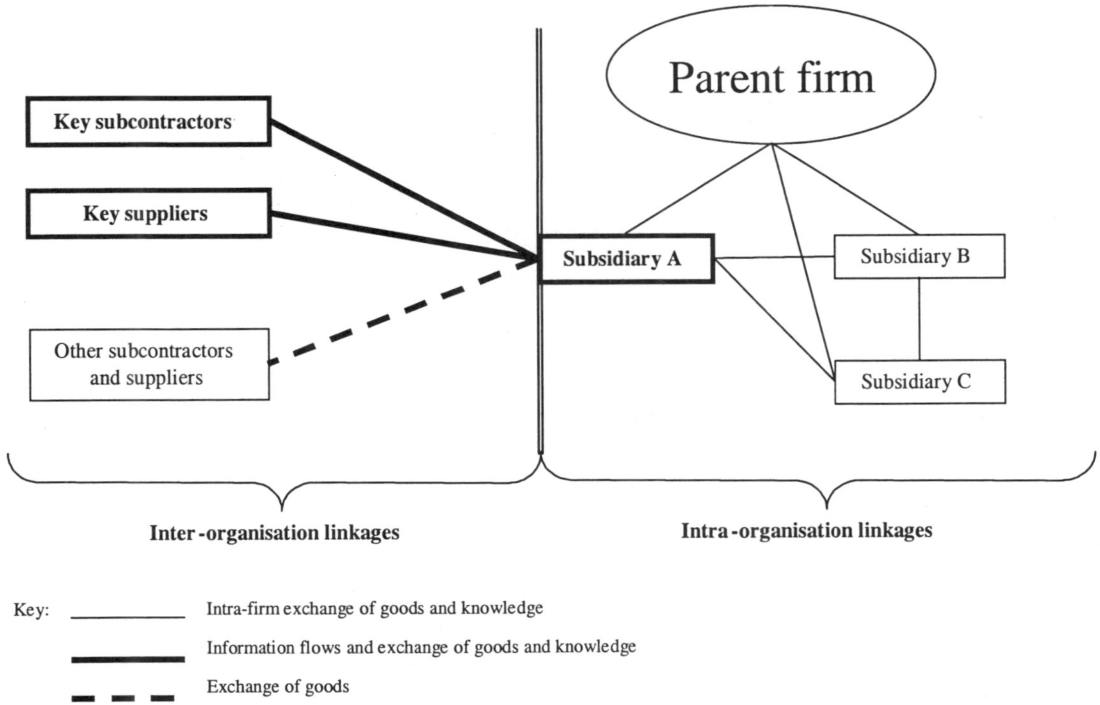

Figure 2.4 TNCs and backward linkages

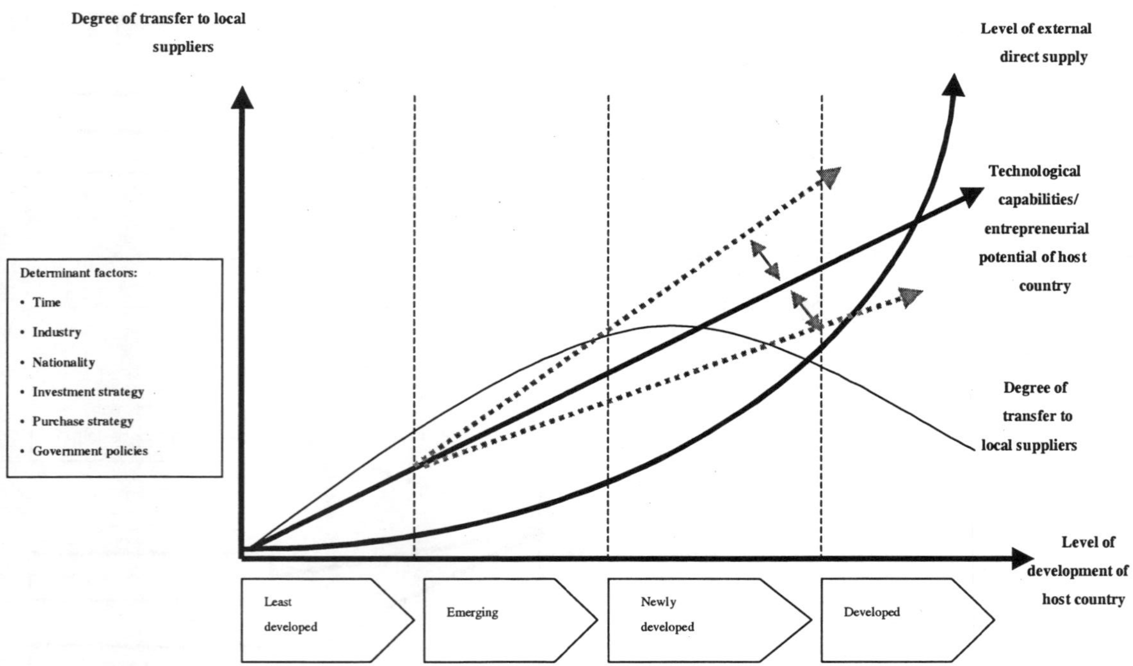

Figure 2.5 Stages of growth and backward linkages

Backward Linkages and Upgrading of Host Country's Production Capabilities

The description of inter-organisational linkages leads us to investigate the impact of the linkage creation on the host country. Figure 2.5 shows how various factors come into consideration when trying to perceive the impact that a TNC may have on the capability development of the host country through backward linkage creation. The vertical scale shows the level of development of the host country from the least developed countries to developed countries. The level of development is closely linked to the level of technological capabilities and the entrepreneurial potential of the host country. The level of technological capabilities in the host country may not follow a straight line, but may vary according to the policies adopted by the government. Similarly, the slope of the line can shift at any point in the development process of the host country, because, for instance, of the government policy, past TNC involvement and previous transfers. The shift in the slope has subsequently a proportional influence on the knowledge and technology transfer line.

The level of external direct supply, that is the supply of key parts and components, also increases with the level of technological capabilities and the entrepreneurial potential of the host country, as greater reliance on the local suppliers' capabilities is possible. The degree of knowledge and technology transfer appears on the vertical axis. The overall level is small, but the degree of transfer varies according to the entrepreneurial potential and the existing technological capabilities of the host country. There are a number of considerations to be taken into account by the TNC in the decision to transfer knowledge and in the implementation of information exchange.

Of course, if the technological potential in the host country is too small, no substantial linkages can be created. However, as the host country's capabilities increase, so does the potential level of information exchange between the subsidiary and its suppliers. The nature of the knowledge and technology transfer and information exchange changes with the level of development. As the technological capabilities of the local companies increase, there is less need for transfer, but instead exchange of ideas and joint projects/designs may occur. Thus, the level of knowledge and technology transfer does not disappear completely, even when the country is developed.

The degree of technology transfer is not a uniform phenomenon. The degree will vary according to the industry concerned, and the extent to which external sourcing is used. The nationality of the firm, or its region of origin, is a key factor, inasmuch as the TNC may also originate from a developing country. The different levels of power and control applied by the parent firm on its subsidiary explain transfer too. Other important factors include the level of trust in the host environment, and the length of time the subsidiary has been operating in the country. Finally, the investment strategy and the purchase strategy of the firm also account for differences in the level of the transfer.

Established Framework of Analysis

While Figures 2.3 and 2.4 give the framework for the analysis of backward linkages and related knowledge and technology transfers, Figure 2.5 helps in drawing preliminary research questions for the empirical analysis of knowledge transfer through backward linkages. As indicated in Figure 2.5, the degree of knowledge transfer varies according to the level of development of the host country. We discussed earlier the different factors likely to influence the slope of technology development and supplying pattern, and their subsequent impact on the host country's development. The determinant factors are time, industry, nationality of the firm, investment strategy and purchase strategy followed by TNCs, and government policies.

In the study of backward linkages, TNCs and impact on host countries, we first focus on the level of local direct sourcing. Second, attention is drawn to the factors likely to influence the subsequent transfers to local suppliers, leading in turn to the improvement in the technology capabilities and entrepreneurial potential of the host country.

CONCLUSION

In this chapter, definitions for key terms used in this research were provided. A comprehensive view of backward linkages was then developed, following the analysis of a wide spectrum of existing theories related to the economics of the firm and the economics of international trade. First, the transaction cost theory, internalisation theory, industrial networks' approach and attempts towards a network theory were looked at. These theories helped to understand TNCs and inter-firm relations. Useful insights were derived about the mechanics through which TNCs operate overseas and create inter-firm relations in their home and host countries, and in third countries. Second, competitive advantage of nations, asset accumulation and restructuring and stages of growth paradigms were covered. These paradigms were useful to grasp the full impact that TNCs have on the upgrading of a country's productive capabilities through backward linkage creation.

From the theories mentioned above, an integrative approach was adopted, and a framework of analysis provided. We looked at the context within which backward linkages are created, and designed a contextual framework for potential transfers between subsidiaries in a host country and local suppliers and subcontractors.

In the next chapter, the analysis of empirical research on the impact of TNCs in the ASEAN countries, and the findings of the previous studies conducted on the interaction between foreign subsidiaries and their local suppliers will provide further evidence on the existence of the backward linkages and their potential links with the upgrading of production capabilities in a host developing country.

NOTE

1. Depending on the author, the firm might be called the flagship firm (D'Cruz and Rugman, 2000), the core firm, or the hub firm (Jarillo, 1988). In this case, the term 'lead' firm is not intended to fit in with the dynamic network description of Miles et al. (1992), but rather is understood as a single term used to include all the above-mentioned words for the 'central' firm in the network.

3. TNCs and their Impact in Host-developing Economies: the Case of ASEAN Countries

INTRODUCTION

Transnational corporations (TNCs) by their activities in host economies have an impact on the local economy. This chapter provides the overall picture of TNCs' impact on host economies, in particular the case of Association of Southeast Asian Nations (ASEAN) countries; hence it does not focus on Malaysia only. The rationale for this is that the TNC concept must be analysed in its broader context for a full understanding of the phenomenon. The requirements for a beneficial environment are described first, as the benefits arising from the presence of TNCs in host countries are likely to vary according to the conditions in which the foreign firms operate. The overall environment includes the political factors surrounding the firms, the increasingly crucial structural factors and the changing profile of the TNCs' global activities and their firm-related factors; each of these groups of factors being very important in the determination of TNCs' impact on the host country. The strategic profile of locational factors is then analysed; this encompasses the host country factors, the external resources and the role of foreign direct investment (FDI) in their promotion. The analysis of the environmental factors precedes the description of the wide variety of potential impacts of TNCs on host countries' production capabilities. This broad contextual framework helps us to understand that the TNCs' impacts, and the backward linkages in particular, depend upon a number of factors external to the firms. These must be understood before studying the impact of TNCs through the creation of backward linkages, because most of these concepts subsequently apply to buyer-supplier relations.

Finally, the upgrading of production capabilities through the creation of backward linkages between foreign firms and their local suppliers is considered. This chapter looks at the definitions of backward linkages, and the ways through which they may be important for the development of local firms. An analysis of the literature on the topic leads the researcher to an evaluation of the impact arising from such relations in the ASEAN region. The specific case of backward linkages and the linkages' creation by TNCs with their local suppliers in the electrical and electronics sectors in Malaysia is taken as a case study in order to show the spillovers arising from inter-firm linkages for the upgrading of production capabilities of local firms. A series of propositions as to which factors influence these impacts is drawn from the literature studied in this chapter.

OVERALL ENVIRONMENT

There are a number of factors that affect both the international flow of investment and the relative transfers from TNCs to the local host economy. This section endeavours to present the overall environment in which foreign firms operate, and the factors that are likely to have an impact on TNCs' operations and decisions. The regional dynamic in ASEAN is strong, and the region as a whole, as opposed to individual countries, thus represents an attractive location for foreign firms. This section therefore covers the ASEAN region, Malaysia being part of this group of countries. The political factors are first analysed, before looking at locational factors and firm-related factors.

Political Factors

Political factors are very important in the explanation of FDI flows and TNCs' activities. These factors include the changing face of global competition, the role of governments and the regional perspective.

Changing face of global competition

Competition. The global environment in which international investment operates has changed dramatically over the past 20 years (Dunning, 1998b; UNCTAD, 1998; Fortanier and Maher, 2001). Whereas initial investment internationally has benefited until the late 1960s from a general *laissez-faire* attitude from host and home government, such a trend has slowly changed, together with the increase in the level of such investment. Drastic changes in government attitude led to a restrictive era, during which host governments closed strategic sectors to foreign investors, imposed strict requirements on foreign firms' operations and in some cases proceeded to the nationalisation of existing foreign enterprises. Such procedures are nowadays rarely used. As the beneficial impacts of international investment are widely emphasised, host governments are now struggling to attract rather than restrict FDI, and compete against each other in a race to become a favoured location for foreign firms. These changes can be noticed both in developing and developed countries, and the political trends, especially in Southeast Asian countries, have been to liberalise requirements towards international investment (Lecraw, 1992; Lall, 1993b; South Centre, 1997; UNCTAD, 1997; Brewer and Young, 1998). In parallel to the multilateral agreements on international trade, it is arguable that the 21st century will bring a multilateral era of international investment (Segafi-Nejad, 1995). The Organisation for Economic Cooperation and Development (OECD) has been acting as a negotiating forum in the effort towards a multilateral agreement on investment (MAI). The negotiations have proved very difficult and did not result in any agreements (http://www.worldbank.org).

Changes in locational advantages. The increasing focus by host countries on the benefits that foreign firms may bring with them has not only influenced the way policies are designed but it has further increased the bargaining power of foreign firms compared to that of the host country (Contractor, 1995; Dunning, 1997). The bargaining power is biased towards the increase in technology advantages of international firms, whereas the locational advantages offered by the host countries, and especially by the developing host countries, are of less importance than in the past (UNCTAD, 1998, 2001). This occurs because a greater number of developing countries offer similar locational advantages which, as a result, increase the location choices open to the investing firms. This trend has been accentuated as a result of the Asian crisis, and governments have become even more lenient (UNCTAD, 1999).

Concentration of international investment. Despite this increased choice in locations, recent trends of international investment show the narrow range of host countries gathering the bulk of this investment. The top 10 developing host countries gathered nearly 75% of total FDI inward stocks in 2000 (calculated from UNCTAD, 2001). Such a high concentration does show the difficulty of other developing countries in trying to attract part of these investments.

Role of governments

Proactive role of host governments. Location advantages traditionally include the value of the natural resource base, the rate of growth of the labour force, the rate of growth of consumption, the perceived risk of the country, the real exchange rate, tariff rates tax rates, and the opennesss of the country's foreign investment incentives system (Lecraw, 1991). Therefore, the host government plays a major role in nurturing and promoting some of these advantages. Nowadays, there is a rise in the competition between firms and countries to produce similar products. Recent trends in the relationship between the host government and international investors have pushed governments to take a more active part in the development of new assets upon which the country can then compete (Narula, 1996). As created assets gain importance compared to natural assets in the value-added process, governments in developing countries need to adapt their policies in order to develop an appropriate environment. Many aspects need consideration such as encouraging local firms' learning capabilities, scale economies, increasing return, and agglomerations. These factors imply market failures, and so a role of policy in developing genuine comparative advantages (Lall, 1998). Specific actions can include the development of the local infrastructure, the emphasis on increasing the overall educational level of the country or even financing projects in high-tech areas.

Facilitated government policies towards inward foreign investment. Another challenge faced by host-developing countries lies in the development of a facilitated framework to invest within their boundaries. Not only do governments need to create new assets, but they also tend to liberalise their policies. In ASEAN, the term 'competition' is rarely used to comment on each country's policies towards inward investment. However, the Philippines, Malaysia, Indonesia and Thailand, for instance, do enter into competition for similar types of foreign firms, insofar as they offer similar locational advantages and attract similar types of investments. In order to facilitate the investment procedures, developing countries such as Mexico, Costa Rica, Venezuela and Honduras have created private investment promotion agencies, whereas others such as Malaysia or Singapore resort to quasi-independent investment promotion bodies (Contractor, 1995). Such liberalisation is often coupled with special incentives, such as inducements towards export-oriented activities, towards high-tech-intensive industries or towards the development of targeted underdeveloped geographical areas. It remains to be proved whether such incentives really attract foreign firms or not (UNCTAD, 1996, 1997). It was shown, for instance, in the case of the UK that specific research and development (R&D) incentives were not primary reasons for the foreign firm's choice of location, but could influence on a secondary basis, when coupled with other types of advantages (Cantwell and Mudambi, 2000). As regards tariff protection, Ariff and Lim (1987) mention that companies regard it as a more important incentive than tax holidays. With increasing favourable governmental policies for foreign firms, such incentives may not indeed be the primary reason for choosing an investment location; however, government incentives play a substantial role in determining the final investment location decision.

Reasons for being welcoming. This massive change in host governments, attitude towards FDI is primarily the result of the fact that ownership advantages of firms embodied in their technology, the human capital, learning experience and organisation structure and knowledge have become easier to transfer to subsidiaries around the world. That way, the host country is more likely to benefit from the experience of the TNCs through its subsidiary (Ariff, 1991; Dunning, 1994). Host countries actively seek to modernise their economy, diversify their production structure and industrial base, create employment and generate technology transfer by welcoming foreign firms, in trying to encourage sufficient spillover and linkage effect on the host economy.

Regional perspective

Geographical Situation of ASEAN. On top of the rising competition between developing countries, inter- and intra-regional competition has become increasingly important with the opening of China to the market economy. On the one hand, the role of China within the world economy is potentially complementary to that of other countries. As a country with a lower level of economic development, China enters the Akamatsu model of the 'Flying Geese' pattern at a lower level in the hierarchy of countries in the region, and as such is attracting foreign firms, mainly in labour-intensive activities (Kojima, 2000). On the other hand, the entry of China is also forcing other countries into higher value-added activities. Competition does occur in the short run insofar as the other countries have not adapted yet to higher value-added activities. Moreover, the scale of China means that the pace of adaptation needs to be fast.

Such concepts as competition versus complementarity are also crucial for intra-regional relationships in ASEAN. Singapore is the only country in the region with a substantially different set of economic activities to be considered complementary to the other countries (Ariff, 1991). Natarajan and Miang (1992) surveyed foreign firms' activities in Thailand, Malaysia and Singapore in order to see the way in which these firms used their regional links. Seventy-one per cent of the firms started operations in the region in Singapore and then opened other subsidiaries in one of the other two countries. Twenty-three per cent of the TNCs in the sample started operations in either Malaysia or Thailand first, whereas only 6% started operations in the three countries at the same time. Singapore clearly had a demonstration effect on the other two countries. Product lines were transferred out and the Singaporean operations served as the technical support centre of their operations in Malaysia and/or Thailand. Transfers occurred through sub-assembling or full product line transfer. Such results show the product mix between each country, most of the product lines being transferred from the

Singaporean operations. Malaysia shows an intermediary position in which the best strategy is to develop niches in targeted industries.

The results found by Natarajan and Miang (1992) highlight the existence of both complementary and competitive activities between countries in the ASEAN. They also underline companies' strategies in the region. The product mix gives place to localised segments of production in different countries. TNCs are thus creating production networks on a regional basis (Dobson, 1993; Dobson and Chia, 1997). Such strategic behaviour is particularly clear in the electronics industry. In some instances, where foreign investors have been required to meet local content requirements, they have built large-scale facilities for the manufacture of particular components with a view to reaping mass production economies and exporting that portion of the output not required for locally made end products. For example, a TNC manufacturing TV sets in Malaysia now supplies flyback transformers to its other TV manufacturing plants in the ASEAN region. Similarly, some import-substituting TV manufacturers in the region have begun to export electrolytic capacitors, tuner assemblies and so on (UNCTC, 1986).

These studies emphasise the need for ASEAN countries to work closely together. These inter-country relationships also show the limits of the 'Flying Geese' model in explaining the economic development in the region. Indeed:

> Rather than replicating Japan's development experience in country after country throughout the region as the Flying Geese analogy suggests, the diffusion of manufacturing in East Asia has been increasingly characterised by shifting hierarchical networks of production linked both backward to Japanese innovation and forward to American markets for the export of finished goods (Bernard and Ravenhill, 1995: 172).

Economic integration. The three drives behind the economic integration of East Asia are the globalisation of production networks as seen above, the increased inter-governmental disputes over bilateral economic relations and the rapid pace of technological change (Bernard and Ravenhill, 1995). Economic integration however becomes feasible where there is a high degree of interaction between economies that have achieved an approximately similar level of industrial development. The ASEAN investment area within ASEAN is therefore focusing on harmonisation. This is because integration is more difficult if wage and technological skill levels between Singapore and its neighbours continue to diverge (Chalmers, 1991). These elements show how ASEAN is linked to the rest of the East Asian region, not only through its political actions, but also through trade and investment links, which largely help the region into the acquisition of new technologies. Substantial technology transfer occurs through the relocation of Asian firms within ASEAN (Giroud and Mirza, 1998).

FDI as agents of increased regional trade and investment. Overall intra-ASEAN direct investment flows (balance-of-payments basis) have totalled over US$15.4 billion over the 1995-2000 period (ASEAN, 2001: 175). Many Southeast Asian firms invest in nearby countries, but increasingly many TNCs from outside the region perceive ASEAN as a production base. Those firms adopt a regional strategy, so that:

> Improvements in communication and transportation links due to technological development have enabled Japanese electronics producers to adopt a regional strategy. High-technology products (such as semiconductors) are produced in Japan or other subsidiaries in developed countries or those in the NIEs where high technological capability exists. The components are then shipped to the subsidiaries in ASEAN where final products such as TVs or refrigerators are assembled by local labour. Another production arrangement is a production differentiation strategy, where a production of a particular item is located where it can be in high demand (ex: standard colour TVs are produced by affiliates in ASEAN while the more sophisticated TV sets are still produced in Japan). Such strategies lead to inter-process, intra-firm and intra-industry trade (Pangestu et al., 1992: 341).

Other studies also emphasised the same pattern (Dobson and Chia, 1997; Arndt, 2001). For instance, Dobson and Chia (1997) specifically studied the role of TNCs in the integration of the East Asian region. Their work includes surveys in eight East Asian countries (Singapore, Taiwan, Hong Kong, Thailand, Malaysia, China, the Philippines and Indonesia) on the trade patterns of TNCs. They conclude that:

> Regional integration is influenced by international firms in two ways. First, it is enhanced by NIE producers based in the region who invest defensively in other regional locations and by multinational firms' (or their local affiliates') investments in components, intermediate products, and services that are shipped around the region, both to add value and to reach final customers. Second, regional integration can be reduced by firms from outside the region whose trade is outside the region. Although the relative magnitudes of these offsetting trends are difficult to estimate, the positive influence of investments from within the region is growing (Dobson and Chia, 1997: 250).

Importance of intra-firm trade in the region. There are four groups of factors important in intra-firm trade: government and regional arrangements, product-specific and firm-specific factors (specificity of product design;

quality, quantity and price uncertainties; the degree of use of proprietary technology; excess capacity in some subsidiaries; and economies of joint operations and vertical integration). TNCs, through their role in international trade and production, have played a major role in the steady integration of the global economy that has taken place over the past 30 years (Gray, 1993). Trade-related investment measures (TRIMs) cover policies on concepts such as local content requirements, trade-balancing requirements (permission to import the foreign exchange it generates through exports at some predetermined ratio), and export requirements (as a percentage of domestic output). The best candidate as a TRIM is an export-processing zone, which effectively creates an enclave of modernity, separate from the tax regulations and rates.

The propensity of an industry to engage in intra-firm trade varies with the life cycle of its product. When a product is newly developed, the extent of international trade in components and semi-processed material tends to be small, but the incentive to conduct this trade within the company is high. When the product matures, it becomes standardised, and enters into mass production; components and other intermediate products enter more extensively into trade, but the tendency for trade to be intra-firm is lower. Hence, the share of intra-firm trade in total trade within the industry falls as its products mature (Casson and Pearce, 1988). The level of intra-firm trade depends on the propensity to circulate for a product, the propensity to trade, the propensity to internalise and the composition of the final product.

Intra-firm trade often involves transfer pricing, which refers to the setting of prices on transactions (or transfers) between different parts of the same firm, as distinct from the setting of market prices on transactions between independent producers (Helleiner, 1981; Murray, 1981; Plasschaert, 1996; Moosa, 2002). Transfer pricing allows a firm to consolidate its profits and losses internationally (through the use of tax-free places, or low-tax countries).

Such considerations explain the behaviour of firms in host countries; their tendency to create linkages with local firms will depend on the age of their products and the incentive to resort to intra-firm trade.

Locational Factors

Structural factors

Structural factors are the primary reason behind the high concentration of foreign investment in a few developing areas, mainly Southeast Asia and Mexico (Lall, 1993a). Agglomerations and industrial clusters appeared in these places through the creation of special economic zones. According to the host government policies, these economic zones can take various forms; they can be licensed manufacturing warehouses (LMWs), export processing zones (EPZs) or free trade zones (FTZs), as in the case of ASEAN countries such as Malaysia or Singapore. These special economic zones have attracted TNCs in concentrated areas, and have led to industrial clusters within the host country (Meyanathan, 1994; Athukorala and Menon, 1996). Still, some benefits were highlighted. For instance, FTZs and LMWs in Malaysia have led to employment creation, net gains from conversion of foreign currencies into local currencies, gains from purchase of local materials and capital equipment, gains from tax revenues and sale of electricity (Lim and Pang, 1991). The lack of local linkages is explained by the low quality of supplies, although local suppliers have developed in Singapore, Taiwan and Hong Kong, and recently in Thailand and Malaysia.

Problems of concentration are slowly diminishing. Firms have tried to disperse because of the improvement in infrastructure, the rise in the cost of land in congested areas, the emergence of small-and medium-sized enterprises (SMEs) in the FDI profile, and the policy environment whereby host governments give incentives for locating in less congested areas. Overall, the high concentration of TNCs has not encouraged a substantial creation of linkages with the host economy. However, some studies highlight the increase in flows of information between firms. The increase in product differentiation has helped in increasing the subcontracting relations with suppliers who have specialised knowledge and research. Flows of information remain higher in developed countries than in developing countries; therefore, the main challenge for many developing countries is to turn subcontracting relations into opportunities for skills' upgrade. Such concepts are looked at in the latter part of this chapter.

Local conditions

Attractiveness of ASEAN for foreign investors. There are a number of reasons associated with the high increase in foreign investment in the Southeast Asian region in the 1980s and 1990s. First, Southeast Asia has an important place in corporate plans because the interest in the region is partly market-driven, and ASEAN is a rapidly growing area, or at least it was until the recent crisis. Second, the yen appreciation in the 1980s encouraged Japanese firms to establish overseas operations in the rest of Asia. As the European Community and the USA started to protect their markets from rising imports of Japan products, many Japanese firms relocated to developing countries with favourable market access. Proximity played in favour of ASEAN, as shipping costs are lowered if production sites are located close to the growing Asia-Pacific market. Also, an important motive for investment is the competition aspect. There are a greater variety of products customised for particular markets, so TNCs relocate simple R&D facilities to the region. Finally, encouraging government policies in the mid-1980s played a role in attracting some of these investments. Motives such as factor costs are less important, processes tend to be more and more automated and labour-intensive production to be located in other countries such as China or Vietnam.

Competitiveness of location. The competitive advantage is determined only to a minor extent by location factors. Endowment factors such as economies of scale, product differentiation and location decisions of TNCs tend to be of prime importance. The competitiveness of a location relies on the ability to fulfil firms' requirements and needs. For instance, a country must be able to furnish the minimum levels of skills required when new technologies arise; this, combined with higher levels of training, provides a productive, flexible and low cost labour force. A country also needs to provide a supplier system that can meet the needs of just-in-time (JIT) production techniques and of information-based processes. Finally, the technological and physical infrastructure must be in place for new activities to run efficiently. At each stage, progress is determined by the ability to absorb technology efficiently, and by the nature of government policies (Lall, 1993b, 1998).

Technological mastery. Technological mastery is the effective use of technological knowledge through continuing technological effect to assimilate, adapt and/or create technology (Lindsey, 1986). We analyse later the different aspects of technology transfer. What is important to underline when looking at the location factors is the crucial role of the governments to ensure efficient technological mastery. If no such effort is produced, a possible outcome may be the inability of developing countries to fill the technological gap they have with the developed countries (Kosai and Tho, 1994). This is why indigenous efforts and technology absorption are crucial. In the case of Korea and Taiwan, local restrictive policies on FDI and import-substituting policies have played a role in stimulating the growth of local firms in the electronics industry (Hobday, 1994).

Market orientation. The market orientation of the TNC located in a host country is influenced by the host government policies in developing countries. In the 1960s and early 1970s, most Asian countries adopted import-substituting policies. Such policies were not found to be successful, and therefore led to a change in policy orientation and to the development of export-oriented policies. The effects of both types of policies varied in the local economy. Thus, linkages by export-oriented industries were found less important than those of import-substituting industries (Riedel, 1975; Casson and Pearce, 1987; Athukorala and Menon, 1996). For instance, it has been found that export-oriented firms are more import-dependent and show a lack of substantial linkages, with limited external effects on economic development.

Riedel (1975) carried out the largest survey on the topic, gathering information from 445 manufacturing firms. The results showed that foreign firms did not tend to be more or less capital-intensive; on average, they employed significantly more labour per unit capital in apparel, plastic and metal products, they had a strong demonstration effect, were more import-dependent and they created few local linkages. In other words, the external effects of export-oriented FDI on the economic development of the host country was limited. Such an extensive study, however, has not been done since. Although indicative, the results cannot be considered applicable in today's situation in the ASEAN region, where countries have already reached a higher level of economic development. The benefits of export orientation were shown by Hobday (1994) in the case of the electronics industry development in selected East Asian 'dragons'. The export orientation of local firms was beneficial in helping to launch a local electronics industry. In Korea, for example, the acquisition of appropriate technology has helped in the development of the three largest conglomerates, Samsung, Goldstar and Daewoo.

Time scale. These patterns vary with the time spent by the TNC in the host country or with the host country concerned. While Riedel's (1975) study applied in Taiwan in the 1970s, which is mostly a subcontracting success story, Premachandra and Jayart (1996) in the 1990s find in Malaysia that export-oriented FDI has brought significant returns, principally because the general economic climate has been favourable for the internationalisation of production for a considerable period of time. In the newly industrialising economies (NIEs), each export-led growth forced the pace of local technological learning among firms (Hobday, 1994). Thus, results from studies vary according to when they are carried out, which indicate that time is an important factor.

Strategic orientation

Need for strategic policies. In the light of the importance of the government policies when choosing between the type of external relations for the country and its industry, the role of policies in adopting adapted policies to maximise linkage effects also comes to the fore. The government needs to allow market forces to create linkages, to use market-friendly incentives and to force linkages through state intervention, for example, local content requirements. Hence, the liberalisation trends of government policies should be coupled with specific actions towards the development of linkages between foreign firms and the local economy, especially in the case of export-oriented firms, which will tend to generate fewer relations with local firms. Other necessary interventions by the government include the investment in human capital and policies on technology transfer, as we have already discussed. Well laid-out guidelines for technology agreements between local and foreign counterparts are needed, as well as protection for intellectual property rights. The specific actions taken by the Malaysian government are analysed in the next chapter.

Industrialisation. The need for indigenous technological development is imperative. Developing countries cannot entirely rely on foreign firms to provide the technology necessary to catch up with the technological level of developed countries. Some industries in ASEAN extensively rely on foreign technology, the electronics sector being the most striking example. The micro-electronic revolution did not take place. Technological diffusion has been partial. Foreign firms primarily resort to intra-regional hierarchy of production. Export manufacturing in the region relies on foreign technology and components.

Therefore, it becomes important that external support for the development of indigenous firms is provided. This can occur through the supply of generic and specific skills, the provision of physical infrastructure, the development of input suppliers, subcontractors and service firms, technology infrastructure, financial system and administration (Lall, 1993a). FDI has an impact on capability development; however, if dependence on import of know-how is passive, it can have deleterious effects. The cost of absorbing and developing a technology is higher if local capabilities are not developed at all. Local firms must undertake technological tasks otherwise mastery, adaptation and development of a technology are less likely to take place and beneficial externalities are reduced. The quality of FDI cannot be raised if the local economy is not able to offer higher levels of complementary skills and technological efforts, and reverse, high-quality FDI is more likely to reduce the technology gap (Glass and Kamal, 1998; Sjöholm, 1999).

Critical role of FDI in industrialisation. Nowadays, FDI is perceived as a safer source of finance in many developing countries than other investment flows (Lipsey, 1999; Mallampally and Sauvant, 1999; Fernandez-Arias and Hausmann, 2001). This trend has proved particularly relevant in the case of ASEAN countries, following the Asian crisis in 1997 (Loungani and Razin, 2001). As an example, following the crisis, many Japanese and European firms chose to retain their presence in ASEAN, and adapt their management strategies to the new environment (Legewie and Meyer-Ohle, 2000; Edgington and Hayter, 2001). Governmental help is not always sufficient. In Singapore, industrialisation took place with foreign firms; other countries in the region heavily used foreign technology and skills to develop specific sectors in the economy. Similarly, the Malaysian industrialisation programme would not have been possible without foreign participation (Ariff, 1991).

The importance of FDI in capital formation can be high in ASEAN. Inward FDI averages more than 11% as a share of gross fixed capital formation (GFCF) in the region. Table 3.1 exemplifies the impact of FDI inflows on the GFCF of Indonesia, Malaysia, the Philippines, Singapore and Thailand. Malaysia and Singapore show the highest share of inward FDI in GFCF, while foreign influence is less pronounced, but on the increase for Thailand until 1999. The Philippines shows low dependence on foreign capital, while the foreign investment role in Indonesia has decreased sharply since the Asian crisis.

Through their study, Alburo et al. (1992) show that FDI in ASEAN remains a small but significant fraction of total capital formation. They show through a regression analysis for eight countries from 1986 to 1988 a positive influence of the US and Japanese FDI. Results show that relative to domestic capital formation, the annual movements of US and Japanese FDI were either a complement or substitute for local investment. Overall, FDI plays a significant role in creating a critical mass of overall investment growth.

Table 3.1 Inward FDI flows as a percentage of gross fixed capital formation in ASEAN-5 1987-99

	1987-92	1993	1994	1995	1996	1997	1998	1999
Indonesia	2.7	4.3	3.8	7.6	9.2	7.7	-1.6	-11.0
Malaysia	18.1	20.3	14.9	15.0	17.0	15.1	13.9	20.1
Philippines	6.0	9.6	10.5	8.9	7.8	6.2	12.7	5.1
Singapore	32.2	23.1	36.1	31.2	29.7	35.3	20.6	26.1
Thailand	5.6	3.6	2.4	2.9	3.0	7.2	20.7	13.7

Source: UNCTAD (1999, 2001).

Firm-related Factors

Changing face of global firms

Change in firms' behaviour. One of the features of the global economy is that an increasing proportion of economic activity is footloose in its location nowadays, so that firms are more flexible in their location and tend to be much more mobile than ever before (Dunning, 1997). This resulted in the fact that companies take a systemic approach to their global activities and are pursuing more integrated production and marketing strategies (Dunning, 1994). The reasons and implications of this trend in footloose location are numerous. They include the need to cope with increasing costs of R&D, the rise in the number of countries with similar levels of skilled labour and technological capacities, the need for firms to cope with global integration on the one hand, to be more sensitive to localised supply capabilities and consumer needs on the other hand and the increasing difficulties to draw the line of firms' boundaries among their numerous networking agreements.

Advancing technologies. New production techniques used by corporations have an impact on the way firms interact with each other. Substantial changes in technologies notably led to small batch production or JIT techniques. Similarly, modular designs and the increased interchangeability of parts in sophisticated products have also improved efficiency in manufacturing (Meyanathan, 1994). Improvements in technology further influence firms' behaviour and strategic choices by allowing easier international transport and communications. Such improvements have made it possible for firms to decentralise decision-making departments and activities. As firms can communicate more easily, they have a wider choice in locating their subsidiaries wherever it is most beneficial to do so.

Investment forms and types

Forms of investment. FDI has a growing role in the world economy. With the change in government attitude towards FDI and the reasons behind the liberalisation of government policies, the new forms of international investment have also played an important role. Indeed, traditional forms of foreign entry, such as wholly owned subsidiaries or market entry through exports, have been complemented by a wide variety of new forms of investment (Oman, 1989), such as minority joint venture, licensing, turnkey projects, management contracts or subcontracting contracts and more recently strategic alliances.

Impact on technology transfer. Traditionally, technology transfer related to the presence of TNCs involved the provision of capital with technical know-how, equipment, management, marketing and other skills. But new forms of investment have brought a new range of technology transfer to the host country since the 1960s (Lall, 1993b). Altogether, investment flows by TNCs provide technical skills and access to foreign markets quickly and efficiently. In LDCs, firms often opt for in-house production/vertical integration to decrease transaction costs and inter-firm linkages; this in turn limits the linkage creation in the host economy.

Type of investment. In the 1980s and 1990s, the majority of FDI was undertaken by already established TNCs. It is likely that such investment will generate its own costs and benefits for the host country (Kogut, 1983; Buckley and Casson, 1985).

Table 3.2 FDI and country/region of origin in the ASEAN countries (balance-of-payments basis, US$ million)

Investing Economy	1995-2000*		1995		1997		1999		2000	
	Value	Share of total FDI	Value	Share of total FDI	Value	Share of total FDI	Value	Share of total FDI	Value	Share of total FDI
EU	26,022.4	21.7	3,648.9	17.3	4,370.3	16.0	4,564.8	27.4	2,905.8	27.9
Japan	19,213.7	16.0	4,238.0	20.1	6,002.8	22.0	769.3	4.6	-55.5	DIV**
USA	17,994.8	15.0	3,262.1	15.5	2,685.9	9.8	2,977.7	17.9	2,320.4	22.3
Triad total	63,231.0	52.7	11,244.0	53.4	13,156.1	48.2	8,410.8	50.5	5,170.7	49.7
All Countries	120,088.0	100.0	21,062.8	100.0	27,301.5	100.0	16,641.9	100.0	10,408.4	100

Source: Adapted from ASEAN, 2001: 22-35.

Note:

* Total cumulative figure, excludes data on FDI in Cambodia.

** This corresponds to disinvestment. Japanese firms have been disinvesting massively from Indonesia. Japanese investments to other ASEAN countries have continued.

Firms and their origin

Competing with new investors. Depending on the subject studied, it has proved very difficult to draw a clear line between the different types of firms and their nationality of origin. Traditional investing countries are developed countries in Europe and North America, while Japan has been gaining a second rank as an international investor over the past two decades. Despite its second rank as an international investor, Japan can still be considered as a new investing country.

The late 1980s and early 1990s have witnessed a substantial outward flow of investment from Asian and selected South American NIEs (Pangarkar, 1998; Mirza, 2000). Third World TNCs have increased in size and scope of activity, which now range from labour-intensive agricultural enterprises to highly capital-intensive computer production. On looking at the ASEAN countries, the role of traditional investing countries is of somewhat less importance. The Triad countries' share in FDI flows to the region substantially decreased in the early 1990s from 74.6% in 1985-87 to 44.7% in 1990-93 (UNCTAD, 1996). Such a decrease reflects the heavy investment conducted by Asian NIEs within ASEAN; in particular, investment from Taiwan and Hong Kong. Flows from Singapore to other ASEAN countries have been particularly substantial, and as part of intra-regional investment flows (Chia, 1993; ASEAN, 2000). Overall, the increase in FDI from new investors in ASEAN slowed down after the Asian crisis in 1997. As shown on Table 3.2, the share of Triad countries decreased to 48.1% in 1997, but increased slightly thereafter to 50.3% in 2000. Importantly, Japan was a key source of investment in the first half of the 1990s, and remains a key investor in most ASEAN countries in the second half of the 1990s, except for Indonesia from which Japanese firms have been divesting.

Difference between firms according to origin. Various authors have found distinctions between TNCs originating from NIEs and those originating from developed countries (Chen, 1990; Ramstetter, 1994; Ulgado et al., 1994; Yeung, 1996, 1998). Marked differences in the 1990s are unclear from the results. Ulgado et al. (1994) carried out a survey among TNCs from the Asian NIEs and compared the results with previous studies. They found that Asian NIEs TNCs do not hire as many local top managers as US firms. In this respect, those firms had a similar profile to Japanese TNCs. Clear results as to the level of centralisation of the firms were not found. Centralisation of US firms seemed to have been higher than that of Asian NIE firms, which in turn was higher than that of Japanese and European firms. Among the Asian TNCs surveyed, decisions concerning new product, chief executive selection and expatriate use were highly centralised, while matters relating to training, credit-giving, maintenance, servicing and advertising were decentralised.

Clear distinctions were therefore not found between different TNCs and their country of origin. However, it was found that Asian TNCs did employ a lesser degree of formalisation than their developed country counterparts, as 85% of them did not rely on written policies from their headquarters. Finally, 74% of Asian TNCs adopted regional or non-global operational structures similar to those of US and Japanese firms, and in contrast with European TNCs. Nevertheless, from ASEAN's viewpoint, it can be argued that the proximity of Asian TNCs could mean that they are more likely to be responsive in Asia and work with local ASEAN suppliers. Proximity would need to be coupled with the length of the international experience and the specific company's strategy.

Theoretical Implications

Prior to the study of the various benefits that can arise from FDI in a host-developing country, we have described the overall environment that influences greatly such benefits. As the international scene has become more favourable for the increase of FDI world-wide, government policies have been liberalised, and improvements in technology have facilitated a change in companies' strategies. The ASEAN region also offers a conducive environment for international investment. Political factors have improved up until the Asian crisis and are still stable in most ASEAN countries, so have locational factors, which offer a wide choice for location of production to TNCs. TNCs have therefore changed their international operations' strategy and operate on a more global basis. Each of these sets of factors lead to a variety of effects of FDI in the host countries and need careful attention.

The increasing importance of the host government in capability development is undeniable; it is determined at the national level by policy regime on trade and industry, by its investment in skills, information flows, infrastructure and supporting institutions. However, comparative advantages still determine capital flows, especially as TNCs benefit from better production, transport and communication technologies and can spread their activities further than was previously possible, and can locate where the benefits are the greatest. The traditional theory relating FDI to industrialisation emphasised the benefits in terms of growth, employment and balance-of-payments' effects. Recently, the experience of the NIEs has drawn attention to the technological and entrepreneurial development, the provision of strategic market access and the creation of international business linkages and expertise and experience (Lim and Pang, 1991).

POTENTIAL IMPACT OF TNCs ON HOST COUNTRIES' PRODUCTION CAPABILITIES

General Impacts

There is a wide existing literature on the impacts of FDI on the host-developing country (Lall, 1996; UNCTAD, 1996; IFC, 1997; Wells, 1998; De Mello, 1999; Moosa, 2002). The general impacts include the change in the business environment, and the creation of links between the host country and the rest of the world. Subsequently, foreign firms generate trade effects through their import of materials and export of production. The effect on the economic structure includes the modernisation of the industry, a shift towards larger scale, more concentrated industries, the increase in demand for TNCs' products and process technology, the TNCs' requirements for high and uniform quality in their products and the entry barriers with product differentiation through advertising and product innovation (Lecraw, 1992). Another impact of FDI is the effect on labour through the increase in wages and the training of employees. The effects on technical efficiency depend on whether the TNC employs overly capital-intensive technologies or not. The main problem for the developing countries lies in the creation of the right economic climate. Government policy makers seem to be more concerned with the amount of 'high-tech' and the cost of this technology than whether or not it is appropriate to factor market conditions. The effects on investment include the discussion between substitutability or complementarity of foreign investment with domestic investment, the time path for inflow and outflow of foreign investment and the gross amount of outflow of foreign exchange compared to inflows. Finally, the social and political effects of inward investment can be substantial.

Immediate impact of FDI

Real impact of FDI. It is difficult to evaluate the real impact of TNCs, because such an evaluation depends on a number of factors. The main problem is the lack of comparison. It is often impossible to evaluate the economic situation in which the host country would be if the investment had not taken place. Therefore, there is no point of comparison on which to base the evaluation of the impacts. Furthermore, impacts are likely to differ widely according to the sector or the industry under study. It is often underlined, for example, that subsidiaries in the electronics sector do not bring as many benefits to the host country as TNCs in other sector, because much of the technology inputs and technology decisions come from the parent firms (Ariff, 1991).

After many TNCs were concentrated in special economic zones in developing countries, there has been growing concern about the creation of modern enclaves. There has been an increase in the number of large TNCs, but a notable lack of corresponding growth in the SMEs employing modern and up-to-date technologies in the urban informal sectors. Therefore, the presence of foreign firms has not led to the expected spread of SMEs in the host countries, leading to the modern enclaves. In a study of large firms in Malaysia among 71 large firms and 56 foreign joint ventures (JVs) Fong (1990) shows that only 25% of large firms and 30% of JVs offer subcontracting jobs to local SMEs. The main reasons provided for this lack of linkages include the lack of quality of the products of the subcontractors (16.9% of respondents), the self-sufficiency of production capacity (25.4%), and the delay in delivery, shortage of adequate small firms and instability of management in small firms. Other important reasons behind the lack of linkages included the financial resources, the financial shortages, the problems with marketing, labour and production technology, the lack of marketing knowledge, the bias of suppliers (shortage of raw materials), the low quality of inputs, the lack of skilled labour, the high turnover and the rigid structure (family-type, individual-dependent, obsolete machinery and technology).

Enclaves are further created by the high export propensity of foreign firms, which indicates a low forward linkage effect.

Regional agglomerations. Countries like Malaysia have witnessed a fast development in selected geographical areas such as Selangor, Johore or Penang. Foreign investors have aggravated industrial congestion by clustering around areas where infrastructural facilities are most developed (Ariff, 1991).

Linkage creation in the host economy

Types of inter-firm linkages. TNCs have various means through which they can intervene in the local economy. There are several kinds of inter-firm linkages that may enter through this process; such linkages vary from complete internal organisational control to complete independence on the market, going through a hierarchy to a market control. These include outright equity control (with 100% ownership), JV (majority, 50:50, or minority), industrial cooperation agreement, licensing and franchising or arm's-length market relationships. Arm's-length relationships involve either 'spot' market transactions or 'longer-term', recurrent contractual supplier-buyer relationships (Wong, 1991).

Creation of local industries. The high concentration within host countries has aggravated the enclave creations. In FTZs, domestic linkages are weak. TNCs maintain strong external links, given the enclave nature of their assembly operations, and the duty-free access to imported raw materials and inputs. Still, some local economic activity arises from these enclaves. Most of the time, such economic activity is centred around services, such as insurance and freight, or around light manufacturing activities, such as stamping, mould and die-making, and packaging activities. These are the local industries that develop to serve the FTZs.

Original machinery and equipment used by TNCs. One of the impacts can be created when the foreign firms first move into the host country through the purchase of machinery and equipment. Such benefits, however, are often limited. Hence, in his study of 27 Korean firms with offshore production facilities in Southeast Asia, Chung (1994) found that in the electronics sector, 77% of the machinery and equipment was imported from Korea, of which 54% was used previously by the parent firm. In the apparel industry, 71.4% cent was imported from Korea, of which 24.3% was used formerly by the parent firm, and similar results were found in the footwear industry. Such results clearly indicate the preference of Korean firms in purchasing original equipment from the home country, and depending on the sector concerned, the tendency of firms to recycle their machinery and equipment in offshore locations.

Influence of foreign firms on local firms. In his study of foreign manufacturing investment in Malaysia, Rasiah (1992) shows that, between 1979-88 labour productivity in foreign-dominated branches grew stronger, rising capital intensity occurring in the 1980s, due to the automation in export-oriented industries. Direct sales and purchases by these branches in the domestic economy have grown strongly, even if the linkages are still weak. The import-substituting and export-oriented sectors are still unintegrated in the domestic economy. However, one side-linkage results from the demonstration effect created by the presence of TNCs in modern technology-intensive sectors.

Autonomy from parent firm. Sivalingam and Yong (1993) carried out a survey in Malaysia among 24 TNCs, 4 resource-based, 7 electrical and electronics, 6 consumer electronics, 2 pharmaceuticals and 5 chemical-based. six firms were from the USA, 12 from Japan, 3 from Europe and 3 from developing countries. On the production management side, they found that 50% of the subsidiaries were autonomous in their choice of inputs, but not in their choice of technology. The management style was similar to the parent firm; for the employment practices, recruitment was done locally, except for senior managers. Only 20% of the subsidiaries had autonomy for decisions related to promotion and remuneration. Concerning the marketing and management style, there was a very low level of independence. Financial management style was determined for 50% of the firms on a joint decision for major new investment, return on investment and payback period.

Knowledge Transfer

Conditions for knowledge transfer

Means of knowledge technology transfer. Although knowledge transfer may occur through several means, FDI-related technology transfer is often considered the most efficient. In a survey of FDI in 69 developing countries, Borensztein et al. (1998) find that over the last two decades FDI has been an important vehicle for the transfer of knowledge, contributing relatively more to growth than domestic investment. First, the foreign firm is linked to its local subsidiary, which thus creates a long-term commitment. Second, the foreign firm is more inclined to provide many of the elements of new technology and the access to world class technology and market. Third, although technology can be purchased, such a method is uncommon. Its implementation and success will widely depend on the existing and evolving levels of local skill and capabilities, on the type of technology involved and, very importantly, on the alternative means of gaining the technology, if any.

Role of host policies. As mentioned in the locational factors, one of the key elements in knowledge transfer is the ability of the host country to absorb technological expertise. Technology absorption depends on the existing educational and skill levels of the labour force, the availability of local entrepreneurship and the government's capability to maintain a stable political and macroeconomic environment (Lall, 1994). The higher the absorption capacity of a location, the more likely it is to be competitive. The government must put in place the appropriate technological and physical infrastructure. The two keys to technological progress rest in the absorption capacity and the nature of government policies. In cases where host countries primarily depend on TNCs for technology transfer, governments must develop local research skills and capabilities to upgrade TNC operations in their economies.

In addition to providing a favourable environment for the dissemination of the technology, it is important that the host government encourages the TNCs to proceed to some kind of transfer of technology. Until the late 1980s, there was a lack of such political incentives towards technology transfer in the ASEAN countries. Some authors believe that it is because companies were not required to transfer technology or to create linkages with local businesses that the dualistic structure in the host economies was reinforced (Chalmers, 1991). Policies can interact with economic forces to act against the creation of a dualistic structure.

Specificity of industry concerned. It is important to differentiate between the type of industry and the resulting technology transfer. In his analysis of the semiconductor industry, Rasiah (1989) notes the relationship between the changes in the dynamics of production in the mid-1980s and the technology transferred by TNCs. Changes included large investments in R&D, which resulted in shorter product cycles (leading to an increase in mergers since 1984), an infusion of automation since the 1970s and a systematic planning and reorganisation of production. The latter took a dynamic turn with the incorporation of JIT techniques and material requirement planning (MRP) (Rasiah, 1989: 43). This is the change that provided the basis for some form of technology transfer in Malaysia, through increases in mergers and commercial alliances.

Types of knowledge transferred

Knowledge, in a broad sense, includes production technology, management expertise, marketing skills and other intangible corporate assets. Additionally it includes knowledge of goods and services' production and distribution methods. It can be divided into two types, the hard technology embodied in the product technology, in the machinery and equipment, and the soft technology embodied in the know-how of one firm that may be transferred to the other firm, in the people who operate, maintain, adapt and develop the machines. Physical transfer of machinery and equipment is the first step; it includes transfer of information and skills from low levels to R&D scientists and engineers (Natarajan and Miang, 1992). In many cases, positive transfer is found through the training of local employees and local suppliers of intermediate products (Pangestu, 1991). Lall (1994) identifies three simple indicators of the indigenous technological content of technology export: the diversity, complexity and sophistication of the industries where the export of technology takes place, the enterprises and countries concerned (Did they provide basic process or product technology? What plant and equipment are utilised?) and the local engineering and manufacturing content of the technology sold overseas.

Knowledge transfer and country of origin

Transfer of knowledge by firms from NIEs. TNCs from NIEs usually establish JVs with local partners and produce mature products using standardised technology with high price elasticity of demand. They operate mostly in the region where they belong, and they often use less capital-intensive technologies (Tolentino, 1996; Pangarkar, 1998; Kumar, 1998; Mirza, 2000). Firms from the NIEs are therefore less reluctant to transfer standardised technology. Such technology is also, in many cases, more appropriate for host countries. As host-developing countries' technological capabilities remain limited, the ease of transfer of standardised technology is greater.

The ownership-specific advantage of TNCs from NIEs is the transfer of small-scale, labour-intensive technology. Technology diffusion in a less developed country occurs through the managerial and technical training that TNCs offer to local employees, through the presence of TNCs in those industries where there are proportionately more TNCs, and where local firms then have a higher propensity to adopt new technology (Chen, 1990). TNCs from Asian NIEs view appropriate and flexible technology, skilled and low-cost management, low overheads and access to market information as their primary strengths and sources of competitive advantage. Other strengths, such as home government support and ethnic linkages, are not considered important (Ulgado et al., 1994).

Overall, an appropriate description of technology transfer occurring through these firms is that:

- the transfer takes primarily the form of training offered to employees,
- many TNCs have introduced new operational strategies,
- there is little technological collaboration, sharing of market information or technical assistance between TNCs and local enterprises (Sivalingam and Yong, 1993).

Among NIEs, Korea and Taiwan have developed capital goods production, while Hong Kong has primarily developed managerial, marketing and production skills (Lall, 1984). Taiwanese firms create little technology export, insofar as they have a limited technological capability, and they are relatively small-sized. Therefore, their technology advantage lies more in the product technology, and their managerial and marketing skills than in the process technology (Chen, 1983). It has been emphasised that Taiwanese firms are smaller, which is why their advantage lies in their flexibility and rapid response in market niches for non-standardised products, rather than competitiveness through price. They have been oriented towards innovation, trying to compete in the international market relatively early in the life cycle of individual products (Levy, 1988).

Transfer of knowledge by firms from developed countries. Focusing on the automobile industry, Fujita and Hill (1997) find that the Japanese firm

Toyota Motor Corporation carries out an ongoing incremental transfer of products and processes, which underpins a growing ASEAN middle class, and derives from a post-*keiretsu* network among Japanese and indigenous companies in the region. Thus, it is shown that the transfer of technology depends on the transnational production organisation, on the national industrial strategies and on the local development on the Southeast Asian automobile industry as a whole.

From a survey of 16 companies, Lee (1991) found that the most important technology transfer was in training, expatriate consultant visits and presence. For the transfer of management techniques, the Japanese systems seem to have greater impact, with new innovation in management, more flexibility in approach and less bureaucratic control.

Of a different nature, Japanese and US firms seem to emphasise training on 'show-how' rather than 'know-how'. On the other hand, NIE firms are more willing to transfer their technological skills (Ariff, 1991). Sivalingam and Yong (1993) find that the most common form of technology transfer for the companies in the study is training. Such training occurs in the parent company (19 firms out of 24), in the Malaysian subsidiary (15 firms), through foreign experts (13 firms), or resident expatriate consultants/experts (12 firms), through technical assistance contract (7 firms), licensing agreement (6 firms), turnkey agreement (2 firms) and only rarely through selling technology or through seminars.

Implications for host country. The existence of various linkages between foreign firms and the existing firms in the host country shows in terms of immediate impacts and through the process of technology transfer. A specific type of linkages less often studied is that of backward linkages, which we study in the next part of this chapter.

UPGRADING PRODUCTION CAPABILITIES THROUGH THE CREATION OF BACKWARD LINKAGES

Linkages generally encompass all the existing forms of economic relationships between firms operating within an economy. A linkage usually implies a continual relationship involving recurrent transactions (Wong, 1991). Backward linkages include all the upstream relationships a firm creates with other firms, whether they are one-off suppliers, key suppliers or subcontractors (UNCTAD, 2001). As seen in Chapter 2, linkages between firms vary from complete internal organisation to complete dependence on the market.

The bulk of the research done on backward linkages focuses on the sheer amount of local materials purchased by foreign firms, as a share in total materials used in the production. In developing countries, the mere pecuniary benefit of buying a local input is far from being the only benefit that TNCs can bring to their local suppliers. Humphrey et al. (2000) found very little positive spillover by Japanese firms to their local suppliers in the automotive industry. On the other hand, Kelegama and Foley (1999) note that even though few backward linkages exist in the garment industry in Sri Lanka, these contribute significantly to foreign exchange earnings, in terms of employment creation, and by providing opportunities for local entrepreneurs to gain new marketing and managerial skills. Thus, although local supply by TNCs is often limited in developing countries, TNCs can bring other potential benefits to their local suppliers by creating close relationships with them. For instance, local companies can take advantage of special technological and management links otherwise non-existent in their country, or gain recognition in the business environment for having contractual relations with foreign-owned customers. As a result, backward linkages with local suppliers are known to be a potentially very important channel of technology transfer from foreign to domestic firms, as well as a major instrument for building up and developing the technological competence of domestic enterprises (ILO/UNCTC, 1988).

Even though the benefits from the creation of relationships between foreign subsidiaries and local firms in the host country are widely recognised, there is still a substantial lack of research on the topic (Wong, 1991). Unlike customer-supplier relations in the developed countries, backward linkages in developing countries between TNC and local firms tend to be less substantial and less strategic. Yet, we argue that since major TNCs apply the same strategy world-wide, they will create strong relations with a few selected firms in their host-developing countries. It is those that lead to an increase in the local technological and production capabilities.

Existing Backward Linkages

The extent to which a foreign firm may wish to use local suppliers, as opposed to sourcing internally, from within the firm, or import parts and components, depends on a number of factors. This section aims to determine the factors that influence backward linkages' creation. The analysis of the set of factors important for supplier support system comes first, followed by the factors related to the firms' strategy, and finally, government requirements are tackled. This analysis helps to establish a list of propositions to understand further the way backward linkages are created, and under which conditions they are likely to lead to buyer-supplier transfers.

Set of factors important for supplier support system

Importance of supplier support system. Purchasing has had an increasing place in today's company strategy. It is to be expected that there may be various purchasing patterns according to the country in which the TNC operates, and according to the level of economic development of the host economy (ILO/UNCTC, 1988; Halbach, 1989). TNC subsidiaries create backward linkages with local firms, if the local firms are developed enough. The cultural similarities between firms also bring firms to cooperate more easily with local partners, if the psychic distance is not too substantial (Hallén and Wiedersheim-Paul, 1979). It should be added that backward linkages and linked bonds depend to a large extent on the choice of products manufactured in the subsidiary (ILO/UNCTC, 1984). The life cycle of the product is a powerful explanation element in the propensity of a firm to resort to external sources of supplies.

Overall, Lall (1993a) considers that as in the case of technology transfer, there are four sets of factors that are important in the supplier support system. These include the nature of the technology involved by the company, the strategy of the seller, the capability of the buyer and the host government policies. We further argue that there exist other important factors, such as, for instance, the buyer's characteristics or the host country's characteristics.

Favourable economic climate in host country. There seems to be a tendency for foreign firms, especially export-oriented firms, to increase their local business purchase as their operations in the given host country mature. This takes place if the local business environment is conducive to such behaviour through a rapid economic growth and a healthy investment climate (Lim and Pang, 1982, 1991). To show the increase by TNCs of local and ASEAN content, Pangestu et al. (1992) conducted a series of interviews among six US firms, three Japanese firms and four European firms. All electronics firms were export-oriented, while in other sectors they were domestic-oriented. Most TNCs indicated they were beginning to look at the domestic and regional market for sourcing of parts and components. This trend was less important in the case of export-oriented companies, for which the access to imported inputs did not pose a problem, as they were located in duty-free zones or were eligible for duty exemptions or drawback for their imports. These infrastructural facilities made the use of imported inputs in the production easier, and incentive for local purchasing was less important. Belderbos et al. (2001) also highlight the importance of the quality of infrastructure and the size of the local components supply industry as key determinants for the promotion of backward linkages.

Table 3.3 Examples of local purchasing behaviour by Japanese and North American firms in the Singaporean electronics sector

	Firm A (Japan)	Firm B (Japan)	Firm C (USA)	Firm D (USA)
Industry	Consumer electronics	Consumer electronics	Industrial electronics	Industrial electronics
First investment	Early-1970s	1987	Early-1970s	Early-1950s in Japan
Use of local suppliers	1. Own production highly automated 2 24% parts obtained from local suppliers 3 20% parts are key components from Japanese parent 4. 56% parts obtained from local Japanese suppliers	1. 80% parts from local Japanese suppliers 2. 20% parts from Japan	1. For new products, 100% US components for first 12 months 2. After 12 months production, 30-50% parts from suppliers in region 3 After 24 months, 50-60% parts from local or regional suppliers	1. 20% from locally controlled companies 2. 20% from Japanese controlled local suppliers 3. 60% from other US TNCs
Products supplied by locally controlled firms	1. Paper and plastic packaging, plastic parts, metal parts	1. Same	1. Same	1 Keyboards 2 Mechanical assemblies
Selection criteria applied	1. Quality standards met 2 Priced competitively 3 Timely	1. Timely 2. Quality standards met 3. Price competitive 4. Information on product readily available 5 Reliability as suppliers	1. Quality 2. Proximity	1. Quality 2. Proximity

Source: Dobson (1993: 55).

> *Proposition 1* As the environment in the host-developing economy matures, buyer-supplier links increase and become more effective.

Backward linkages and industry concerned. Combining time and sector of activity in their study, Sivalingam and Yong (1993) described the impacts of FDI in two broad industry groups. In the resource-based industry, there had been a slight increase in local input, from 74% of all inputs being purchased locally in 1978, to 76% in 1983. In the non-resource-based industries, shares of local inputs decreased slightly from 43% in 1978 to 41% in 1983. These figures show that the share of local input in total input in the ASEAN did not vary greatly between the 1970s and the early 1980s. The situation has now changed however, due to a large inflow of foreign investment that was made in the region in the early 1990s. In addition to these considerations, there are wide industry differences to be underlined.

> *Proposition 2* Buyer-supplier links vary according to whether the industry is resource-based or non-resource-based.

Type of product manufactured and industry. Looking at the electronics industry in Singapore, Dobson (1993) interviewed four firms, two from Japan and two from the USA. Results from the interview concerning local purchase are shown in Table 3.3. and show the share of local supplies and the type of supplies purchased from locally controlled firms. All firms have local supplies, but the ownership of the local suppliers, the quantity of local supplies or the types of products differ. None of the firms relied on locally controlled suppliers for more than 20% of parts or components by volume. The products supplied locally included packaging materials, moulds, plastic and metal parts and mechanical assemblies used in computers, whereas the key components came from Japan or the USA, or from locally based foreign firms. Other considerations clearly appear in Table 3.3. The Japanese subsidiaries tend to purchase primarily from other Japanese subsidiaries, although the US firms also rely to some extent on Japanese firms' supplies (firm D purchases 20% of its inputs from Japanese-controlled local suppliers).

In line with Dobson's study, Halbach (1989) emphasises the specific nature of the electronics industry. The electrical and electronics industry is often cited as being based largely on assembly work, which often leads to the creation of international subcontractors. It is not only the industry, but also the nature of the product manufactured and/or the production stages transferred to the developing country, which lead to various linkages. Firms were found to subcontract 10-20% of their activities in high-tech electronics, but only 20-40% in consumer electronics and 50-80% in household electrical appliances. It can therefore be said that the industry concerned and the type of product manufactured are important factors in the determination of local backward linkages (Rasiah, 1989; O'Brien, 1993).

Proposition 3 Buyer-supplier links vary according to the level of technology of the product manufactured and the industry concerned.

Ownership and industry. The major survey carried out to distinguish between various backward links created according to different sectors was that of O'Brien (1993). The survey targets all types of companies, whether foreign or local. Out of 86 companies in the survey, 66% of the sample sourced production equipment and instrumentation from abroad, while only 33% had some Malaysian inputs. Malaysian-owned firms or JVs had more local links, which led to the conclusions that local ownership and control did aid the development of a more self-sufficient and interconnected manufacturing sector. As shown in Table 3.4, essentially resource-based food and beverage companies in the study were highly reliant on local raw materials, the same as rubber and non-metallic mineral companies. The electrical and electronics and chemicals were least likely to have local inputs. Finally, firms that sourced the majority of their inputs domestically also tended to sell their goods on the local market.

Table 3.4 Source of input/output by industry in Malaysia

	Share of local sourcing		Share of local sales	
	Less than 50% (*N*=23)	More than 50% (*N*=49)	Less than 50% (*N*=44)	More than 50% (*N*=28)
Food and beverage	20	81	31	71
Non-metallic minerals	17	83	17	83
Chemicals	9	91	73	27
Fabricated metals	5	95	68	32
Electrical and electronics	69	31	73	27
Electrical	33	67	44	56
Electronics	88	12	88	12
Total	32	68	61	39

Source: O'Brien (1993).

O'Brien reaches the conclusion that locally owned firms create more local links because they are often in low-tech products, they are more in resource-based activities and the establishment of networks and the cultural factor of familiarity in the supply of inputs are important. Similarly, McAleese and McDonald (1978) identified two main reasons why foreign firms purchased less local input in Ireland than local firms did. First, within the same industry group, foreign-owned firms tend to be concentrated in more specialised activities than domestic firms. Second, overseas firms may have a higher import propensity than domestic firms because of such factors as familiarity with foreign suppliers, and real or imagined inadequacies of local producers. Therefore, foreign firms have a higher import propensity than domestic firms because of factors such as familiarity with foreign suppliers, and sometimes inadequacies of local producers.

Proposition 4 Locally owned buyers create more backward linkages than foreign-owned firms do.

Backward linkages and firms' strategy

Market orientation of firm. In Thailand, Supapol (1995) shows that JVs with a primary interest in the domestic market tend to source more locally than export-oriented TNCs, in line with the results of Pangestu et al. (1992), discussed earlier. Local sourcing generally takes place with other foreign firms, or is mostly 'off-the-shelf' sourcing of non-specialised and low-to-medium technology parts and components.

Proposition 5 The export orientation of the firm affects the buyer-supplier links' creation.

Time spent in host countries. The first time consideration is the age of the plant. There is a strong relationship between the age of the plant and the backward linkages in the host country (McAleese and McDonald, 1978). This shows that there is a learning process at work for the foreign firm in its first years of operation in the host country.

Second, the age of the production of the product is looked at, together with the type of production carried out by the firm. The foreign firm may change its output mix and production process for various reasons. Variations in input prices may trigger such a change, or as mentioned above, the learning curve of the foreign firm in the host country. This pattern appears for firm C in Table 3.3. This North American firm uses local supplies, but only after the new product has been manufactured in the host country for some time, has proved successful and if some adequate local suppliers have developed.

Local sourcing will further depend on the stage of development of the local industry. The firm needs to establish credibility as a producer. Therefore, local inputs may be included after the operation has proved successful. Home country suppliers may themselves be encouraged to locate in the host country. Alternatively, the subsidiary may choose to expand its operations vertically. It may however occur that an indigenous supplying industry develops. This phenomenon is directly related to the importance of the supplier support system and to the favourable economic climate in the host country. Consideration of the industrial composition of new small industries suggests that the scope for indigenous backward linkages is most pronounced in the metals, engineering and electronics sectors (McAleese and McDonald, 1978).

Proposition 6 The time spent by the firm in the host economy affects the buyer-supplier links creation.

Reasons for not supplying locally. Dobson (1993) identifies some reasons for the low level of local supply. First, the needed components were high in technology (firm B in Table 3.3); second, the use of local suppliers depended on the length of time the company had spent producing the product locally (firm C). As a conclusion, intense cost competition, rigid quality standards and short product cycles are the driving factors in procurement arrangements. In this environment, internalisation within the group, if not within the firm, minimises material and transaction costs. Dobson (1993) concludes that these are the reasons why Americans rely just as heavily on existing local Japanese suppliers as Japanese TNCs do. Halbach (1989) identifies three major reasons for limited subcontracting relationships. Initially, these include the fact that firms are unaccustomed to the infrastructure and the socio-cultural environment; then, the problems related to the technical and qualitative capabilities of local firms; and finally, the difficulties of planning and management deficiencies of subcontractors. In her study in Indonesia, Gultom-Siregar (1995) explains why, sometimes, local suppliers were not chosen. Reasons provided included (1) incapability of modification whenever needed, which was mainly due to the length of time needed to import machinery, (2) very limited capacity to fulfil a big order and (3) lack of technological and design capability to provide qualified parts and components (Gultom-Siregar, 1995: 163). Rasiah (1995) showed that SMEs in Malaysia lack in technological capability. It appears that efforts to raise further local sourcing would require improvements in product technology, quality and process technologies utilised in local SMEs. Two TNCs suggested that local SMEs require capacity expansion to increase local supply capability.

Proposition 7 The less developed the host country's technological knowledge and production capabilities, the fewer buyer-supplier links are created.

Types of firms and linkages. Fukagwa (1995) identifies three types of TNCs in East Asia. The first is the 'traditional type' with strong market control based on its monopolistic advantage, which is common among US and European firms (for example, Unilever, Nestlé, Coca-Cola and IBM). These firms have a wide experience of international operations. They usually keep a strong control through ownership, a limited number of subsidiaries in each market, a tight network within the region in management resources and minimum R&D activities in local areas based on a policy of concentrating development resources at the headquarters. They tend to favour vertical integration and supply internally for scale economies.

The second type concentrates on offshore production, seeking cost reduction. They usually transfer little technology and repatriate profits. They are largely export-oriented, and subsidiaries are controlled either by the headquarters or by the regional headquarters. They have tried to create a dynamic division of labour among regional facilities to exploit differences in locational advantages. This group shows increasing market integration in its procurement behaviour. For instance, through their export business in ASEAN, Japanese home appliances' firms are still mainly offshore production with a large share of in-house procurement. However, local and regional procurement has been promoted rapidly, and parts' procurement for colour TVs in NIEs has already reached 70-80% at the local level and 90% at the regional level. In Indonesia, the difference expands between less than 30% and 80%. These regional supplies are taken by Japanese firms, but in terms of plastic moulded products by Taiwanese firms, and TV tubes and metal moulds by Korean firms regionally, which means international intermediate goods' exchange amongst NIEs has promoted a regional network, leading to an integrated market for parts (Fukagwa, 1995: 18).

The third group is less oriented towards monopolistic control and is positive about making alliances (such as JVs, technical cooperation, original equipment manufacturing (OEM) contracts and so on) with both local companies and other transnationals. They are more oriented towards the host domestic market, with a product line management that can be localised. This type is common among Japanese firms, especially home appliances' manufacturers such as Matsushita. Still, some of these firms resemble the second group, insofar as they often have wholly owned subsidiaries, with a tendency for greater procurement by intra-firm import (Fukagwa, 1995: 29). Their cooperative nature makes them dependent for their materials on local partners.

Such a division of firms is useful. Instead of differentiating firms by their country of origin, or by the product they manufacture, their international, regional and local strategic orientation is shown in this case to have an impact on their local purchasing strategic orientation.

Proposition 8 The company's overall international and locational strategy affects the buyer-supplier links' creation.

Dependence on parent firm. Although no study comes out with a clear conclusion on potential differences between nationality of firms and backward linkages, it is worth comparing various results. Hence, Chen (1990) argues that developed country firms seem to be more dependent on their parent firm and on intra-firm trade, and have a smaller proportion of local supplies than firms from the NIEs do. This may be true insofar as these firms tend to produce more labour-intensive products, for which the use of high-quality inputs may be of less importance. However, this further depends on other factors, the most important one being the length of time spent in the host country. As the subsidiary will first try and establish its image as a producer, it is likely to import a greater share of its inputs at the beginning of the production.

Proposition 9 Dependence on the parent firm and high levels of intra-firms' trade have an impact on the buyer-supplier links creation.

Backward linkages and origin of firm. For Japanese firms, although they are expected to exhibit the highest degree of subcontracting, they are the least aggressive in promoting the establishment of locally owned supplier firms, since they already rely on well-established Japanese suppliers (Lim and Pang, 1982). Yet, it is sometimes argued that Japanese firms provide important information to their subcontractors when they create subcontracting relationships, as well as technological and managerial assistance. These can benefit the local economy when local firms are hired as subcontractors (IDE, 1994).

In the Singaporean experience in the 1970s, Lim and Pang (1977) reported that Japanese subsidiaries in the electronics sector were more likely to purchase inputs from other Japanese subsidiaries in preference to local Singaporean producers, precisely because of the latter's inability to supply goods of satisfactory quality. Local producers were criticised for failing to appreciate the overwhelming importance of consistency in quality. This situation has changed markedly, but it remains interesting to compare Singapore in the 1970s and today, as a wide base of local electronics' producers has developed. Despite this trend, it is still noticeable that Japanese firms generally rely heavily on other Japanese subsidiaries (Dobson, 1993; Dobson and Chia, 1997). This trend is more pronounced for firms that belong to Japanese vertical industrial groups (*keiretsu*); these firms show higher procurement from local clusters of affiliated Japanese suppliers (Belderbos et al., 2001). As a result, Japanese networks in the East Asian economies may be less open than US firms, because of reliance on Japanese managers, who lack local networks, indifference towards non-Japanese products and that they are more recent investors. Japanese firms have a familiar alternative to fall back on, that is, suppliers from the same country of origin. The relative transaction costs of using non-Japanese suppliers are higher than they are to non-Japanese firms (Dobson, 1993).

Although US firms do not imitate Japanese supplier networks (they are more open to outside suppliers, including those used by the Japanese), increasingly they are investing in long-term relationships in order to achieve exacting quality standards. Interview comments include that the US practice of recruiting local executives is helpful in developing such linkages. These recruits tend to have local knowledge and contacts that are of considerable assistance in creating supplier relationships (Dobson, 1993: 69).

European firms resemble US firms inasmuch as they started investing in ASEAN at an early stage. The time factor, the industry and the type of product manufactured may be of more importance in determining their backward linkages. Studies rarely include European firms as a major study group.

As argued in the comment on technology transfer, these firms are less reluctant to share their knowledge. It remains that they are new investors and their local supplying networks are still at an infant stage.

Proposition 10 The country/region of origin of the foreign firm explains to some extent the buyer-supplier links' creation.

Backward linkages and contractual relationships. Considering the limited amount of local purchase and the type of products purchased locally, for example, low-tech-intensive products, the most common type of contractual relationship between TNCs and their local suppliers is the off-the-shelf contract. Subcontracting is said to bring the highest benefits; however, the tendency towards increased long-term contractual relations makes suppliers special partners, and such partnerships also lead to substantial benefits (Supapol, 1995).

Backward linkages and government requirements

Masterman (1991) carried out a case study of the Japanese firm KGD Indonesia. The subsidiary is a joint venture with Indo Auto Parts' Makers, as the government requirements are strict on ownership. Within five years, the local procurement was supposed to reach 60-70%. In the first year, from 1990 to 1991, the company was only purchasing 10% of its local inputs locally. The reason for it was the low standard of quality of the local suppliers, who did not meet the Japanese requirements. The firm had however to consider ways of improving local supply, because of the agreed local content requirements with the host government. Similarly, Gultom-Siregar (1995) finds that local sourcing was limited to less than 20% of the total inputs in Indonesia for non-critical components, and this was mainly the result of local content requirements and costly import procedures. Gultom-Siregar (1995) concludes that government requirements do not cover for the poor state of local suppliers' technological capabilities and finds no transfer of technology. Thus, even when government policies are in place in terms of local content requirements, their efficiency is not shown. The efficiency may further depend on the type of firms. As an example, Hackett and Srinivasan (1998) study FDI in host economies, and find that strict local content regulations impact on FDI, with a somewhat stronger negative effect on Japanese FDI than on US FDI. Finally, in a study of the determinants of backward vertical linkages by 272 Japanese electronics manufacturing affiliates in 24 countries, Belderbos et al. (2001) find that local content regulations have a modest positive impact on linkages and they do not stimulate procurement from locally owned suppliers.

> *Proposition 11* Government policies may influence firms to supply locally, but they have not been found to have a substantial impact on buyer-supplier links' creation.

Developing Local Suppliers with Backward Linkages

Concepts related to backward linkages

The UN distinguishes three types of linkages by foreign firms in host countries (UNCTC, 1981). The first link implies the relationships that TNCs develop with suppliers or buyers in the same chain of production. The second type of links is embodied in the relations TNCs may have with competing local enterprises in the same industry. This can lead to a change in infrastructure, firms' take-overs, the training of workers or to the introduction of new products and technologies. The third type of links includes the effects of TNCs on local corporations in other sectors through their impact on the economy in general, by affecting the net income, the balance of payments, the government revenue or the industrial infrastructure as a whole.

Scale and scope of knowledge transfer. The analysis of the relationships existing between foreign subsidiaries and their local suppliers in less developed countries has recently been the object of specific research. While some authors analysed these relationships in the case of subcontracting only (UNCTC, 1981; Halbach, 1989; Wong, 1991), studies often omit the increasing tendency of firms to treat their suppliers as close business partners. Thus, it is likely that the information exchange between foreign subsidiaries and their key suppliers may lead to close relationships, as in the case of subcontracting contractual agreements.

> A review of the international subcontracting literature indicates that characteristics of the TNC affiliates and of the indigenous subcontractors (technical and absorption capabilities of subcontractors' activities of foreign affiliates and the type of products sources from local suppliers (i.e. technological complexity and specificity idiosyncratic investment), as well as the nature of the sourcing arrangement, are important factors in determining the scope of local sourcing and technology spillovers. New technologies and knowledge that are embodied and uncodified and are difficult to transfer through arm's length are more likely to be made available to long-term suppliers with established relations with the TNC rather than to occasional suppliers with no long-term commitment to the TNC (Supapol, 1995: 25).

Thus, by focusing only on subcontracting relationships, it is impossible to obtain the overall picture on buyer-supplier links' creation. However, it is acknowledged that greater links will exist between a foreign firm and its local subcontractors. Stories of success include Acer, which grew as a supplier of IBM convertible machines, but uniquely developed technology to upgrade the speed of ordinary machines twice by adapting a 64-bit base, and provided CPU technology even to Intel. At present, even the import agent of Acer of Indonesia, the Metropolitan group, is trying to expand the business to manufacturing in the same way.

> *Proposition 12* Greater buyer-supplier links are created with subcontractors, but other long-term suppliers, considered as business partners, may also benefit from substantial relations with TNCs.

Benefits of linkages' creation. Technology and knowledge transfer through backward linkages (that is purchasing of goods and services from locally owned suppliers by companies controlled by foreign firms) have long been known as beneficial to the host country (ILO/UNCTC, 1984, 1988; Supapol 1995), especially in the host-developing countries. These transfers comprise transfer of technology and know-how from one firm to another, or possible benefits through the long-term relations and the exchange of information. Channels of technology and knowledge transfer include FDI, turnkey operation, purchase of machinery and equipment, licensing and management agreements, technical assistance contracts, direct employment of foreign experts and training of local staff. Through the specific case of subcontracting, backward linkages include information flows, technical, financial, procurement, locational, managerial, pricing flows and flows related to sales arrangements or export help (UNCTC, 1981). Such transfers were summarised in the World Investment Report (UNCTAD, 1995) as follows:

> The crucial contribution of linkages created by foreign affiliates involved in production activities with host countries, lies not only in the business generated for local firms, but in the knowledge and technology flow from TNC systems to local firms. Such flows could be in the forms of designs, drawings, specifications, manufacturing knowledge, process know-how, quality control, productivity enhancing techniques, management know-how, training and the like. Linkages with subcontractors often involve joint preparation of specifications, designs and drawings. Often, personnel from foreign affiliates are deputed to train personnel in supplier firms in the production of required components or parts, leading to the upgrading of skills in local firms. As an increasingly common practice in recent years, whenever products or process changes are effected in foreign affiliates, they involve their suppliers and subcontractors in the process from an early stage so that the latter group can gain the know-how to effect changes in their products supplied to the affiliate (UNCTAD, 1995: 159).

Knowledge transfer can therefore include financial help provided to another firm, product and process technology information exchange, or transfer of managerial, technical and marketing skills through various transfer methods.

Evaluation of beneficial transfers through backward linkages

The literature review discussed above emphasises the factors likely to explain why foreign firms might use local suppliers, as opposed to imported supplies. This analysis provides a good idea of the important factors when studying buyer-supplier relations in a developing country. The literature on overall transfers between buyers and suppliers helps further in the provision of a clear picture of the factors influencing knowledge transfer from the foreign affiliate to its local suppliers.

The most common source of transfer from foreign subsidiaries is linked to the technical support related to the production process and the product. In the great majority of cases, the foreign firm will supply blueprints to its suppliers. It may further provide special tools. In the case of subcontracting, the subsidiary may take part in the capacity and production planning of the subcontractors (Halbach, 1989). Support is frequent for the learning of specific process technologies, but upgrading of technological know-how to product design is limited (Wong, 1991).

Even in the transfer of business and organisational know-how, buyer-supplier links are related to the technical management or involves support by the subsidiary in technical production and quality control. Such findings confirm the endeavour of foreign firms to try and improve the technical level of local suppliers in the host developing countries, and consequently the quality of the purchased items, so that they fit the required specifications. Still, the extent of technology absorption will depend on both the potential scope for absorption created by the supplying relationship and the internal absorption capacity of the supplier. TNCs will only devote resources to transfer technological know-how to their suppliers if the expected returns from such efforts outweigh the costs. In most studies, foreign subsidiaries favour the creation of long-term relationships with their local suppliers so that trust is created between the two parties, and the level of assistance provided remains beneficial to the initiator.

The level of technology transferred is often difficult to evaluate. Out of 20 SMEs interviewed, Rasiah (1995) found 8 that considered the transfer of technology through subcontracting was extremely important, 10 that it was quite important and 2 that it was mildly important.

Some firms emphasise management transfer. Hence, suppliers of Japanese firms were found to adopt management restructuring inspired by Japanese firms, such as quality control centres, multi-task jobs, group work, employee suggestion schemes and other human resource-related management techniques (Fleury et al., 1994; IDE, 1994).

Although financial assistance is often mentioned as one type of assistance provided by foreign firms to their local partners, nearly all the studies found that no financial assistance was provided (Wong, 1991; Gultom-Siregar, 1995). For only a very few companies, Halbach (1989) found that firms had pre-financed materials, tools and facilities or proceeded to prepayment of orders placed.

Buyer-supplier transfers were shown to differ according to the nature of the products and components bought from local suppliers (Gultom-Siregar, 1995; Supapol, 1995). In his study on subcontracting relationships in Thailand, Supapol found that among the cases with low-tech parts and components, 38% realised spin-offs, whereas for cases with medium- and high-technological components, respectively, 71% and 50% realised technology spin-offs. Therefore, when the nature of the component is general and the production involved old and well-known technology, there is less need for the foreign subsidiary to provide assistance to its suppliers.

A study on technological development through subcontracting linkages with SMEs in the electronics industry in Singapore, with a comparison with Japan, South Korea and Malaysia, led Wong (1991) to some striking conclusions as to the conditions behind the extent of transfers:

1. Direct intended technology transfer is likely to be less important than the indirect mechanisms of learning.

2. The biggest scope of technology transfer through subcontractors is by means of learning facilitation.
3. The major forms of technology absorption by SME suppliers depend upon (i) the TNC's strategic emphasis on subcontractors, and (ii) whether the subcontracted items are standard or specialised.
4. Scope and form of technology absorption through subcontracting is influenced by the extent of the TNC in-house knowledge of the process technology of the supplier.
5. The actual speed/magnitude of technology absorption by an SME supplier is determined jointly by its own internal technology absorption capacity and by the scope for indirect technology transfer. The internal technology absorption capacity of SMEs, in turn, as argued earlier, is influenced by their business vision and technology strategy of the entrepreneurs as well as their managerial/technical/financial capacity (Wong, 1991).

These points show that learning facilitation is the most important buyer-supplier link in the study carried out by Wong. For a maximised benefit for both partners, Wong further advises that it may be necessary to reduce the number of suppliers and to seek to maintain a reasonably long-term relationship. The key issue in the buyer-supplier relationships for a TNC in a developing country is to ensure that the suppliers' outputs perform according to specifications. Specifications are followed more easily if the buyer places the emphasis on quality assurance technology; such a process is made easier if the foreign affiliates keep a limited number of suppliers.

Proposition 13 The purchasing strategy of the buyer, in terms of number of suppliers, is indicative of the potential links created.

The case of electronics in Malaysia. The share of domestic sourcing can be high in Malaysia, but it is often the result of purchase from other foreign firms in the country. Competition and the use of state-of-the-art technology from TNCs may have increased high-precision and quality control technologies from the subcontractors. Rasiah (1995) finds the transfer of product and process technology important. Such a transfer was only beneficial for technologically advanced suppliers. Hence, the need to promote the technological capabilities of local suppliers, local R&D and information about existing suppliers to the TNCs was underlined.

The Malaysian electrical industry is mainly concerned with the assembly of medium-tech consumer durables (air-conditioners, radios, TVs and low-to-medium-tech intermediate goods, such as switchgear). This industry has created backward links with both local producers and importers. The electronics industry is involved with the production/assembly of semiconductor devices, themselves components for consumer durables and investment goods. In this case, most of the input is imported in the first stage of production, but secondary inputs such as metal lead frames are purchased locally. Most of the transaction takes place at the subsidiary-to-subsidiary level (O'Brien, 1993). Some ancillary firms still benefit from the activities of foreign firms, notably in the semiconductor industry, by specialising in auto-precision turning work, precision engineering, plastic mould and parts' fabrication, manufacture of automation systems, precision plastic moulding products, plastic extrusion products, making moulds and stamping, tooling, die-making and packaging (Rasiah, 1989).

Transfers through backward linkages: what does it involve?

A description of the types of transfers operated by foreign affiliates to their local suppliers has been given and discussed above, but without giving a priority to the key transfers operated. Tables 3.5 and 3.6 are compiled from two major studies on the creation of backward linkages in developing countries. Table 3.5 indicates the results obtained by Halbach (1989) for four sectors, showing examples of support of TNCs to their subcontractors, as seen by the contractor. Within the 'technical support and decision aid' category, the most frequent transfers are the supply of blueprints, cooperation with the subcontractors, supply of tools and assistance in the procurement of raw materials and intermediate products. These results vary according to the sub-industry concerned. The second group of transfers describes the general business-organisational assistance. In this category, the business and technical management assistance are the most frequent transfers. The third category describes financial assistance. Compared to the other two categories, there is little frequent transfer of financial assistance. It might be noted, however, that the pre-financing of materials, tools, facilities, and the prepayment of orders placed are used by some TNCs.

Table 3.6 shows the perceived importance of various processes of technology transfer/absorption through subcontracting relationships, as mentioned by SME suppliers as an important source. Wong (1991) collected these results through interviews. None of the direct transfers seems to have a high influence on the suppliers. Three transfers scored 'high', namely 'product design specification and performance requirements', 'exposure to the multinational enterprise (MNE) systems' and 'testing and diagnostic feedback'. These are either spillover transfer or learning facilitation categories. In the same study, Wong (1991) provides an application of transfer per country of origin and host country. In the case of Malaysia, suppliers to North American firms follow specific quality control methods, receive technical assistance and work with buyers on specific products. Suppliers to Japanese firms receive no training, but receive technical assistance, and their quality performance is closely monitored.

Table 3.5 Support of TNCs to their subcontractors, as seen by contractors (percentage of firms with frequent aid support)

Percentage of firms with frequent aid	Automobile	Electrical industry	Machinery and precision instruments	Food processing industry
1. *Technical support and decision aid*				
Development of initiatives in establishing subcontracting enterprises	27	15	8	57
Cooperation in capacity- and production-planning of the subcontractors	43	20	38	37
Consultation on the technical organisation and the purchasing of machinery	13	22	21	50
Supply of blueprints	80	83	83	37
Supply of (used) machinery	10	2	13	17
Supply of (special) tools	27	49	46	17
Assistance in the procurement of raw materials and intermediate products	13	32	38	43
Other technical support	47	27	21	-
2. *Transfer of business/organisational know-how including training*				
General business-organisational assistance	3	2	8	33
Support for accounting and book keeping	7	0	4	33
Business management assistance	7	2	0	67
Technical management assistance	27	32	29	-
Support in technical production and quality control	67	78	63	100
3 *Financial assistance*				
Contribution to the risk-capital of the subcontractor	0	0	4	16
Granting of repayable loans at usual commercial rates	3	0	4	0
Granting of repayable loans on concessional terms	0	0	0	16
Giving of non-repayable financial grants	0	2	0	0
Pre-financing of materials, tools, facilities etc.	20	20	38	16
Pre payment of orders placed	15	27	38	50
Other financial support	0	0	8	-
Total number of firms	30	41	24	8

Source: compiled from Table 2 in Halbach (1989: 13).

Table 3.6 Perceived importance of various processes of technology transfer/absorption through subcontracting relationships, as mentioned by SME suppliers as an important source

Direct transfer	
Advice on plant layout, equipment selection and operations planning	Low
Advice/training on quality management system and other 'good manufacturing practices' (GMP)	Medium
On-site audit of plant operation and troubleshooting of specific productivity problems	Low
Loaning of equipment and machinery, either temporarily or permanently	Low
Training of supplier staff through formal courses/seminars or informal consultations/visitations	Low
Spillover transfer	
Product design specification and performance requirements	High
Early supplier involvement in prototype development and value engineering stage	Medium
Access to technical and market information on competitors' products	Low
Informal sharing of technical information and ideas among the technical staff of both companies	Low
Exposure to MNE system of managing and organising manufacturing activities and observation of MNE corporate culture (organisational and management technology)	High
Exposure to ongoing R&D activities and other technological developments of MNE	Low
Recruitment of MNE technical/managerial staff	Low
Learning facilitation	
Testing and diagnostic feedback on quality and other dimensions of performance of supplier's products	High
Sourcing of technical experts to solve specific technical problems encountered by the supplier	Low
Early feedback on prototype design	Low
Advanced indications on future quality/performance/feature requirements and targets	Medium
Introducing potential technology supplier or JV partner	Medium
Inducement	
Reduction of the perceived risk of technology investment decision by implicit MNE procurement commitment	Medium
Provision of a stable source of income to finance the investment	Medium
Access to superior market demand information that improves investment decision	Low

Source: Wong (1991: 42-3).

Although some financial flows may exist and take place, the two major categories of buyer-supplier transfers are the knowledge and technology transfer and the transfer through training (the latter often being technical in nature and content). These two categories of transfer are a means for TNCs to ensure greater quality performance from their suppliers in developing countries, and they account for two of the specific information flows categories presented in Table 2.1. Thus, the emphasis, when studying buyer-supplier transfers in a developing country, falls on the technical bonds described by Håkanson and Johanson (1988) and Holmlund and Kock (1995).

Framework of Important factors in Transfers through Backward Linkages

Figure 3.1 is a graphical representation of the discussion on backward linkages, transfers and key factors in the analysis of the two concepts. On the left-hand side of the figure, the factors affecting the level of transfers have been grouped by major categories. Thus, the four major categories of factors would include 'the investment environment' provided by the host country, the 'firm-related factors', the 'purchase-related factors' and 'time factor'. The 'investment environment' covers propositions 1, 2, 7 and 11. The 'firm-related factors' cover propositions 3, 4, 5, 8 and 10. The 'purchase-related factors' include propositions 9, 12 and 13. The 'time factor' relates to proposition 6. This framework of key factors varies in content according to the specific industry and the country studied.

The left-hand side of Figure 3.1 shows the technology and knowledge transfer categories. For the purpose of discussion, we mention financial flows although previous studies indicate these flows tend to be minimum. The technology transfer and the training categories are the two key categories.

CONCLUSION

The discussion presented above emphasised different factors influencing the creation and level of beneficial linkages between foreign subsidiaries in host developing countries and local firms. In the ASEAN countries, many factors have contributed to facilitation of the development of such linkages. The description of the overall environment, comprising political factors, locational factors and firm-related factors led us to the theoretical implications that the study of traditional benefits generated by TNCs in host-developing countries was incomplete, and attention should be drawn towards the managerial, entrepreneurial and technological linkages in the host economy.

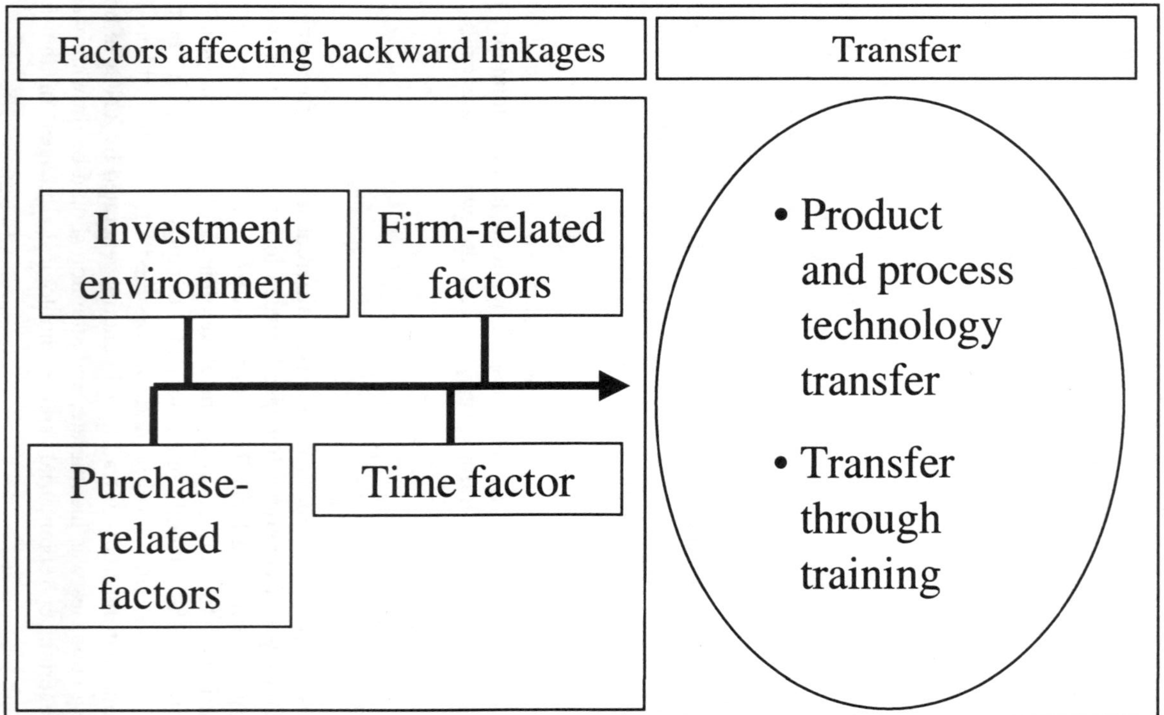

Figure 3.1 Framework of analysis

After a description of the potential impact of TNCs, the specific upgrading of production capabilities through the creation of backward linkages was developed. It is argued in this chapter that the process of local purchasing by TNCs in developing countries is not the most important benefit. Local purchasing by foreign firms is often limited, and its extent depends on various factors, such as the level of economic development of the host country, the foreign firm's strategy and the host government requirements. These are shown in our theoretical framework in Chapter 2. It is the special relationship arising from the purchase of local goods and equipment that generates substantial benefits and leads to the upgrading of production capabilities in the host-developing countries.

The literature review of the transfers occurring through backward linkages shows how foreign firms engage in close interactions with their local suppliers, often in view of long-term benefits through a better quality of supply and services. Transfers include both technological and managerial criteria, and take place either directly through the provision of product specification or intervention in the suppliers' production process, or indirectly through training. Such transfers are provided to the foreign firms' subcontractors, as well as close suppliers with which the subsidiary has entered into a long-term partnership.

Thirteen propositions were drawn from the literature and tackle the conditions under which TNCs may have an impact on host countries' indigenous production capabilities through backward linkage creation. A framework for the analysis of backward linkages encourages paying attention to the investment environment, when studying TNCs and backward linkages. Chapter 4 looks at this issue and provides an in-depth analysis of the Malaysian investment economic environment, so as to assess for this particular developing host country the conditions within which backward linkages take place.

4. Developing a Favourable Environment for Backward Linkages: Malaysia, Foreign Direct Investment and the Electrical and Electronics Sector

INTRODUCTION

Literature on the impact of foreign investment in the economic development of a developing country is extensive. Impacts are likely to vary according to the situation, but also to a large extent according to the amount of foreign investment in the country. Malaysia has been among the 10 largest developing country recipients of world investment since the 1980s, a large proportion of its gross domestic capital formation is generated by foreign capital, some of its sectors of activity depend almost entirely on foreign firms, a large share of its exports are generated by foreign firms, and in some sectors, such as the electrical and electronics sector, nearly all exports are generated by foreign firms. This shows a high dependence of economic performance on foreign firms. Therefore, the government has been trying to diversify its economic base to encourage the development of indigenous firms, to encourage foreign firms to move into higher value-added production and to create more links with local firms.

Pushed by the rise of lower-wage economies in the neighbouring countries (primarily Vietnam and China), rising wage rates at home, and at times, near full employment, Malaysia needs to ensure future economic prosperity. It is unlikely that foreign firms already in Malaysia will relocate or disinvest, considering the existing sites they have created in the country. However, it is crucial to ensure that future waves of investment continue to pour into the country, and also to make sure that local firms become more competitive, both nationally and internationally. The development of supporting industries at home may be crucial to the future inflow of foreign capital. Such a development will also lead to a decrease in import-dependence of foreign firms for their input of parts and components, and may improve the balance-of-payments' situation. Backward linkages and supporting industries are increasingly important for host-developing countries, which endeavour to generate high potential benefits from the presence of foreign firms within their national boundaries.

This chapter aims to evaluate the current situation of Malaysia's past and current dependence on foreign capital and to give a picture of the trends in foreign investment in the country. By describing the various economic development stages Malaysia has undergone, we show that the country has long been welcoming foreign investors, and has always tried to create a favourable and sound economic environment for them to operate in. The example of mixed government policies, combining both import-substituting policies with export-oriented ones, is unique. Since the late 1980s, the government has opted for a more monolistic approach, favouring export-orientation as a means for economic development. Since the New Development Policy was launched in 1986, the government has managed to attract increasing amounts of foreign direct investment (FDI). Nowadays, concerns are not solely related to export promotion and investment attraction, but very importantly they turn towards the creation of substantial linkages in the economy. The government has realised that a comprehensive promotional approach for the ancillary industries is needed to create industrial linkages and to further widen and strengthen the industrial structure. After providing a historical background to Malaysia's development path and a detailed analysis of government policies related to investment, current trends in FDI are described. The last section of this chapter is dedicated to the analysis of the electrical and electronics sector, and the special place it takes within the Malaysian economy, notably in terms of foreign firms' operations in the country.

HISTORY OF FOREIGN INVESTMENT TRENDS AND GOVERNMENT POLICIES

Over the past few decades, Malaysia has gone through a series of economic 'stages' characterised by a strong government involvement in business, a substantial reliance on foreign firms for the development of key economic sectors and the growth of indigenous companies, some within the public arena. Malaysia has a long merchant tradition; its major ports have facilitated trade from the East and the West since the 14th century. The concept of an open economy is deeply embedded in the local culture, and the external orientation of the economy means that exports and imports have nearly always and will continue to constitute a high percentage of the country's income.

The colonial era has been much criticised, especially for its role in specialising the Malaysian economy in a handful of natural resources activities. It is said that the country was used by the British Empire as a major supplier of rubber and tin, and, as a result, became almost wholly dependent upon those two primary commodities for its export earnings by the early 20th century (see, for example, Fong, 1990a). Regardless of criticism, it is true that indigenous firms have a strong foothold in natural resource-related industries, and Malaysia currently remains today a major world producer of rubber, palm oil and tin. The country does benefit from a well-developed resource-based sector providing a substantial share of the export earning. From the colonial era, Malaysia has also inherited a liberal economic system with a well-functioning bureaucracy, a tradition of stable macro-economic policies and a good infrastructure.

Over the last decades, successive governments have adapted national policies, but have always followed a trade-oriented stance and kept open policies towards foreign investment in the productive sectors. We can perceive four stages that arise in the Malaysian economic development pattern since its independence. These broadly follow the Malaysian Master Plans. Phases I and II are described in-depth in other publications (Fukushima and Kwan, 1995; Lall, 1995). We intend therefore to build on these, and develop models for Phases III and IV to indicate the relatively 'conducive' economic environment in which foreign firms have operated. As the international and national conditions have changed, so has the government's perception of the potentials of FDI. In recent years, the Malaysian government has strongly emphasised the possibilities of spillover generated by foreign firms. The policies adopted were adapted accordingly in order to maximise the positive impact of transnational corporations (TNCs) in various sectors. Phases III and IV show clear signs in the government's change of attitude towards the effort to encourage inter-firm linkages throughout the country and across sectors of activity, with particular emphasis on the manufacturing sector.

PHASE I: 1957-70 First Steps after Independence

Malaysia became independent from the UK in 1957, and the need for an independent political stance started after independence. Over the first phase, government policies showed moderate protection for import-substituting activities, but there was no pursuit of protectionist policy that would have boosted manufacturing at the expense of agriculture. The primary aim of this policy was to reduce dependence on imported consumer goods, promoting the use of domestic natural resources, and creating employment opportunities. Foreign investment is a very old phenomenon in Malaysia; in the 1920s, there was already substantial British investment, mostly concentrated in the plantation and mining sectors (90% of the total). In the 1960s, the government established measures to attract FDI into export activities. It is very important to note that FDI in export-oriented activities has thus been promoted ever since the 1950s. Major government interventions were found in the promotion of rural development and in the development of infrastructure. A *laissez-faire* attitude was adopted in the industry, yet with the endeavour to create a favourable investment climate.

The *Pioneer Industries Ordinance* was promulgated in 1958. As is reflected in Table 4.1, the aim was to provide financial inducements in the form of tax holidays,[1] industrial estates and supporting services and industries. The *Pioneer Industries Ordinance* was replaced in 1968 by the *Investment Incentives Act* (IIA), as Malaysia made a policy shift away from import substitution, and the government showed a clear commitment to encourage investment, regardless of its origin. This change was coupled with the creation of the *Federal Industrial Development Authority* in 1965 in an attempt to promote and regulate industrial development. This body became the *Malaysian Industrial Development Authority* in 1968, a governmental body that has gained in importance ever since, and to this day serves as a central body for investment governance. In the 1960s, growth was primarily the result of import-substituting industries and industry was essentially Chinese-owned and concentrated in light first-stage manufactures such as textiles, footwear and garments. Close links between the state and major Chinese business leaders existed, showing a strong importance of interpersonal relationship between the Malay political and bureaucratic élite and the business community. Searle (1999) refers to the Chinese business groups prominent in the 1960s and 1970s as the Old Chinese Wealth in his classification of Malaysian capitalist groups. The place of minority ethnic groups in the overall share of economic wealth in Malaysia became particularly important in the late 1960s, as is emphasised in Phase II.

PHASE II: 1971-85 Combining Import-substituting Policies with Export promotion

Although the second stage in economic development of Malaysia is assimilated to the launch of the *New Economic Policy* (NEP) by the government, it could be argued that this reversal in political stance originally started with the social conflict that arose in 1969. The *Bumiputeras* (native Malays) showed increasing discontent in their relatively poor economic conditions compared to people from minority ethnic groups: the Chinese and the Indians. In an attempt to level out differences in wealth between various ethnic groups, the government launched the NEP, to eradicate poverty and restructure society by redressing economic imbalances between the ethnic groups (Table 4.2). The ethnic crisis had a substantial influence on the government's attitude towards the manufacturing sectors. The government embarked on new domestic industrial interventions, by launching state-owned firms in all activities, thus increasing *bumiputera* employment rates. Education was used to increase the qualification of the latter group.

In parallel, foreign investment continued to be encouraged. With the Free Trade Zones Act, the objective was to attract TNCs within certain industrial estates. Considerable effort was devoted to generate a favourable investment environment, especially through the granting of incentives. The incentives granted to firms willing to set up in free trade zones (FTZs) included duty-free imports of raw material and capital equipment in the zones, streamlined customs formalities with reduced administrative formalities, infrastructure facilities within the zone (the government rented factory buildings but encouraged foreign firms to build their own) and company income tax incentives (including *pioneer status, labour utilisation relief, investment tax credit, export promotion, deduction and locational incentives).* The domestic policies aimed at increasing the share of wealth owned by the *bumiputera*-affected foreign firms to some extent, because they were encouraged to create joint ventures with *bumiputera*-owned companies, and employ a specified share of *bumiputera* employees.

One of the promotional approaches by the government to encourage foreign investment included targeted sector policy. In this respect, MIDA (Malaysian Industrial Development Authority) saw an opportunity for the development of the electronics industry in Malaysia. In the 1970s, Singapore was trying to move up to higher value-added products (Mirza, 1986). MIDA targeted its efforts on attracting firms that would previously have chosen Singapore as a location for their electronic product manufacturing. Particular effort was made on attracting TNCs from the USA. MIDA offered generous incentives packages, as defined within the Investment Incentives Act, a favourable investment climate, and it emphasised the disciplined educated workforce. The fact that this workforce could speak English was perceived as an advantage for the US TNCs. All these factors led to foreign firms setting up semiconductor manufacturing plants within the FTZs in the first half of the 1970s. The increase in inward foreign investment in the first half of the 1970s should have been further facilitated with the amendment of the Customs Act in 1975 (see Table 4.2), by which firms that were designated as Licensed Manufacturing Warehouses were granted similar advantages as those located in FTZs. However, two factors played against this upward trend. The first one resulted from the reinforcement of the objectives of the NEP, through the enactment of the Industrial Coordination Act (ICA).

Table 4.1 Government bodies and policies related to investment, 1950s-60s*

Year	Government body/law	Aim/Role
1951	Employee Provident Fund	Provides for a compulsory contributory provident fund that is payable for employees in full on reaching the age of 55 years. All employers and employees are required to contribute to the fund at the rates of 12% and 10%, respectively, of the employees' monthly wages
1955	Employment Act	Main legislation regulating terms and conditions of employment
1958	Pioneer Industries Ordinance (PIO)	Provides financial inducements in the form of tax holidays, industrial estates and supporting services and industries. Replaced in 1968 by the Investment Incentives Act, as Malaysia made a policy shift away from import substitution, and currently known as the Promotion of Investment Act, set up in 1986
1965	Companies Act	Provides guidelines and registration procedures for all companies conducting business in Malaysia
1966	Action Committee on Tariff and Industrial Development (ACTID)	Provides regulation and ensures application of tariff scheme. Considers applications for pioneer status, thus has control over part of industrial development
1967	Industrial Relations Act	Provides for regulation of relations between employers and workers and their trade unions, and the prevention and settlement of trade disputes
1967	Tariff Advisory Board	Grants tariff protection
1967	Income Tax Act	Contains tax law and incentives for some sectors
1968	Malaysian Industrial Development Authority	Government body responsible for promoting and regulating industrial development, originally created in 1965, first named the Federation of Industrial Development Authority
1968	Investment Incentives Act (IIA)	Replaces the Pioneer Industries Ordinance, set up in 1958. Emphasis drawn away from import substitution. Incentives administered by ACTID. Measures include accelerated depreciation, reinvestment allowances and investment credit for non-pioneer industries. Important incentives also include export incentives, with export allowance (to provide deduction of taxable income according to export performance and domestic input content), tax deductible promotional expenses for exports, accelerated depreciation allowance (if exports represent more than 20% of output), and export financing and insurance facilities at preferential rates
1969	Employees' Social Security Act	Administers Employment Injury Insurance Scheme and Invalidity Pensions Scheme

Source: compiled by the author, information provided by MIDA.
Note: * legislation in Malaysia is not specific to foreign investors.

Table 4.2 Government bodies and policies related to investment, 1970s-80s*

Year	Government body/law	Aim/Role
1971	New Economic Policy	To eradicate poverty and restructure society by redressing economic imbalances between ethnic groups
1971	Free Trade Zone Act	To attract TNCs within certain industrial estates. Incentives include duty-free imports of raw material and capital equipment in the FTZs, streamlined customs formalities, with reduced administrative formalities, infrastructure facilities within zone (government rents factory buildings but encourages foreign firms to build their own) and company income tax incentives (including Pioneer Status, Labour Utilisation Relief, Investment Tax Credit, Export Promotion Deduction and Locational Incentives)
early 1970s	Regional Development Authorities	To integrate agricultural, commercial and rural development in selected regions
1972	Sales Tax Act	
1975	Customs Act	Originally created in 1967, amended in 1975 to allow firms to be designated as Licensed Manufacturing Warehouses (LMW) so as to allow them to operate with similar advantages as those located in FTZs
1975	Industrial Coordination Act	Regulated investment activities in Malaysia in accordance with NEP. Initial requirements were amended in the 1980s. The ICA requires manufacturing companies with shareholders' funds of RM2.5 million and above or engaging 75 or more full-time employees to apply for a manufacturing licence for approval by the Ministry of International Trade and Industry (MITI).
1976	Excise Act	
1986	New Development Policy	Change in policies. Creation of first Industrial Master Plan to target specific activities so as to strengthen comparative advantages
1986	Promotion of Investment Act	Provides incentives for manufacturing, agriculture, tourism and hotel projects, replaces IIA
1990	Free Zones Act	To enable operations in zones to enjoy minimum customs' control and formalities in import of raw materials, parts, machinery and equipment, as well as in export of finished goods

Source: compiled by the author, information provided by MIDA.

Note: * legislation in Malaysia is not specific to foreign investors.

This Act started to regulate investment activities in Malaysia in accordance with the NEP. It required all manufacturing enterprises with 25 or more employees (or with paid-up capital greater than M$250,000) to obtain a licence to manufacture. This made administrative procedures more troublesome and slowed down the granting of authorisation through the licensing process. Obstacles also lay in the way by which the licence was granted.

By the late 1970s and early 1980s, inward foreign investment flows started to decrease as a result of several factors. The government strengthened measures to increase *bumiputera* participation, both as a share of employees, and as a share in companies' ownership structure. Restrictive features of the ICA, such as the emphasis on joint ventures-linked projects also accounted for a sharp decrease of foreign investors' interest in Malaysia. Additionally, FTZs grew into enclave economies. Exports generated by TNCs concentrated in a narrow range of economic activity, and linkages between foreign firms and the local economy were virtually non-existent (UN-ILO, 1991). As a result, TNCs were not seen as beneficial to the economy, apart from their generating income from exports.

The development of a dualistic economy between the industries in the FTZs producing for export and the local industries producing for the domestic market was the result of a dual governmental strategy. The government used state interventions through licensing and quotas, and regulated prices in some sectors, and promoted foreign investment in export-oriented industries.

The reversal in policies reached its peak in the early 1980s, when the Mahathir government launched tougher industrial policy measures. The model followed was that of the Korean policy measures with the heavy industry drive. Korea promoted economic development by promoting *chaebols* in key economic sectors (Amsden, 1989). In the Malaysian context, however, this policy was not successful for a number of reasons. The Malaysian government launched the Heavy Industries Corporation of Malaysia (HICOM). The best known project carried out under HICOM is the Proton national car (Searle, 1999). As presented by Lall (1995), this policy was not successful for a number of reasons. Contrary to the Korean heavy and chemical industry, HICOM was domestic-oriented, the Malaysian interventions were led by 'soft' budgets, without a previous 'learning' base for absorbing complex organisational and production technologies. Similarly, whereas exports in the newly industrialising economies (NIEs) came from domestic companies, it was nearly entirely due to foreign firms in Malaysia, which strongly lacked interventions to promote local research and development (R&D) and procurement-led technological development and linkages.

HICOM did give rise to some successful ventures. In 1995, HICOM Holdings was one of Malaysia's biggest listed conglomerates. It had functioned as the government's holding for heavy industrial enterprises. Listed companies in the HICOM include the Perusahaan Otomobil Nasional (whose main product is the Proton automobile), and the Edaran Otomobil Nasional or Eon, in charge of distribution for Proton (*Financial Times*, 2 November 1995).

Table 4.3 Ownership of share capital in corporate sector (percentage)

	1971	1980	1988
Foreign	61.7	42.9	24.6
Share in Malaysian companies	32.9	24.0	15.8
Net assets of local branches	28.8	18.9	8.8
Malaysian	38.3	57.1	75.4
Bumiputera individuals	2.6	5.8	13.0
Bumiputera trust agencies	1.7	6.7	6.4
Other Malaysian residents	34.0	44.6	56.0
Total (RM million)	6,564	32,420	97,971

Source: Ariff (1991: 109).

The result of government policies is striking when one looks at the share of foreign firms in the corporate sector over the period of Phase II. The latter has decreased continuously during the 1970s and 1980s. Since the enactment of the NEP, the share of *bumiputera* in the corporate sector has increased substantially, from 4.3% in 1971 to 19.4% in 1988 (Table 4.3). Despite the emphasis given by the government to the share of *bumiputera*, the most noticeable change in the corporate sector's ownership structure has taken place among 'other Malaysian residents', whose share in the corporate sector was boosted from 34% in 1971 to 56% in 1988.

PHASE III: 1986-95 Creating Favourable Investment Environment

Phase III's clear starting point is that of another major change in government policies. The New Development Policy and the Promotion of Investment Act were both enacted in 1986. The former led to the design of the first 10-year Industrial Master Plan (IMP, 1986-95). The government reinforced the outward-oriented development strategy and, as part of it, the import liberalisation policy was implemented again. This occurred because the early 1980s' policies did not generate enough economic performance, and a number of infant industries never grew, despite being protected for a long time. This prevented the export industry from strengthening its competitiveness in the international market. Malaysia therefore opted for a trade neutrality policy, that is, the coexistence of export promotion policy and import liberalisation policy (Okamoto, 1994). Twelve sub-sectors were given high priority status for long-term development.[2] Some of the products that were promoted were no longer attractive to producers in Asian NIEs owing to relatively low value-added technology and skill content; however, they could help home producers to control products and process technologies. The aim of the Promotion of Investment Act (see Table 4.2) was to target specific activities in an attempt to strengthen the existing comparative advantages of the country, while the aim of the latter was to provide incentives for manufacturing, agriculture and tourism. Selective strategies included targeted skills, training, technical support, finance and quality improvement, together with more import protection.

A Technology Action Plan was drawn up, and the Industrial Restructuring Plan was devised to provide cheap finance for wood, textiles and engineering. Trade was liberalised while infant industry remained protected. Finally, just as Singapore had done a decade before, MIDA applied selective policies to draw export-oriented TNCs into higher value-added activities. These efforts show the emphasis put by the government on upgrading local capabilities. Supplies were targeted in particular. Foreign suppliers that had invested in Malaysian export processing zones to supply their principal customers were denied some privileges received by wholly export-oriented firms, and were treated as local firms.

The New Development Policy and the Promotion of Investment Act in particular resulted in the surge of private investment, both domestic and foreign, which has continued until the recent financial crisis in the region. In the late 1980s and early 1990s, new investors from NIE countries started to invest heavily in Malaysia. These trends are covered in this chapter.

PHASE IV: 1995 onwards towards 2020, from 'Tiger' to 'Dragon', or the Bumpy Road to Economic Recovery

The second Industrial Master Plan was launched in 1995 for the period 1996-2005. There has not been any major reversal in the government policy orientation, continuing the promotion of reinvestments, industrial linkages, export and training. However the New Industrial Master Plan takes a cluster-based approach, whereby strategies focus on the development of specific industry clusters. In this connection, the clusters at various levels of evolution are:

- naturally evolving clusters: these are resource-based industries such as wood-based, rubber-based, palm-based, petroleum-based or chemical products;
- policy-driven clusters: these are the automotive, aerospace, machinery and equipment industries, which constitute the strategic sectors;
- clusters with international linkages: these are the electronics and electrical appliances, and textiles industries (MIDA, 1996).

Such policy orientation does show a more proactive approach on the part of the Malaysian government. It appears clearly that investment incentives are not seen as the only way towards development any longer. The new cluster-based approach and a stronger concern for local linkage creation show the willingness of the government to encourage further benefits from investment. In line with this idea, the eighth Malaysia Plan, 2001-05 aims at deepening the knowledge content of the economy, and pays particular attention to the human resource enhancement, as well as the development of R&D facilities (MIDA, 2001).

Despite these improvements in policies, in the mid-1990s, Lall (1995) stressed five remaining worries for the future development of the economy. First, there is still a high degree of concentration of the activities. Second, TNCs continue to dominate export, and there is still a low participation of local private firms, except for garments. Third, local content of most manufacturing exports remains low, and its increase has been driven by foreign investment in the component sectors. Fourth, both local and foreign firms have few high value-added and technologically demanding tasks like design and development. The general level of the Malaysian technological capabilities is low, and the underlying supply of high-level technical skills is inadequate. Fifth, the Malaysian exporters have not developed independent marketing capabilities, constraining their ability to upgrade into high value-added products and markets. There were therefore still many challenges that the government needed to address. Arguably, these problems have persisted or worsened as a result of the Asian crisis.

Following the devaluation of the Thai baht, a wave of speculation hit Malaysia, and with its foreign exchange reserves down to US$28 billion, Malaysia let the ringgit float on 14 July 1997. The devaluation brought the ringgit down from US$1 for RM2.525 before the crisis to $1 for RM4.15 six months later. There were several reasons why Malaysia should be brought into the speculative attacks after Thailand. Reasons include a combination of excess investment, high borrowings much of it in dollar-denominated debt, and a deteriorating balance-of-payments' position. Malaysia was however able to stop the slide in its currency and stock markets without the help of the International Monetary Fund (IMF). The crisis has prompted the Malaysian government to make some changes in its policies towards investment. These changes will be covered in the following section.

To conclude comments presented on Phase IV, one can argue that despite a strong emphasis on investment, there has been little development of locally owned firms in key sectors of the economy. Training and skills' development have been emphasised since the mid-1980s by the government, but the positive effect of this political strategy is still to come. First of all, it takes time for these objectives to generate sufficient beneficial spillovers in the economy; second, the country is only slowly recovering from the events of the Asian crisis. The government has only recently started to promote specific links between TNCs and local firms, the cluster-based approach, and the need to help in the generation of inter-firm linkages and to push towards higher-tech industries that only started in the 1990s.

NATIONAL INVESTMENT STRATEGY AND GOVERNMENT POLICIES

The historical background above helps to understand why the government and the private sector are closely interlinked in Malaysia. The review of policies and government attitude towards foreign firms has also shown the effect of pursued policies on FDI inflows in Malaysia. To understand fully the current trends and prospects for foreign firms in the Malaysian context, and their subsequent linkage creation in the economy, it is essential to study the national investment strategy and government policies. There exist several sources to study government policies carried out in Malaysia; most of the information in this chapter was compiled from the MIDA itself (http://www.mida.gov.my), but also from the Asia Pacific Economic Cooperation (APEC) *Guide to Foreign Investment* updated regularly (http://www.apec.org), and the newly published *Investing in ASEAN: a Guide to Foreign Investors* (ASEAN, 1999). This section provides an overview of government policies, showing the development strategy followed by the current Malaysian government; the FDI procedures and requirements to understand the process through which foreign firms have to go to localise their operations in Malaysia; and finally, the particular initiatives taken by the government to encourage backward linkage creation.

Overview of Government Policies

As was developed earlier, Malaysia has adopted an export-oriented industrialisation strategy since the early 1970s. The policy aimed at reaching a higher level of international competitiveness in the 1970s through the creation of an enclave economy. The government established FTZs and special incentives were offered to investors.[3] Other incentives to promote exports and employment were offered through the Investment Incentives Act of 1968. The government has long relied on foreign firms to generate the country's exports. The current policies towards foreign investment still reflect this policy orientation.

In 1999, growth rates relied heavily on two sectors, which represent the Malaysian strength after the Asian crisis. These two sectors are the manufacturing sector and the services sector; they showed respective growth rates of 13.5% and 2.9% in 1999. The manufacturing sector was the fastest-growing sector throughout the 1990s. This sector contributed to 33.1% of the gross domestic product (GDP) in 1995 and 30.1% in 1999; and the share of manufactured exports to total export earnings was 79.6% in 1995, and 84.6% in 1999 (MIDA, 1996, http://www.mida.gov.my). Currently, Malaysia is among the world's leading exporters of manufactured goods such as semiconductors, room air-conditioners, rubber gloves, catheters and audio-visual equipment. Additionally, the importance of natural resources needs to be cared for. Malaysia continues to be one of the world's largest producers of natural rubber and palm oil. The country is also a leading producer of tropical hardwoods, pepper and cocoa and a net exporter of crude petroleum and natural gas. Another emphasis of the government is on the development of the infrastructure. There are over 200 industrial estates throughout the country. Some were developed by government agencies, namely, the State Economic Development Corporations (SEDCs), Regional Development Authorities (RDAs), port authorities and municipalities. Others have been developed and are being established by private developers in certain states. Additionally, 14 FTZs[4] (also called free industrial zones), exist where companies that export more than 80% of their production, and mainly import raw materials may be eligible to locate. Three specialised parks were launched in the 1990s, namely Kulim High-Tech Park, Technology Park Malaysia and JSEDC Technopark located, respectively, in the states of Kedah, Selangor and Johor. These parks are equipped with modern and up-to-date facilities for high-tech manufacturing activity and R&D companies. A new Biotechnology Park has also been developed in the state of Sarawak. Last, considerable efforts concentrate on the enhancement of a favourable investment environment. Phase IV showed how the Malaysian government increasingly recognises the contribution of foreign investors to technology and knowledge transfer, capital inflow and export market expansion. More than 4,000 international manufacturing companies from over 50 countries have operations in Malaysia (MIDA, 2002).

Keeping in mind the Malaysian strengths, the importance of natural resources and the investment environment, the following section introduces the strategy pursued by the government to promote exports, to try and attract a higher value-added production range within its national boundaries, and to try and influence the creation of local linkages that would benefit the improvement of indigenous firms, while at the same time trying to maintain a favourable and attractive environment for foreign firms to operate in. First, we look at the way the government tries to encourage investment through incentives, then we look at selective industrial policies and finally at the way the government tries to influence shifts in comparative advantages.

Policies: encouraging investment through incentives

The major investment incentives offered by the government are included in the Promotion of Investment Act (PIA, launched in 1986), and are listed in Table 4.4. Those incentives are available to both foreign and domestic firms. Other incentives are provided for in the Income Tax Act 1967, the Customs Act 1967, the Sales Tax Act 1972 and the Excise Act 1976. Tax incentives are provided mainly for the manufacturing sector, but also for agricultural and tourism activities. There are three different lists of activities and products promoted by the government. The first list is general, whereas the second one targets high-technology companies, and the third small-scale companies. A list of specific products promoted was also published under the PIA in 1986.

The majority of the incentives listed in Table 4.4 apply to the manufacturing sector, while additional incentives are aimed at the agricultural sector and the tourism sector. Similarly, specific incentives for the protection of the environment are applied. Investment is encouraged from both foreign and local firms. The government's objectives come out clearly from the analysis of the incentives.

Selective industrial policies

Selective industrial policies are carried out by the promotion of chosen sectors of activity and products, as appears in the lists of promoted products and activities. Emphasis is strong on infant industry protection, or through investment tax allowance, and on infrastructure development, as mentioned within the 'General incentives' in Table 4.4. Alongside these selective industrial policies, the government is actively promoting the move towards higher value-added production locally. To facilitate manufacturing development, the government has created over 200 industrial estates and 14 FTZs.

Table 4.4a Incentives offered to investors

1. ***General incentives***
Company not entitled to benefit from pioneer status and investment tax allowance simultaneously.

- *Pioneer status:* Granted for companies that take part in promoted activity or produce promoted products. Incentive takes form of abatement of 70% of statutory income for 5 years. Full income tax exemption granted for 5 years to high-tech companies, and for 10 years for projects of national and strategic importance involving heavy capital investment and high-tech. Abatement of 85% granted for companies located in Eastern Corridor state of Peninsular Malaysia.
- *Investment tax allowance:* Allowance given on 60% of capital expenditure incurred on factory or plant and machinery used for purposes of approved activity for period of 5 years. Allowance can be used to offset 70% of statutory income in year of assessment. A 100% allowance permitted for project of national and strategic importance for period of 5 years, 80% for project located in the Eastern Corridor.
- *Reinvestment allowance:* For manufacturing companies that incur qualifying capital expenditure for expansion of production, capacity, modernisation and upgrading of production facilities and diversification into related products. Allowance can be used to offset 70% of statutory income in year of assessment. There again, further allowances provided for firms in Eastern Corridor.
- *Incentives for industrial adjustment:* Companies in operation before 31 December 1990 in the wood-based, textile, machinery and engineering sectors eligible for allowance of up to 100% in respect of qualifying expenditure incurred by manufacturing company in efforts at undertaking industrial adjustment.

2. ***Export incentives***

- *Double deduction for promotion of export:* Includes first, double deduction of export credit insurance premiums paid by product exporters to approved local insurance companies. Second, double deduction for promotion of exports. Expenses qualified include wide range of expenses incurred by presence of company abroad such as overseas advertising, supply of free samples abroad, export market research, expenses of personnel abroad, or cost of maintaining sales' offices overseas.
- *Export credit refinancing* (ECR) Provides short-term credit at preferential rates of interest. Facility operated by commercial banks. Central Bank Negara Malaysia will refinance those commercial banks that have extended export credit to eligible exporters. Scheme includes either pre-shipment or post-shipment refinancing for goods that have minimum value added of 20% and minimum domestic resource content of 30%, for companies that have secured ECR credit facility with any commercial bank. Pre-shipment refinancing granted for companies whose exports' volume is at least RM3 million per year, to fund inventory and raw materials prior to receipt of export orders. Maximum period of financing: 4 months for pre-shipment ECR, 6 months for post-shipment ECR. Minimum amount for ECR financing RM10,000, minimum drawdown RM2,000.
- *Industrial building allowance*: Granted in respect of buildings used as warehouses for storing goods for export, initial allowance is 10%, followed by 2% annually.

3. ***Incentives for high tech and strategic industries:*** High-tech companies are defined companies engaged in promoted activities or production of promoted products in area of new and emerging technologies. Local R&D expenditure to gross sales should be at least 1% on annual basis and percentage of science and technical graduates to total workforce should be at least 7%. Granted pioneer status with full tax exemption or investment tax allowance (ITA) of 60% of qualifying capital expenditure incurred within period of 5 years.

Table 4.4b Incentives offered to investors

4. ***Incentives for research and development:*** 'R&D means any systematic or intensive study carried out in the field of science or technology with the object of using the results of the study for the production or improvement of materials, devices, products, produce or processes'. Contract R&D company or R&D company can benefit from the pioneer status with full income tax exemption for 5 years or from investment tax allowance of 100% on qualifying capital expenditure incurred within 10 years. ITA can be used to offset 70% of statutory income in year of assessment. Companies carrying out in-house research in Malaysia eligible to apply for ITA of 50% on qualifying capital expenditure incurred within 10 years. ITA can be used to offset 70% of statutory income in year of assessment. Additional incentives provided, such as double deduction on revenue expenditure, on payment for use of services, industrial building allowance, capital allowance on capital expenditure, and exemption from the duty/tax for machinery/equipment, materials, raw material/component parts and samples for R&D purposes.
5. ***Incentives for training:*** Investment tax allowance of 100% for 10 years given to companies that establish technical or vocational training institutions. Single deduction for contributions in cash to technical or vocational training institution established and maintained by a statutory body, exemption from import duty, sales' tax and excise duty on machinery, equipment and materials, double deduction for expenses incurred for approved training and industrial building allowance may also be granted.
6. ***Incentives for operational headquarters:*** Approved OHQ refers to locally incorporated company, whether locally owned or foreign-owned, which carries on business in Malaysia of providing qualifying services to offices or related companies outside Malaysia. Paid-up capital of the company should be of RM0.5 million and total business spending should be at least RM1.5 million per annum and should carry out some particular activities (management and administrative services, treasury and fund management services, etc). Company can then benefit from two types of incentives:

- *Fiscal incentives:* Concessionary tax rates of 10% granted: (1) for income arising from qualifying services rendered by OHQ company to its offices or related companies outside Malaysia; (2) for income derived from interest on foreign currency loans extended by OHQ to offices or related companies outside Malaysia; and (3) for royalties received from R&D work carried out in Malaysia by OHQ on behalf of offices or related companies outside Malaysia. Concessions last for 5 years from the date of approval of OHQ status, and can be renewed for 5 years.
- *Non-fiscal Incentives:* Expatriate posts facilitated for period of 3-5 years, obtaining credit facilities in foreign currency without approval of Bank Negara in Malaysia, possibility to borrow in ringgit up to RM10 million without approval from domestic sources for use in Malaysia and thereon invest freely in foreign securities and lend to related companies outside Malaysia, open one foreign currency account or one multi-currency account with any designated bank in Malaysia to retain export proceeds, with any designated bank, licensed offshore bank in Labuan, or overseas bank for crediting foreign currency receivables other than export proceeds.

Table 4.4c Incentives offered to investors

7.	***Non-fiscal incentives:***
•	*Tariff protection*: Can be considered for infant industries, which are in position to supply major portion of domestic market. Degree of use of domestic raw materials, level of local value added and level of technology of industry taken into consideration.
•	*Exemption from import duty on direct raw materials/components:* Full exemption normally granted for manufacture of goods for export, provided raw materials/components not manufactured locally or not at acceptable quality and price. Full exemption will also be considered for manufacture of goods for domestic market.
•	*Drawback of excise duty on parts, ingredients or packaging materials*
•	*Drawback of sales tax on materials used in manufacture*
•	*Exemption of import duty and sales tax on machinery and equipment:* Exemption given for machinery and equipment used directly in manufacturing process, and equipment used for environmental control, recycling, maintenance and quality control.
•	*Drawback of import duty:* Drawback in full granted for all duty-paid goods used as parts or ingredients or as packaging materials in the manufacture of other goods subsequently exported.
8.	***Incentives for strategic projects***
9.	***Incentives for small-scale companies***
10.	***Incentives for agricultural sector***
11.	***Incentives for tourism industry***
12.	***Incentives for computers and information technology assets***
13.	***Infrastructure allowance***
14.	***Incentives for the storage, treatment and disposal of toxic and hazardous wastes***

Source: compiled by the author, from the information provided by MIDA.

Influencing shifts in comparative advantages

Over the past decade, the Malaysian government has paid particular attention to the shift of the country's comparative advantages, towards high-tech intensive sectors of activity. Action is taken in five key areas: promoting high-tech industries through incentives, developing infrastructure for hi-tech industries, encouraging training to develop local skills, attracting headquarters in Malaysia and export orientation.

Promoting hi-tech industries through incentives. With the rise of new competitors, attracting investment in labour-intensive industries, the Malaysian government has undertaken efforts to encourage the development of more capital technology and skill-intensive industries that have a competitive edge in global markets. High-tech products and activities are heavily promoted and can easily benefit from the above 'general incentives'. High-tech companies are defined companies engaged in promoted activities or the production of promoted products in area of new and emerging technologies. R&D is also actively encouraged by the government. R&D means 'any systematic or intensive study carried out in the field of science or technology with the object of using the results of the study for the production or improvement of materials, devices, products, produce or processes. But it does not include the quality control of products or routine testing of materials, devices, products or produce, the research in the social sciences or humanities, the routine data collections, the efficiency surveys or management studies, or the market research or sales' promotion (from the PIA, launched in 1986). The government recognises that the involvement of private companies in R&D activities is crucial to the nation's industrialisation drive (MITI, 1996).

Developing infrastructure for hi-tech industries. Hi-tech industries and R&D activities are not only encouraged through incentives, they are also facilitated by improved infrastructure, such as science parks. Kulim Hi-Tech Park (KHTP) was Malaysia's first science park, and was inaugurated on 25 June 1995. It was developed to attract hi-tech investors to the still poor, largely agricultural state of Kedah, only 26 kilometres from the island city of Penang, where there is a high concentration of electronic industries. Companies that set up in this park benefit from subsidised land prices. Additionally, KHTP builds and rents premises to selected companies according to specifications, with options to purchase, and selected companies may set up joint ventures (KHTP, Inaugural Issue, 1995). Similarly, the Perak Science Park was designed to attract hi-tech manufacturing and R&D, and it was developed by the Seri Iskandar Development Corporation Sdn Bhd, a joint venture (JV) between the Perak State Development Corporation and the Teras Cemara Sdn Bhd (a subsidiary of the Malaysian Resources Corporation Berhad). The science park consists of productivity centres, a local commercial centre, R&D centres, a resource centre and a sports and recreation centre (Perak Science Park: A Symbiosis of Minds and Hearts). Johor Technology Park also aims at attracting hi-tech manufacturing and R&D. The latest attempt to facilitate hi-tech development has been the conception of the Multimedia Super Corridor (MSC). The MSC is 15 kilometres' wide and 50 kilometres' long. It starts from Kuala Lumpur city centre, through the Kuala Lumpur International Airport, to two newly developed smart cities: Putrajaya, new host to the government and Cyberjaya, an intelligent city that aims to attract multimedia industries, R&D centres, a multimedia university and operational headquarters for multinationals (http://www.mdc.com.my). This ambitious project has not yet attracted a substantial amount of high-tech companies (*The Economist*, 05/02/2000: 91), but it indicates the strong will of the Malaysian government to modernise the country.

Encouraging training to develop local skills. Training is actively encouraged to provide hi-tech and R&D activities. An example of successful training delivered privately is that of the US electronics firm Harris Corp. The firm has invested M$500 million (US$198 million) in state-of-the-art technology, and has trained a team of skilled workers who have helped turn Malaysia into the world's third largest producer of semiconductors (*FEER*, 27/14/95). To multiply and ease these initiatives, an investment tax allowance of 100% for 10 years is given to companies that establish technical or vocational training institutions. In addition, the Human Resource Development Fund (HRDF) is aimed at encouraging direct private sector participation in skills' development. It was launched in 1993 with a grant from the government. One of the most successful training centres arose from the collaboration between several TNCs in Penang. The Penang Training Centre provides off-the-job training for foreign and local firms' employees in the area.

Attracting headquarters in Malaysia. Finally, the government is actively trying to encourage firms to set up their operational headquarters (OHQ) in Malaysia. An approved OHQ refers to a locally incorporated company, whether locally owned or foreign-owned, which carries on a business in Malaysia, providing qualifying services to its offices or its related companies outside Malaysia. Approved OHQs can benefit from substantial tax incentives.

Export orientation. An important policy stance appears through intensive promotion of export activities. Although exports are primarily generated by foreign firms in Malaysia, the government encourages all types of firms to export, regardless of their origin. Major incentives are export credit refinancing, which provides short-term credit at preferential rates of interest, double deduction for promotion of exports, which comprises a double deduction of export credit insurance premiums paid by product exporters to approved local insurance companies, and a double deduction for promotion of exports, and finally, building tax allowances, for companies that need to build warehouses for storing goods for export.

The analysis above shows how the Malaysian government is using policies to try and reach its objectives, with fiscal as well as non-fiscal incentives. Whereas only 9.8% of approved investment in the manufacturing sector was conducted without any tax incentives in 1988, this figure rose to 46% in 1995, and was 34.5% in 1998 (MIDA, 1996, http://www.mida.gov.my). Other tools include equity guidelines through which the government also encourages specific sectors of activity and exports; these are developed in the next section.

FDI Procedures and Requirements

A strong indicator of a country's commitment to attracting foreign investment lies in its ability to facilitate investment procedures from application to implementation. This section endeavours to show how Malaysia has greatly improved its investment procedures to ease foreign investors' entry into the country. The three topics covered include foreign equity guidelines, the foreign investment process and the protection of foreign investment.

Foreign equity guidelines

Although the government is encouraging projects undertaken on a JV basis, equity facilities are awarded to export-oriented companies.[5] Until the recent events of the Asian crisis, the Malaysian government used the level of exports as a determinant for foreign equity participation in manufacturing projects.

Up until July 1998, foreigners were permitted 100% equity ownership of their companies if they exported 80% or more of their products; indeed, the government considered that FDI and enterprises that were export-oriented bring in foreign exchange earnings, technology and heavy equipment and capital goods. Foreign investors could obtain a majority of 51-79% if the export share of production stood between 51 and 79%; an equity share of 30-51% when the export share stood between 20 and 51% of production; or else an equity share of up to 30% when the export share in production stood below 20%. Exceptions to equity constraints included such cases as when no partner could be found, or if the firm employs more than 350 full-time Malaysian workers; 100% ownership may then be allowed. Majority foreign ownership of up to 100% was also accepted for projects involving the extraction or mining and processing of mineral ores.

Additional guidelines in terms of JV partners were and still are provided, as for the distribution of Malaysian equity. In line with continuous attempts to increase the local Malay participation in the private sector, foreign firms are also concerned in a number of ways. If no local partner has been identified for the establishment of the JV, then the balance of equity will be reserved for *bumiputera* if 70% or more of the equity is held by foreigners. If less than 70% of the equity is held by foreigners, 30% will be reserved for *bumiputeras* and the balance for non-*bumiputeras*. If the project was initiated by *bumiputeras*, similar guidelines are followed when the foreign partner holds 70% of the equity, but *bumiputeras* account for all participation if the foreign partner holds less than 70% of the venture. Finally, if the project was initiated by non-*bumiputeras* and 70% is owned by the foreign partner, the remaining equity is attributed to the non-*bumiputera* partner. But if less than

70% of the equity is held by foreigners, 30% cent will be allocated to non-*bumiputeras* and the balance to *bumiputeras*. If no *bumiputera* partner can be found, the balance guidelines may be altered by the Ministry of International Trade and Industry (MITI).

The equity guidelines described above have been momentarily suspended by the Malaysian government, to overcome a short-term threat of a decrease in FDI inflows, as a result of the Asian crisis. The Malaysian government has loosened its requirements on export performance to allow generally foreign equity ownership of up to 100% for all manufacturing projects received between 31 July 1998 and 31 December 2003. This relaxation of export requirements will be reviewed after December 2003. Similar relaxation in guidelines applies to the telecommunication sector, the insurance sector and the stockbroking industry (Table 4.5).

Foreign investment process

MIDA gives all the approval for the investment projects; MIDA has been the Coordination Centre for Investment (CCI) since October 1988. For the purpose of implementing the regulations of acquisition of assets, mergers and take-overs, the Foreign Investment Committee is responsible for major issues of foreign investment. It formulates policy guidelines on foreign investment in all sectors, solves problems regarding foreign private investments, advises on policies towards FDI, coordinates and regulates the acquisition of any assets or interests, mergers and take-overs of companies and businesses in Malaysia and monitors and evaluates FDI to keep up-to-date information.

These new regulations have led to a substantial increase in foreign investment in Malaysia, especially in the electrical and electronics sector in the late 1980s and in the early 1990s, as the latter part of this chapter shows.

Protection of foreign investment

Once equity participation has been approved, a company will not be required to restructure at any time as long as it complies with the original conditions of approval. Moreover, the authorities signed investment guarantee agreements with major investor countries,[6] and bilateral agreements on double taxation, so as to secure multinational companies against non-commercial risk, such as expropriation and nationalisation, and to allow for the remittance and repatriation of capital. These agreements also include permission to remit or repatriate profits or equity capital on investments in any convertible currency. Finally, the Malaysian government in 1966 ratified the provisions of the Convention on the Settlement of Investment Disputes established under the auspices of the International Bank for Reconstruction and Development. In addition, the Kuala Lumpur Regional Centre for Arbitration was established in 1978 under the auspices of the Asian-African

Legal Consultative Committee (AALCC). The Centre is a non-profit organisation and serves the Asian and Pacific region.

Table 4.5 Main changes in FDI policies and measures after Asian crisis

Date	Sector	Policies and measures
With effect from 31 July 1998	Manufacturing	Prior to 31 July 1998, foreign equity participation in the manufacturing sector was governed by the level of exports, up to 100% in manufacturing sector companies is generally permitted irrespective of the level of exports (with the exception of specific activities and products where Malaysians and Malaysian companies have the capabilities and expertise). The current policy is applicable to all project applications received between 31 July 1998 and 31 December 2000. Projects approved under this policy will not be required to restructure their equity after that period, provided that the companies continue to comply with the original conditions of approval and retain the original features of the projects. Companies licensed before 31 July 1998 must comply with the equity condition stated in their respective licences. If such companies undertake expansion or diversification, however, the current foreign equity policy will apply to such expansion and diversification projects
With effect from March 1998	Telecommunication	Raised the maximum level of foreign equity investment in locally licensed basic telecommunication companies from 30% to 49%
June 1998		The Government announced that it was prepared to consider applications to raise the 49% ceiling on foreign equity holdings in locally licensed basic telecommunication companies up to a maximum of 61%, provided that the companies concerned reduce their foreign equity holdings to a maximum of 49% within five years
With effect from March 1998	Insurance	Raised the maximum level of foreign equity investment from 49% to 51% in the insurance sector under specific conditions
With effect from 1 July 2000	Stockbroking industry	The maximum level of foreign equity investment in financial leasing and stockbroking companies will be increased from 30% to 49%

Source: ASEAN (1999: 88).

Undoubtedly, the Malaysian government has created a favourable investment environment for foreign firms, and substantial inflows of FDI are the result of this policy. However, these policies have not led to substantial amounts of local investment, especially in the high-tech sectors. Local firms are mostly composed of big conglomerate groups in most manufacturing activities. These conglomerates are successful, but to date have been heavily subsidised by the Malaysian government, often to the detriment of real managerial and technical capability development. Small firms prevail in the resource-based sector and have less developed capabilities. Government policies offer a wide range of incentives to encourage small and medium sized enterprise (SME) development, but these policies are fairly recent, and still have not led to a real spurring of these firms. Additionally, real emphasis on managerial and technical capability development was only recently initiated as a clear government strategy. The Malaysian government actively promotes export orientation, as opposed to domestic-oriented FDI. This, as a result, leads to fewer inter-firm networks and linkages between TNCs and local firms. More inter-firm links might be created when and if foreign firms start operating in Malaysia to serve the local market itself.

Government Initiatives for Local Linkages: Soft Incentives towards Backward Linkages

The Malaysian government has recognised the contribution of FDI to industrial and economic development and international competitiveness of the nation. It has thus encouraged investment in high-tech industries, and the siting of R&D facilities and headquarters within its national boundaries. Additionally, the latest master plan emphasised the need for inter-firm linkage and cluster development. Following this policy stance, soft incentives have been launched to try and encourage TNC-local supplier linkages. This section gives an account of the way the Malaysian government encourages TNCs to create partnerships with their local suppliers, in particular, soft incentives and existing initiatives.

Soft Incentives: seeking to increase backward linkages

In recent years, governments have recognised the lack of efficiency of protectionist measures and have taken steps towards economic liberalisation. As we have seen, Malaysia has made special efforts to maintain strong market incentives and to integrate public and private resources in alleviating constraints in technology acquisition, worker training and financing. Another major step towards liberalisation has been to promote the creation of backward linkages using soft 'voluntary' incentives rather than compulsory policies, often cumbersome for foreign investors and that may simply keep them away from investing in the country. Additionally, Malaysia has initiated special programmes that work closely with both foreign investors and domestic suppliers in promoting mutually beneficial linkages between the two parties. In most of the successful developing countries, government policies regarding local purchasing of foreign firms have recently been modified, abandoning traditional types of import restrictions and content requirements in favour of more interactive policies, where the government intervenes in the upgrading process of the local supporting industries (Battat et al., 1996). Many countries have come to realise that increased institutional support to promote backward linkages could lead to greater benefits for local suppliers; in this sense, it can be argued that regulatory policies based on protectionism are not a sound approach to improving backward linkage.

Existing initiatives in Malaysia

The Malaysian government has chosen to encourage inter-firm linkages by using market-friendly incentives (Meyanathan, 1994). Such incentives include tax incentives, specific institutional arrangements, vendor development schemes, local content requirements, subcontracting exchange schemes and information provision and exchange.

Incentives for international procurement centres (IPC). Companies that undertake procurement and sale of raw materials, components and finished products for its group of companies in Malaysia and abroad may qualify as IPCs (http://www.miti.gov.my/inddev-ipc.html). As specified in the PIA (launched in 1986), the benefits of this status encompass foreign equity ownership in wholesale and retail trade, and permission to open foreign currency accounts to retain export proceeds (http://www.mida.gov.my). Companies such as Sony, Aiwa or JVC all enjoy the IPC status in Malaysia. Applications must be made to the MITI. While this policy does not directly encourage local purchasing, it does point to the importance of the purchasing function in the firm, and the many advantages of hosting the IPC in Malaysia.

Incentives through the Industrial Linkages Programme (ILP). This programme aims at both large and small companies in Malaysia. On the one hand, it provides incentives for large companies to undertake actions to enhance the quality of local vendors. Such actions may include training of employees, product development and testing and factory auditing to ensure the quality of vendors' products. In exchange for such efforts, the government offers reductions in income tax. On the other hand, local vendors that participate in an ILP-related project may receive incentives.[7] Applications must be made to MIDA.

Further incentives for inter-firm relationships. Institutional arrangements are composed of a set of agreements launched by the government together with private firms; for instance, the 'umbrella strategy' implies that an 'umbrella' firm that possesses the necessary financial resources and expertise helps to coordinate the production and marketing of another firm. This arrangement includes skills and management training.

In the same way, the vendor development scheme was launched in the mid-1990s to encourage a large firm to provide assistance to a local vendor. The Malaysian government benchmarks the existing potential locally owned suppliers, and organises events whereby TNCs and local suppliers can meet and interact. During these events, the government will provide advice for local suppliers on how to improve the quality and efficiency of their services. A series of tax incentives and financial packages are available for SMEs to try and obtain ISO (International Organization for Standardization) certifications. TNCs also provide advice on the type of relationships that they wish to create with their local suppliers. In the case of agreements between two parties, a memorandum of understanding is signed, leading to the creation of a close relationship between a foreign firm and a locally owned SME. The primary aim of this programme is to enhance close beneficial relationships for local firms. TNCs may not derive so much benefit from this relationship, but it helps them to contribute to the economic upgrading of the economy they are in. The Small and Medium Industries Corporation replaced this scheme in 1999 by the Global Supplier Programme (UNCTAD, 2001: 187-8; MITI;[8] SMIDEC[9]). The programme aims at linking TNCs with local small and medium industries (SMIs), with the TNC committing to strengthen the capacity of selected SMIs through skills enhancement, technology upgrading and mentoring activities. One key objective of the programme is to develop local vendors into global vendors. Penang Skills Development Corporation (PSDC) was the first organisation to organise the supplier development programme (*Malaysian News*[10]). Training is organised for SMIs and run by staff from participating TNCs. This constitutes the first initiative of the programme, that of training in critical skills to acquire competencies to adopt and use new technologies, with a focus on quality and productivity. The second initiative of the programme is the linkage programme where TNCs adopt local companies and commit to their development. Local companies are also encouraged to prioritise R&D.

Effectiveness of these initiatives

These market-friendly incentives concern only a very small number of firms in Malaysia. They have produced a few spillover effects, but successful ones (Meyanathan, 1994; Battat et al., 1996; Jomo et al., 1999; UNCTAD, 2001). Indeed, the few agreements between a foreign firm and a locally owned SME have led to positive links for the local firm. These programmes have been successful in other developing countries. For the moment, they are still new and the full benefits that can arise remain to be seen, but the latter are likely to increase as the programmes gain importance, in Malaysia and in other Asian countries.

TRENDS IN FOREIGN INVESTMENT

The historical background of the Malaysian environment and the account of government policies towards investment allow us to explain fluctuations in FDI flows over the past two decades. The impact and importance of government policies in attracting foreign firms was emphasised. Government policies and strategy help create a favourable investment environment. In addition to internal factors, external factors have had substantial effects on investment distribution within such an open economy. The recession in the early 1980s was important in explaining the huge decrease in exports from Malaysia, and therefore in partly explaining a decrease in inward FDI, since this was pushed towards export orientation by the Malaysian government. The decrease in firms' profitability, the local recession and lack of domestic investment led to the decrease in foreign investment. Similarly, the change in government attitude in the mid-1980s showed tremendous results in terms of attracting foreign investment.

As we have seen, the government has dramatically deregulated the rules with respect to FDI since 1986, in the hope that foreign companies would revitalise the economy. These government policies have led to a massive increase in inward FDI in Malaysia, but also in the structure of this investment. In the 1960s and 1970s, foreign capital was predominant in the agricultural (75.4%), mining (72.5%), and trade (63.6%) sectors, and to a lesser extent in the manufacturing sector (59.6%) (Ariff, 1991: 101). In the 1990s, foreign capital primarily originated from other Asian countries, and predominated in the manufacturing sector, in particular in the electronics industry.

Background to Inward Investment in Malaysia

Before looking at FDI trends, it is useful to look into the available data on FDI in Malaysia, and to analyse the factors that explain investment flows to the country.

Inadequacies of investment data in Malaysia

Table 4.6 shows the speed of project implementation, whether foreign or local, after the Malaysian government has approved the project. The share of projects implemented over the first two periods of time has roughly remained equal, at three-fifths of all projects. During 1992-95, as at 31 December 1995, 33.4% of approved projects were still in the process of negotiation, and 3.8% were unlikely to be implemented. A noticeable slowdown in projects' implementation took place during the Asian crisis. The share of projects implemented in total approved manufacturing projects declined from 70.5%

in 1996 to 63.1% in 1997, and to 51.9% in 1998. This indicates the cautious behaviour of some investors, and the incapacity of others to proceed with planned investment.

From these figures, it can be argued that data on approved FDI exaggerate foreign equity participation in the sectors covered, and therefore interpretation of these data has to be taken with caution. The problem lies in the lack of other means of data gathering. It is too demanding to look at foreign investment from the countries of origin, not to mention that such data are often too aggregated for analysis. Furthermore, capital exports are sometimes channelled through particular countries for certain tax reasons, thus hiding the actual destination of capital transfers. Also, foreign investment data by international organisations, such as the IMF and the UN, are aggregated, and international comparisons are difficult because of dissimilar standards of compilation of data between countries and differences in accounting procedures and depreciation rules. Since there are no standard international classifications for investment, unlike trade (Standard Industrial Trade Classification, SITC) and industry (International Standard Industrial Classification, ISIC), international comparisons of investment are made more difficult. Recent efforts in gathering data have been made by Asia Pacific Economic Cooperation (APEC) (see *APEC Investment Regime Guidebook*, APEC, 1999), as well as by the ASEAN, which has been publishing an ASEAN Investment Report since 1999. However, core inadequacies in the data persist. This, therefore, helps explain why the analysis and conclusions regarding inward foreign investment in Malaysia need to be taken with caution (Ariff, 1991).

Factors motivating FDI flows in Malaysia

There are different kinds of factors in Malaysia that are attractive to foreign investors. Economic factors included, first, continued rapid economic growth in the late 1980s and early 1990s. The slowdown in growth during the Asian crisis has already led to higher growth rates in 1999 and 2000. Second, Malaysia benefits from good infrastructural facilities, with notably a good road network connecting all the major towns, and adequate water and electricity supplies, port facilities and several airports, both national and international. Industrial and technology parks also attract foreign firms. This is strength in the region, as other major recipients such as Thailand, Taiwan or China have yet to overcome infrastructure bottlenecks. Third, Malaysia attracts investment from Asian-based TNCs driven by cost factors or market access, such as the NIE countries. Fourth, labour costs in Malaysia are low, even though other countries like China, Thailand, Indonesia, India, Burma and Sri Lanka have cheaper labour.

Table 4.6 Progress of implementation of approved manufacturing projects, 1988-98 (frequency and percentage)

	1988-91*		1992-95**		1996		1997		1998***	
	No.	%	No.	%	No.	%	No.	%	No.	%
Projects implemented										
a- Production	1,809	53.1	1,856	55.8	551	70.5	479	63.1	438	51.9
b- Machinery installation or factory construction	213	6.3	154	4.6	16	2.0	23	3.0	40	4.7
Sub-total	2,022	59.4	2,010	60.4	567	72.5	502	66.1	478	56.6
Projects in initial stages of implementation										
a- Site acquired	115	3.4	76	2.3	11	1.4	18	2.4	19	2.3
b- Active planning or negotiation	963	28.3	1,114	33.4	135	17.3	155	20.4	266	31.5
Sub-total	1,078	31.7	1,190	35.7	146	18.7	173	22.8	285	33.8
Projects not implemented										
a-No action taken	47	1.4	23	0.7	1	0.1	12	1.6	9	1.1
b- Temporarily shelved or abandoned	213	6.2	101	3.0	61	7.8	66	8.7	70	8.3
c- Withdrawn	43	1.3	4	0.1	7	0.9	6	0.8	2	0.2
Sub-total	303	8.9	128	3.8	69	8.8	84	11.1	81	9.6
Total	3,403	100	3,328	100	782	100.0	759	100.0	844	100.0

Source: MIDA, *Statistics on the Manufacturing Sector* 1988-92, 1991-95 and 1995-99.

Notes:
* as at 30 June 1992.
** as at 31 December 1995.
*** as at 30 June 1999.

The quality of labour in Malaysia was and is still relatively high. Among non-economic factors, the stability of the Malaysian government, its currency and economy were very attractive to foreign investors, at least until the outbreak of the Asian crisis. Recent liberalisation of FDI and outward-oriented trade policies play in favour of increased FDI. Malaysia further benefits from its strategic location in Southeast Asia and the strengthening regional economic integration (with notably ASEAN, the ASEAN Free Trade Area (AFTA) and growth triangles).

Negative factors in attracting FDI

With the changes in government regulations in the late 1980s (away from restrictive policies, such as the ICA and the Guidelines for Acquisition, Mergers and Take-overs), Malaysia has substantially decreased the negative factors against inward FDI. Other negative factors persist; those include intensive competition in a limited domestic market, bureaucracy, shortage of skilled workers and a high turnover of personnel. Unemployment was at 3.1% in 2000 (http://www.statistics.gov.my) and there are serious labour shortages in almost every sector. Wages increased rapidly in the 1990s, and problems of and expanding current account deficit sent negative signals to foreign investors in the mid-1990s; this situation has changed however with a positive current account in the late 1990s. The import of consumer goods and overall consumer spending are rising fast, despite import duties, but overall exports have been grown at a faster rate since 1998 (http://www.malaysiaexports.com). Recently, factors external to the Malaysian economy have been affecting FDI inflows negatively, such as a global downturn in some sectors of the electronics industry, and the stiff competition of China as an investment location.

FDI Trends in Malaysia

The review of inward FDI in Malaysia includes a description of overall trends in inward flows, drawing on the origin of the investment and the sectors of activities into which it is directed. Starting with the overall trends and pattern of FDI, we then discuss the sources of FDI in Malaysia and the sectors of activity concerned, before paying particular attention to the manufacturing sector.

Overall trends and pattern of FDI

Figure 4.1 illustrates the trends in inward investment in Malaysia from the late 1970s to the late 1990s. FDI and portfolio investment flows are reported. The major trends noticeable are that of a sharp increase in inward FDI and a sharp decrease in portfolio investment in the late 1980s and early 1990s.

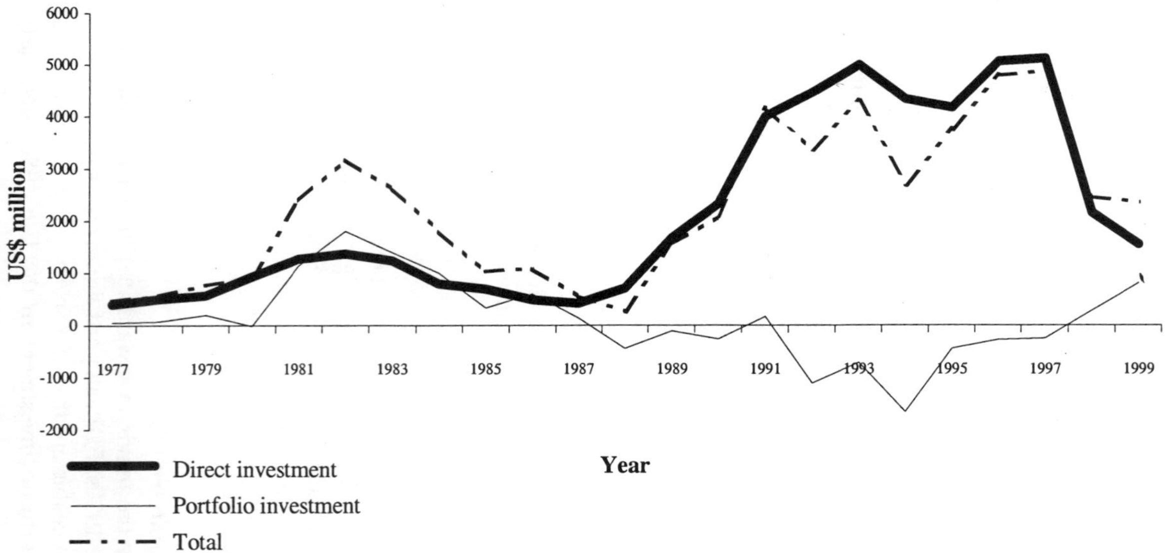

Figure 4.1 Inward investment flows in Malaysia, 1965-99 (Million US$)

Source: IMF, *International Financial Statistics Yearbook*, various issues (1985-2002).

It is very interesting to see how efficient the change in government policies has been in attracting FDI after 1986. FDI underwent a sharp increase from 1987 onwards, reaching a peak in 1992. The discussion in the earlier part of this chapter led us to separate Phase II in the economic development process that ended in 1986, and Phase III with the new government policy orientation. Of course, as mentioned in the factors motivating FDI in Malaysia, other factors have led to this sharp increase, such as the entry of the NIEs in the international investment field, or a generally favourable international economic environment. Lower investment entry in the mid-1990s may have resulted from the entry of new host competitors, such as China or Vietnam, and the shift in the Malaysian comparative advantages, away from low-intensity activities.

The sharp increase in inward flows slowed down in 1997, with the outburst of the Asian crisis. Although Malaysia suffered less economic turbulence than its neighbours, Thailand and Indonesia, inward flows decreased from over US$6.5 billion in 1997 to US$2.7 billion in 1998, before picking up again, reaching US$5.5 billion in 2000 (UNCTAD, 2001). Rather than a loss in confidence in the prospects of the Malaysian economy, this decline reflects the domestic problems in the major investing countries, global excess capacity and continued uncertainty in the region (ASEAN, 1999).

Inward investments have a steady role in the Malaysian economy. Inward FDI stocks' share in GDP increased from 21.1% in 1980, to 24.1% 1990, and to 65.3% in 1999. Such figures show the key role of TNCs in the economy. Whereas foreign firms account for over a third of domestic production, their share in local investment generation is decreasing. Inwards FDI flows' share of gross fixed capital formation gradually decreased from 19.4% over the 1989-94 period, to 15.1% in 1997 and 20.1% in 1999 (UNCTAD, 2001). While the share of foreign investment in total GDCF remains constant, when considering total foreign investment stocks as a share of GDP, Malaysia increased its dependence on foreign firms.

Sources of FDI

Sources of foreign investment in Malaysia have changed widely since the 1970s and the late 1980s, with the emergence of Japan as a major international investor and new investors from the neighbouring NIEs. In 1988, the key country/region investor in Malaysia was Japan with 25% of total inward investment, followed by the NIEs with 24%, Europe and the USA. The third position of Europe was unusual, with 20% of the total, whereas the normal level in previous years was 10%. The NIEs and ASEAN countries have considerably increased their inflow of investment in Malaysia. Taiwan became the first inward investor in the early 1990s, with 36% of total investment in 1990. This position reversed however in 1995, when Japan recovered its position as first investor, with 23% of all investment, against 16% for Taiwan. Investment from the European Community has been decreasing since the late 1980s, while that of the USA has increased substantially, rising from 11% in 1988 to 15% in 1995. Inward investment reached a peak in 1990 when Taiwan invested RM6,339.1 million in Malaysia. In the late 1990s, Singapore became the main foreign investor, accounting for 26% of inflows during 1997-98. The USA came second, and European countries increased their investment in the country in the aftermath of the Asian crisis.

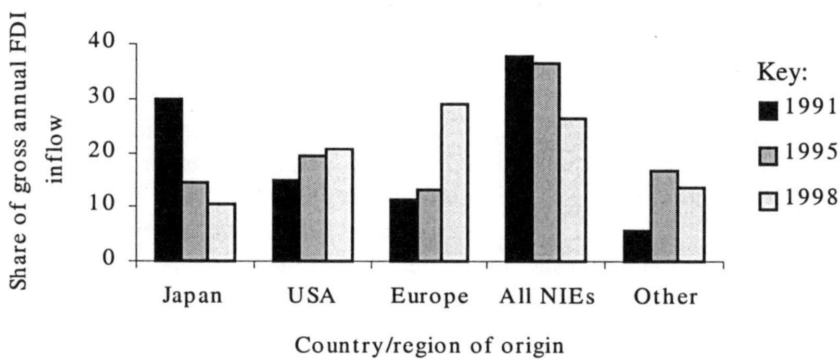

Figure 4.2 Malaysia: share of total gross annual inflow of FDI (BOP data) by country of origin, 1995-98

Source: *ASEAN Investment Report* (1999: 90).

As a breakdown of investment and sources of origin, Figure 4.2 shows that during the 1990s, and especially after the Asian crisis, the share of Asian investors (from Japan and the Asian NIEs) in FDI inflows in Malaysia decreased, while that of European and US investors increased. The USA and Europe (with Germany as the leading European investor) were the first and second most important investors, respectively, in Malaysia in 2000, as shown in Figure 4.3.

Sectoral distribution of FDI

Since the 1980s and the early 1990s, two sectors have attracted the bulk of inward FDI: the manufacturing sector, with 45% of gross annual inflow of FDI in 1998, and the services sector, with 37%. Both sectors have attracted an increasing share of FDI as a result of the Asian crisis, and shares were up compared to their level in 1997.

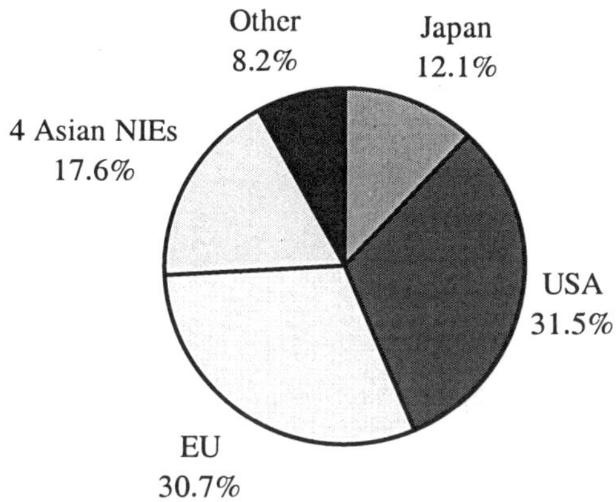

Figure 4.3 Malaysia: share of total gross annual inflow of FDI (BOP data) by country of origin in 2000

Source: *ASEAN Investment Report* (2001: 34-35). Includes retained earnings.

FDI in the manufacturing sector

Approved manufacturing projects increased, both in terms of number of projects and in absolute terms, through the late 1980s and the early 1990s, but the trend reversed in the mid-1990s. Proposed called-up capital has gone up from RM3,469.7 million in 1988 to RM9,679.1 million in 1990, down to RM6,524.1 million in 1995. Overall, the amount of proposed called-up capital in 1995 still remained at nearly twice that of 1988. In 2000, proposed called-up capital in approved manufacturing projects was down to RM5,577.8 million (http://www.mida.gov.my/stats.html) Figures on the potential employment for manufacturing projects indicate that investment projects generated less employment than in the late 1980s, which showed a move towards high value-added projects, and reflected the increase in wage rates in Malaysia. Whereas in 1988 each approved project would have an average of 195 employees, this figure reduced to 131 in 1995 and 109 in 2000. These figures include both Malaysian and foreign investment. A further breakdown between Malaysian and foreign equity is useful to underline the importance of foreign firms in capital generation. While the share of Malaysian investment in total investment increased in the early 1990s, this trend reversed after 1997. Domestic and foreign sources of investment in projects approved by major industry in 2001 appear in Figure

4.4. Foreign sources of investment are major sources of investment in the manufacturing sector, especially in the electrical and electronics sector.

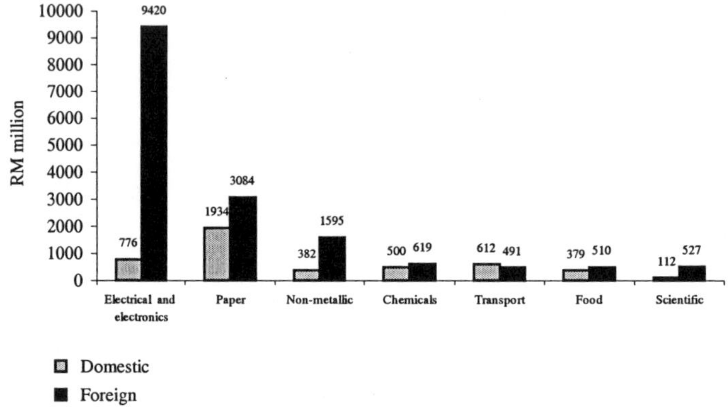

Figure 4.4 Sources of Investment in Projects Approved by Major Industry, 2001

Source: MIDA (2002: 101).

The electrical and electronics products sub-sector has been the main recipient of inward investment, with 33.6% of total foreign fixed assets in 1998 (calculated from http://www.mida.gov.my). Rubber and rubber products accounts for 10% of total foreign fixed assets, chemicals and chemical products 9.5%, non-metallic mineral products 6.7%, and textiles and textile products 5.4%. This very high concentration of foreign investment within the electrical and electronics sector and within the resource-rich sectors shows the high dependence of Malaysia on a few sectors, and the urgent need for diversification.

There is a high level of integration of foreign firms in the manufacturing sector. In 1994, 48% of total fixed assets in the manufacturing sector were foreign-owned, and amounted to RM27,782.5 million. In 1998, this share remained at above 47% with RM47,308.2 million. Table 4.7 shows that two sectors with the highest share of foreign capital are 'scientific and measuring equipment' and the 'electrical and electronics products' sectors. Their respective shares of foreign ownership in total fixed assets as at 31 December 1998 were 96% and 83%.

Key foreign partners across all sectors are Japan, with 27.3% of total fixed assets in 1998, the USA (15%), Taiwan (13.4%), Singapore (11.2%) and European countries (15%).

Table 4.7 Capital structure by industry in companies in production, total fixed assets as at 31 December 1998 (RM'000 and percentage)

	Malaysian	Share	Foreign	Share	Total	Share
Food manufacturing	4,109,239	73.5	1,485,078	26.5	5,594,317	5.6
Beverages and tobacco	276,681	31.1	613,376	68.9	890,057	0.9
Textiles and textile products	1,571,995	24.2	4,920,174	75.8	6,492,169	6.5
Leather & leather products	36,944	39.0	57,902	61.0	94,846	0.1
Wood & wood products	3,811,560	72.7	1,428,081	27.3	5,239,641	5.2
Furniture & fixtures	721,958	74.9	242,447	25.1	964,405	1.0
Paper, printing & publishing	4,739,642	90.2	512,979	9.8	5,252,621	5.3
Chemicals & chemical products	3,101,811	54.4	2,599,449	45.6	5,701,260	5.7
Petroleum & coal	9,516,014	65.7	4,969,392	34.3	14,485,406	14.5
Rubber products	1,651,948	49.5	1,684,121	50.5	3,336,069	3.3
Plastic products	1,690,660	61.7	1,049,138	38.3	2,739,798	2.7
Non-metallic mineral products	5,214,889	66.0	2,685,262	34.0	7,900,151	7.9
Basic metal products	6,041,843	69.7	2,628,525	30.3	8,670,368	8.7
Fabricated metal products	2,244,941	64.2	1,250,253	35.8	3,495,194	3.5
Machinery manufacturing	610,399	34.0	1,182,533	66.0	1,792,932	1.8
Electrical and electronics products	3,463,017	16.9	17,063,726	83.1	20,526,743	20.5
Transport equipment	3,569,633	70.9	1,463,958	29.1	5,033,591	5.0
Scientific & measuring equipment	45,034	4.0	1,068,344	96.0	1,113,378	1.1
Miscellaneous	311,359	43.6	403,482	56.4	714,841	0.7
Total	52,729,567	52.7	47,308,220	47.3	100,037,787	100.0

Source: http://www.mida.gov.my, extracted in March 2000

The impressive rise in inward investment aggravates the increasing dependence of the country on foreign firms for both exports and for specific sectors of activity, and the lack of linkages through the economy leading to the rise in a sound domestic industrial base. We have shown that in response to these factors, recent policies have been refocused to encourage such linkages within the economy. An additional factor to consider is that of the increasing amount of outward FDI from Malaysia.

Outward FDI from Malaysia

Outward FDI stocks' share in GDP rose from 1.7% per cent in 1980, to 6.2% in 1990 and to 22.6% in 1999 (UNCTAD, 2001). FDI outflows have increased dramatically, from an annual average of US$681 million 1989-94 to US$3,768 million in 1996, before the Asian crisis. As a result of difficulties faced by local firms during the crisis, outward flows decreased to US$785 million in 1998, before picking up again, reaching US$2,919 million in 2000 (UNCTAD, 2001). The downward trend was reversed from 1999 onwards, thanks to the economic recovery of the country and its companies.

Malaysian outward investors are principally conglomerates, often operating with government support, primarily located in the finance sector. Four Malaysian companies are listed in the top 50 TNCs from the developing economies (UNCTAD, 2001). Two are diversified groups, namely Sime Darby Berhad and Berjaya Group Berhad; one is in the petroleum sector, namely Petronas – Petroliam Nasional Berhad; and finally, Telekom Malaysia Berhad is in the telecommunication sector. Those firms prefer to invest in neighbour countries of ASEAN, under the JV form of investment (Mirza, 2000). Singapore, Hong Kong and the USA are the major recipients of Malaysian investment abroad. Finding new markets is the main reason for locating overseas. There is still some way before Malaysia becomes a net international investor, but the rise in internationally oriented Malaysian firms shows the country is gaining competitive advantages over other emerging economies and developing countries.

ELECTRICAL AND ELECTRONICS SECTOR IN MALAYSIA

The Malaysian electrical and electronics sector (MEE) is a large contributor to the country's manufacturing output, employment and exports (Table 4.8). In 1999, there were more than 900 companies employing close to 400,000 workers. Malaysia is among the world's largest exporters of semiconductor devices and audio-visual equipment (http://www.mida.gov.my/eeindustry). Similar to most of Asia's electronics manufacturers, companies located in Malaysia sell primarily to the USA, Europe and Japan. Exports from the sector grew by 36.2% in 1998, and by 22.1% in 1999. In 2000, MEE accounted for 58.8% of all exports from Malaysia. This share increased from 53.3% in 1998 (http://www.miti.gov.my). The sector is a strong income generator, but it also leads to substantial imports. MEE's share of total imports has been steady over the years, with 46.2% in 1998, and 47% in 2001. The balance between exports and imports for the sector is positive, exports totalled RM219.5 billion in 2000, against RM154 billion for imports (see Table 4.8), showing a sharp increase compared to RM185.2 billion and RM118 billion, respectively, in 1999.

Although the manufacture of semiconductor devices and other electronic components still continues to dominate the industry, the output structure of the industry has changed significantly since the mid-1980s. In 1999, the output structure for the electronics industry was 42.4% for electronic components, 12.2% for consumer electronics and 45.4% for industrial electronics. These structural adjustments are reflected in the export performance of the three sub-sectors in recent years. In 1999, the export value of components was RM7.6 billion, consumer electronics RM21.7 billion and industrial electronics RM81.1 billion (http://www.mida.gov.my). Electronics components remain the key attractive sub-sector, with over 60% of total approved manufacturing projects in MEE in 2001.

Table 4.8 Malaysia's electronics industry, 1999

Number of companies	>900
Output	RM129.8 billion (US$34.2 billion)
Electronics components, of which	42%
Consumer electronics	12%
Industrial electronics	46%
Average annual growth, 1995-99	18.4%
Employment	381,000
Exports in 2000, RM million	219,555.5
Imports in 2000, RM million	154,996.5

Source: http://www.mida.gov.my/eeindustry (as of November 2000); http://www.miti.gov.my/trdind/annu4.htm (as of March 2002).

Foreign assets in total fixed assets are high in the MEE. They reached RM9,333.4 million in 1994, and RM17,063.7 million in 1998, that is a share of 83.1% for foreign firms. Foreign firms continued to dominate the industry with 92% of approved manufacturing projects in 2001 (Table 4.9). Table 4.7 shows that MEE accounts for one-fifth of total fixed assets across all manufacturing industries. The importance of the MEE sector will continue. The share of this industry in the total approved capital investment in the manufacturing sector has increased from 36% in 2000 to 41% in 2001 (MIDA, 2002).

Japanese firms dominate the MEE sector, but their share in total fixed assets has decreased during the 1990s. Japan accounts for 38.5% of total fixed assets in 1998, down from 43% in 1994. In both 1994 and 1998, total foreign fixed assets in the electrical and electronics sector originated by order of importance from Japan, the USA, Singapore and Taiwan. However, the share of the USA and that of Europe have increased. The share of US firms in total fixed assets in MEE increased from 18.5% in 1994 to 24.6% in 1998. The share of Europe reached 13.3% in 1998.

Table 4.9 Approved manufacturing projects for electronics and electrical products by sub-sector, 2001 (percentage from RM)

	New			Expansion/Diversification			Total		
	Dom.	Foreign	Total	Dom.	Foreign	Total	Dom.	Foreign	Total
Consumer electronics	1.6	0.1	1.7	0.0	2.5	2.5	1.6	2.7	4.3
Electronics components	2.7	28.4	31.0	0.8	28.7	29.5	3.5	57.1	60.5
Industrial electronics	0.5	14.3	14.8	0.7	12.2	12.9	1.2	26.5	27.7
Electrical products	0.7	1.2	1.8	0.6	5.0	5.7	1.3	6.2	7.5
Total	5.4	43.9	49.3	2.2	48.5	50.7	7.6	92.4	100.0

Source: adapted from MIDA, (2002: 131).

Asian NIEs decreased their share, down from 26.7% in 1994 to 20.7% in 1998 (6.4% Taiwan, 8.8% Singapore, 0.8% Korea and 4.7% Hong Kong). On the whole, the Asian share in the MEE decreased as a consequence of the Asian crisis. US and European firms increased their investment and presence in the sector in the second half of 1990s.

US TNCs located in Malaysia include Intel, Motorola, AMD, Agilent, Seagate, Western Digital, Komag, Dell and Gateway. Japanese corporations are set up predominantly in consumer electronics. Companies in Malaysia include Sony, Matsushita, Sharp, JVC, Hitachi, Pioneer, Kenwood and Yamaha. Others include Korean picture tube maker Samsung, Taiwan's Acer and Europe's Siemens, ST Microelectronics, Alcatel and Ericsson (http://www.mida.gov.my).

CONCLUSION

This chapter has shown how Malaysia has always welcomed foreign investment, adopting for a long time both an export-oriented policy and an import-substituting strategy. Such a mix has resulted in creating a favourable environment for foreign firms to operate in. In 1986, the government chose to favour investment from both local and foreign firms as well as export orientation, especially through TNCs' activities. This trend has led to the concentration of the Malaysian economy into a small number of sectors, some of which are foreign-dominated. This high dependence on foreign firms for certain sectors and for exports currently represents a major threat faced by the Malaysian government. The latter is actively trying to diversify the economy through a cluster-based industrial policy launched in 1995 with the New Industrial Plan.

The intensive use of incentives to attract investment is still the major focus of the government to continue attracting foreign firms. The creation of a favourable infrastructure through the creation of technology or industrial parks is only one example of this policy. The target of the government is to reach high levels of value-added activities through such incentives. In addition to incentives to the technology-intensive sectors, the government is trying to increase the spillover effects of existing firms by encouraging linkages.

A major policy stance has been the creation of non-tax incentives to generate backward linkages. The government encourages foreign firms to come into contact with potential suppliers, notably through the Vendor'Net programme.[11] The electrical and electronics sector has long been favoured by the Malaysian government, since the early 1970s, when MIDA actively lobbied electronics firms to come and manufacture semiconductors in export-processing zones. This policy has been extremely successful, and has led Malaysia to become one of the largest semiconductor exporters in the world. As a technology-intensive sector, the electrical and electronics sector is at the heart of the recent endeavour to pull Malaysia towards high-tech product manufacturing. Considerable attention has thus been given to the creation of backward linkages of foreign firms in this sector with local suppliers. Japan and the four NIEs are the major investors in this sector, followed by the USA and Europe. Firms from these countries/regions are therefore the focus of our research.

Overall, the investment environment in Malaysia is supportive of FDI, and to date it has attracted large amounts of foreign investment. Government policies have recently started to encourage inter-firm linkages, giving incentives for training and technical development. There is little incentive in terms of backward linkages' creation. During the different phases of economic development in the 1970s, Malaysia has seen the development of a dual economy, leading to few networks between TNCs and local firms. More recently, emphasis has been put on these developments, but local firms still lag behind in terms of quality and services requirements. As local firms' capabilities are limited, and TNCs' local sourcing primarily takes place with other foreign firms located in Malaysia, there is scope for more stringent policies and more encouragement both to upgrade local firms' capabilities, and to encourage inter-firm linkages between foreign firms and local counterparts.

Details on the research environment in Malaysia is provided in Chapter 5. The methodology adopted for this research is described, with particular reference to adapting the research tools to the Malaysian environment.

NOTES

1. Companies may benefit from tax holidays when the government grant them the right not to pay certain taxes for a fixed length of time.
2. The seven resource-based industries comprise food-processing, rubber, palm oil, wood-based, chemical and petrochemical, non-ferrous metal products and non-metallic mineral products. The five non-resource based industries comprise electrical and electronics products, transport equipment, machinery and engineering products, ferrous metal and textiles and apparel industries.
3. The electronics and electrical sector benefited from special treatment after 1971, with longer tax holidays.
4. Free industrial estates include Bayan Lepas, Prai, Prai Wharf, Batu Berendam, Tanjung Kling, Sungei Way, Ampang Hulu Klang, Telok Panglima Garang, Johor Port, Jelapang, Kinta Phases I and II, Tanjung Gelang and Sama Jaya.
5. This export-oriented policy proved successful, since in 1980 only 31% of projects were accepted in the manufacturing sector under the export category, while by 1990 this figure had increased to 73%, with a majority of the projects in the electrical and electronics industry.
6. Agreements were passed with most European countries in the 1970s, the USA in 1959 and with the four dragons in the late 1980s/early 1990s.
7. Incentives to local vendors cover pioneer status and investment tax allowances (http://www.mida.gov.my).
8. http://www.miti.gov.my/ucapan21.htm
9. http://www.smidec.gov.my/
10. http://www.malaysiadirectory.com/news/03/040806.html
11. As defined in the soft policies adopted by the Malaysian government, this term refers to the vendor development scheme that encourages large firms to provide assistance to local vendors.

5. Research Methodology and Design

INTRODUCTION

In Chapter 3, the literature review on studies of backward linkages revealed that most authors chose the case study/in-depth interviews approach with a qualitative analysis. Such a methodology provides the necessary depth of information required to understand the complex nature of backward linkages and its impact on the development of production capabilities of the host country's suppliers. However, none of the studies on knowledge transfer through backward linkages has tackled a sufficiently high number of companies to generate generalisable results, except for the study carried out by Halbach (1989) who used a mail survey. This chapter describes the methodology adopted for our research. As we need both in-depth information and a reliable number of companies in our sample, a questionnaire has been designed for use in a mail survey, and adapted to proceed with semi-structured interviews with foreign firms' subsidiaries in the electrical and electronics sector in Malaysia.

The database of firms was established from several directories, the target of the research being companies with a manufacturing or assembling activity. A list of 320 foreign firms in the electrical and electronics sector in Malaysia was compiled from various directories. Several sources were used because most of them presented inaccuracies. A total of 108 questionnaires were returned, which is a response rate of 33.75%. Some 95 questionnaires were found usable for the aim of the survey. From the list of products manufactured by the firms, a classification was determined by using various electrical and electronics products directories. This allows a better understanding of each type of activity. It is particularly useful in defining the level of complexities for each category of products manufactured, which is expected to be a key explanatory factor in foreign firms' behaviour.

A similar classification was kept to order interviews by categories. Eleven interviews were conducted within the desired sample, 4 with firms from the USA, 3 with European firms, 2 with Japanese firms, 1 with a Singaporean firm and 1 with a Taiwanese firm. Two additional interviews were conducted for further comprehension of the research topic. One firm from the USA provided information on its 'indirect supply' strategy. Another North American firm was visited; this was a repair centre, but not a manufacturing/assembling plant.

In this chapter, we introduce the reader to the difficulties of doing research in Malaysia in this area of study, and we discuss our choice of research instrument, before discussing the administration of the mail survey and the interview process. The triangulation method is used in the following chapters for the final analysis of the results obtained in this research; thus, this method is discussed at the end of this chapter.

RESEARCH QUESTIONS AND CONTEXT

This research project started at the end of 1994, with literature search and information gathering. A research procedure was designed and kept throughout the research process.

Research Procedure

One of the major difficulties facing a new researcher is to know how to organise not only the information he or she is gathering, but also to know the steps to be followed during the research process. Later in the chapter, we differentiate between a quantitative and a qualitative approach. This distinction is desirable in the design of a research procedure. The qualitative approach generally uses the literature inductively towards the end of the study, but it is very often also used at the beginning to 'frame' the research. The quantitative approach is always be based on the literature search, from which the definition of the variables to be studied arises. Taking this into account, the research started with a thorough literature search, and a review of the theories relevant to the transnational corporation (TNC), its foreign direct investment (FDI) and to the creation of links in the host economy.

We have already listed the three key research questions for this research. Additionally, in Chapter 3, research propositions were formulated; they pose a series of relationships to be studied. This method was preferred to hypothesis testing, because of the lack of specific literature on the topic under study. Questions can be asked, but existing knowledge and theory did not permit the formulation of reasonable predictions about the relationship of variables. The theory was also important in the formulation of the research questions. We also identified FDI trends; such trends were described in Chapter 4, together with the Malaysian government policies in place towards inward investment. From the determination of the research questions and the research objectives, the methodology to be adopted is described in this chapter. We discuss the reasons why a triangulation method was preferred and the procedures behind the quantitative and qualitative methodological issues to be used in this research project.

Research Questions

In Chapter 1, we formulated research questions to be answered in the process of this research. The three research questions were:

> *Research question 1* Do TNCs have an impact on host developing countries through backward linkages?
>
> *Research question 2* To what extent do backward linkages' creation lead to technology and knowledge transfer?
>
> *Research question 3a* What are the factors that may explain the existence of technology and knowledge transfer through backward linkages?
>
> *Research question 3b* What are the factors that may explain the degree of technology and knowledge transfer through backward linkages?

A series of propositions were additionally devised in Chapter 3, framing the objectives of this research project, and Figure 3.1 provided us with a clear structure of research to tackle our third research question. On the one hand, this study focuses on the transfer of technology, through direct product and process technology transfer, and through training. On the other hand, four groups of factors have been highlighted as influential factors in the existence and level of the transfers. The list of questions that follow apply for both research questions 3a and 3b.

Firm-related factors

Firm-related factors include all the factors related to the strategy and operations of the foreign firm. Propositions 3, 4, 5, 8 and 10 are in this category.

> *Proposition 3* Buyer-supplier links vary according to the level of technology of the product manufactured and the industry concerned.
>
> *Proposition 4* Locally-owned buyers create more links than foreign-owned firms do.
>
> *Proposition 5* The export-orientation of the firm affects the buyer-supplier links creation.

Proposition 8 The company's overall international and locational strategy affects the buyer-supplier links creation. The size of the local subsidiary is an indication of the importance of the affiliate.

Proposition 10 The country/region of origin of the foreign firm explains to some extent the buyer-supplier links' creation.

This study focuses on the activities of foreign firms, and does not aim at a comparison with locally owned firms and their transfers to other local suppliers. However, we do acknowledge that there might exist a difference in the transfers generated by the two types of firms. Instead, the factors related to the foreign firm itself are taken into account. Thus,` the following research questions are devised:

Question 3.1 Does the type of product manufactured by the foreign firm have an impact on knowledge transfer generated to the local suppliers?

Question 3.2 Are export-oriented foreign firms more or less likely to transfer knowledge to their local suppliers?

Question 3.3 Does the size of the company have an impact on the transfers to local suppliers?

Question 3.4 Does the country/region of origin of the firm explain transfers to suppliers?

Purchase-related factors

In Figure 3.1, purchase-related factors referred to Propositions 9, 12 and 13.

Proposition 9 Dependence on the parent firm and high levels of intra-firms trade impact on the buyer-supplier links' creation.

Proposition 12 Greater buyer-supplier links are created with subcontractors, but other long-term suppliers, considered as business partners, may also benefit from substantial relations with TNCs.

Proposition 13 The purchasing strategy of the buyer, in terms of number of suppliers, is indicative of the potential links created.

These propositions are centred on the purchasing strategy of the foreign affiliates. The research questions related to this group of factors are:

Question 3.5 Does the level of intra-firm supply have an impact on knowledge transfers to suppliers?

Question 3.6 Does subcontracting as a contractual method explain knowledge transfer to suppliers?

Question 3.7 Does the purchasing strategy, in terms of number of suppliers per item purchased, explain knowledge transfer to suppliers?

Investment-related factors

This group of factors relates to the overall environment in which the foreign firms operate and the transfers and benefits likely to originate in this environment. Four propositions entered this category:

Proposition 1 As the environment in the host-developing economy matures, buyer-supplier links increase and become more effective.

Proposition 2 Buyer-supplier links vary according to whether the industry is resource-based or non-resource-based.

Proposition 7 The less developed the host country's technological knowledge and production capabilities, the fewer buyer-supplier links are created.

Proposition 11 Government policies may influence firms to supply locally, but have not been found to have a substantial impact on buyer-supplier links' creation.

Propositions 1, 2 and 7 are self-restricted by the focus of this study. In choosing to study the Malaysian electrical and electronics sector (MEE), we choose a sector that is likely to generate substantial external links with suppliers, and an emerging country attracting substantial flows of FDI. Chapter 4 has shown that Malaysia has still not developed sufficient indigenous technological knowledge, and the electronics sector is still dominated by foreign firms. However, the country is going through a rapid phase of economic development, and one of the government's objectives is to encourage linkages between foreign and local firms. Thus, the results of this study will bring interesting information to perceive and understand the current changes in this country and in this sector. Proposition 11 is one that can be directly studied in the context of this research. This leads therefore to research question 8.

Question 3.8 Do government policies help to explain knowledge transfer to suppliers?

Time factor

The miscellaneous time factor is the final research question to be answered in the context of this research.

> *Question 3.9* Does the time spent by the foreign firm in the host country explain the transfers to suppliers?

Research issues

The existing literature has highlighted the various factors likely to influence the transfers occurring between foreign affiliates and their local suppliers in host economies. Because of the nature of this research project, we are not investigating all these factors. Instead, we focus on a series of nine key research questions. The expected signs of relationship between these factors and the technology and knowledge transfer are not clearly defined in the literature.

Problems in Macro-data Collection

The problems arising from the collection of macro-level data are common to all researchers in the field of FDI. The main problem, when dealing with less developed countries, is the lack of detailed data, or their unreliability. Hence, data deficiencies may lead to the use of second-best or truncated methods. In this particular study, we are also confronted with the problem of isolating the effects of FDI from other possible change or influence occurring in the economy. It is very difficult to determine what would have happened if the particular foreign investment had not taken place.

Problems with existing FDI records

Secondary sources usually present data of FDI flows and stocks. Depending on the official source, data are recorded differently, with variable definitions and varying times of recording. For instance, the Malaysian government records foreign investment per applications and acceptances, whereas the International Monetary Fund (IMF) provides a record of existing flows. In this context, the full picture of investment flows and stocks result from the analysis of various publications and sources. As a result, data on FDI from different sources (such as international organisations) are not necessarily comparable. This is one of the major impediments to research on FDI.

As an illustration of these discrepancies in registering foreign investment, world inflows and outflows are rarely balanced (UNCTAD, 1992), even though they should be. For all countries, the difference between inflows and outflows was US$120,861 million in 2000. This gives the impression that trends in FDI available are biased. In fact, there are several reasons for such differences; those that are most often put forward are:

- the various definitions given to inflows and outflows of FDI,
- the way branch profits are treated between inflows and outflows,
- the difference in treatment of profits or losses achieved,
- the numerous ways in reporting subsidiaries' operations or reinvested profits,
- the differences in accounting methods for FDI,
- the differences in the treatment of FDI in the construction industry and in the real estate business (Graham and Krugman, 1993; Dunning, 1993b).

Hence, most statistical sources need to be treated with caution. Some countries do not keep a full account of the number of TNCs or foreign subsidiaries operating within their national territories. Sometimes, such statistics are established only starting from a specific amount of sales or assets, and subsidiaries may not be registered in certain industries. In many countries, the national investment bodies report FDI data on an approval basis, and FDI is denominated in terms of the local currency unit. FDI data based on approval by the host country may not necessarily depict the actual flows of FDI into the country. This problem arose in the analysis of inward investment flows in Malaysia and was discussed in Chapter 4.

Discrepancies could arise when foreign firms that invest in a country do not go through the board of investment if they know that they cannot qualify for any incentives, and for companies that have received approval but do not pursue the investment project.

International organisations show a sample range of differences. The Organisation for Economic Cooperation and Development (OECD) sources report FDI data to and from its member countries. This results in the exclusion of many non-member countries. Although the IMF and OECD data appear to be preferred sources, they do not however provide a complete set of data on sources of FDI and FDI in significant economies (for example, Hong Kong, Taiwan). Some data on FDI include reinvested earnings and take into account the intra-firm borrowings, while other data exclude them. Some include petroleum investment, whereas other sources exclude such investment capital flows. In addition, some sources report FDI flows on a net basis, while others report the gross figure. Some FDI data do not cover every industry. Some FDI data may cover only certain industries (for example, manufacturing), while others may lack such data. Also, some countries are more efficient in collecting and reporting up-to-date FDI data, while others (often less developed countries) are slow in releasing FDI flow data to their countries.

Handling problems related to FDI data

There exist various ways of FDI data collection and definition. Therefore, accurate analysis and comparison of FDI flows and effects are fundamentally tedious tasks. Given the numerous potential problems and difficulties in establishing completeness and comparability of macro FDI data, their interpretation needs to be treated with care. In this research, data from international organisations are used whenever possible. However, for recent and detailed analysis of FDI in a specific host country, national records remain the most detailed sources.

Sources of macro or aggregate level data

Data on FDI used in this research were extracted from various national and international organisations and other studies done in the field. Data were collected from Bank Negara Malaysia, the Malaysian Department of Statistics, the Malaysian Industrial Development Authority and the Malaysian Minsitry of Trade and Industry. In addition to literature review, it is important to identify all the useful sources of reliable statistical data available to the researcher. Many international organisations publish data on international investment, useful sources for Southeast Asia include United Nations Centre for Trade and Development (UNCTAD), the IMF, Asia Pacific Economic Corporation (APEC), the Institute of Southeast Asian Studies (ISEAS), the Asian Development Bank (ADB) and the Association of Southeast Asian Nations (ASEAN) Secretariat.

Choice of Methods

The debate between qualitative and quantitative paradigms needs to focus on particular aspects of various research methodologies. It is important to ensure that the research questions drive the methodology. Research questions thus lead to the choice of methods. The five general standards for educational research are:

1. The fit between research questions and techniques of data collection and analysis.
2. The effective application of chosen techniques of data collection and analysis.
3. Understanding the background assumptions.
4. Overall warrant.
5. Value constraints: importance and rigour of appropriate techniques, rather than methodological purity (Howe and Eisenhart, 1990).

Chapter 3 presented the research propositions and variables to be studied. Considering the nature of this research, note that both qualitative and quantitative methods will bring crucial understanding of the phenomena. The quantitative approach through a mail survey will allow gathering of information on more subsidiaries and therefore testing quantitatively the assumptions described for the research, while the need for qualitative data arises from the importance of obtaining managers' opinions on their relationship with their local suppliers. The majority of books on research methods emphasise the quantitative-qualitative dichotomy, and it is often advised to use only one of the two methods. However, within the scope of this research, we have chosen to focus on the inherent complementary nature of the two approaches. We first describe the reasons that lead to the use of both approaches in independent sections. We then show how a triangulation method provides a better understanding of the phenomena studied in this research, and is therefore the favoured methodology.

Quantitative and qualitative methods

Creswell (1994) distinguishes between the two approaches:

> One approaches a quantitative methodology by using a deductive form of logic wherein theories and hypotheses are tested in a cause-and-effect order. Concepts, variables, and hypotheses are chosen before the study begins and remain fixed throughout the study. [...] Alternatively, in a qualitative methodology, inductive logic prevails. Categories emerge from informants, rather than are identified a priori by the researcher Creswell (1994: 7).

Bearing in mind the inherent nature of each approach, we have scanned all the possible research methods available for each paradigm. Saunders et al. (2000) characterise the two paradigms with the words 'positivism' and 'phenomenology'. This follows the above distinction between the quantitative and the qualitative paradigms. Positivism is deductive in nature; its purpose is to explain causal relationships between variables, usually through the use of quantitative data. One of its main features is the design of a highly structured methodology to facilitate replication. At the other end of the spectrum, phenomenology is not quantitative in nature, but rather it focuses on the way people experience social phenomena within the environment in which they operate.

A clear advantage of the quantitative design is that of credibility of results through a large number of responses organised in a structured way. On the other hand, the qualitative approach is particularly useful to grasp social processes and give the researcher a better insight of the phenomena under study. Each research design presents a set of disadvantages ranging from a lack of social understanding for the quantitative design to the lack of generalisation and difficulty of analysis for the qualitative design. Such advantages and disadvantages are useful to keep in mind for the establishment of any research design. Through a list of concepts to be studied, we endeavour to underline the usefulness of each design to our specific study. Following the Oppenheim (1992) differentiation between the research design and the research techniques, first the strategy of the research and the logic behind are exposed, and then the method used to collect data is described.

Choosing research strategies

The most popular research strategy in business and management research is the survey method. This strategy allows the gathering of large amounts of data, and it is highly economical. Data are collected by means of a structured questionnaire, and the techniques are highly standardised. However, the data collected may not be as wide ranging as those collected by qualitative research methods. There is a limit to the number of questions that questionnaires can contain, and the biggest drawback of a survey method is the capacity to do it badly. To try and minimise these limits, we have also used the survey through structured interviews where standardised questions are put to all interviewees. As a result, the central strategy in this research is that of a survey, using a questionnaire administered in two different ways. The first strategy is that of a mail survey to obtain a bigger number of responses. The second strategy lies in a series of interviews to obtain more in-depth data to be analysed in a qualitative way following a case study format. Table 5.1 summarises the topic to research and the reasons behind the choice of method. The case study strategy involves in this instance intensive knowledge of a small number of related cases. This strategy is particularly advisable to gain a rich understanding of the context of the research and the processes being enacted.

The same questionnaire is used twice, but analysed in different ways. Furthermore, the interviews followed a semi-structured questionnaire, in the sense that the respondent was asked only the general questions, and was not given the multiple answer choices as mail respondents were.

This research is cross-sectional as opposed to longitudinal. The sample has been observed at a specific time, and it would be too expensive and time consuming to try and replicate it in the near future. The strategy was that of a survey and a series of interviews conducted over a short period of time.

Table 5.1 Research design to search for links between foreign firms and local suppliers

Methods	Quantitative	Qualitative	Mix of two methods
Topics to research	Groups of variables 1. firm-specific variables 2. input-specific variables 3. supplier-specific variables 4. special requirements by the relation-specific variables 5. different types of links' variables 4. correlation between each group of variables Some questions are qualitative and coded for ease of analysis; they are included under the 'quantitative' heading, although qualitative in nature	1. relationship concepts: how does the subsidiary perceive the existing suppliers and their supplies? 2. types of links between the two parties as perceived by the subsidiary, through use of semi-structured questions to allow freedom of answers from respondents 3. additional factors not planned within research objectives relevant to understanding of the topic	Results from each method will be compared and discussed in perspective with results of the other
Reasons behind choice of method	1. relationships between variables can be evaluated quantitatively 2. signs of relationship can be provided 3. ensures reliability of results 4. clear conclusion building and verification of theory and existing studies	1. understanding of social concepts in the relations between subsidiaries and local suppliers 2. provide new framework for analysis and allow for pattern not planned previously in research objectives 3. increase in-depth understanding of phenomena through specific examples	1. allows easily comparable data through large number of cases and in-depth understanding through specific cases 2. greater reliability of results through complementary methods

The study is descriptive, it provides a portrait of the types of relationships foreign subsidiaries in a host country maintain with their local suppliers and it describes the purchasing activities and habits of the subsidiaries. Furthermore, the study is explanatory, since we try and explain the factors influencing the level of these relationships.

Sample selection

For ease of the research process, the research will target one particular sector. The reasons behind the choice of this specific sector have been given in earlier chapters. From a methodological point of view, it was the intention to obtain as many cases as possible from one particular sector for the robustness of the results. We argue that a higher response rate within one particular sector leads to validation of the results through the sheer number of respondents. Moreover, it is often preferable to target a smaller sample that allows the researcher to spend more time and effort on this smaller sample, and this method generally increases the chance of a better response rate (Dillman, 1978). In the case of the electronics and electrical sector in Malaysia, we estimated the total number of foreign subsidiaries at between 300 and 350 firms at the time of the data collection in 1996, which was a manageable number for a survey. For the interviews, particular effort was made to obtain interviews with companies from each of the nationalities of origin. This means that we paid particular attention to obtain interviews from Japanese, European, North American companies, and firms from newly industrialising economies (NIEs). Difficulties encountered for each method are described in the following sections.

Designing and formulating questionnaire

It would be over-simplistic to say that the analysis of the questionnaire was to be clearly divided between a quantitative analysis of the mail survey results and a qualitative analysis of the semi-structured interviews. In fact, some of the questions included in the questionnaire are qualitative in nature, since they are open-ended questions, and were coded for analysis purposes after receiving responses. Therefore some of the variables used in the quantitative analysis are qualitative in nature.

Designing variables

The first step in the design of the questionnaire was to identify the main groups of variables to be studied following the research questions developed earlier. Additional questions were included in order to obtain a fuller picture of the relationships between foreign subsidiaries and their suppliers. Several samples of existing questionnaires on the activities of foreign firms in a host country were consulted, some found in published research (for example Buckley et al., 1983; Buckley and Artisien, 1987; Buckley et al., 1988). An important source was that of Halbach (1989) who has studied the linkages' creation between foreign firms and suppliers in all countries and all sectors. Some questions of this particular questionnaire are similar to those used within our questionnaire; however, we have completed the set of questions used by Halbach to obtain a picture more applicable for this particular study. For instance, Halbach did not differentiate between product and process technology, he did not include questions on training, although these are very important links created between foreign firms and their suppliers, nor did he include questions on the management styles' transfer. All these categories were later identified as the most important links created by the respondents themselves during the interviews. Additional material was used, these being specialised books on human resources (Dessler, 1994) and types of training, specialised technology transfer publications (Lindsey, 1986; Schrader, 1991; Chen, 1994) and books on production operation management (England and Leenders, 1975; Farmer, 1985; Muhlemann et al., 1992).

Testing questionnaire

Because of geographical distance, a preliminary pilot study was first carried out in England. Each question in the questionnaire was discussed during this meeting, so as to feel how questions were understood, and to identify the variety of answers that could arise. At the end of this exercise, particular attention was given so that questions would be uniformly understood, that they were all useful (if all answers in one question are identical, there is no value for the question) and that the structure of the questionnaire was both logical and easy/interesting for the respondent to follow.

ADMINISTRATION OF MAIL SURVEY IN MALAYSIA

Before embarking on a field trip, the researcher needs to pay particular attention to the various techniques to be applied for the administration of a mail survey, and very importantly, the specificities of the country in which the survey is to be carried out need to be taken into account. In this section, we cover the research techniques identified for our study, before describing the procedure followed to make a full classification of the products manufactured by the respondents. Finally, we present the results obtained from the mail questionnaire.

Research Techniques

Each research is unique and requires a predetermined set of research techniques to ensure success. For the purpose of this study, we carefully designed a research strategy, using a variety of tools and techniques. This strategy has led to a successful mail survey, with over 30% response rate.[1] Each step taken is described below.

Choice of technique

Because of time, financial and distance constraints, the postal technique is the favoured one for this study. The use of questionnaires enables a widely distributed sample of respondents and ensures anonymity for the respondent. The two major reasons for the extensive use of mail survey methods are, first, there is a much lower cost for completing them; second, procedures for mail surveys are simple enough for a researcher to conduct one on his or her own.

Problems likely to arise from this method

- It is difficult to be sure that a substitution of respondent has not taken place.
- The questionnaire might be prevented from reaching the respondent.
- Various problems might arise while the respondent answers the questions, and the presence of the interviewer could be needed.
- The response rates tend to be low.

Actions to be taken to minimise potential problems

- Questions are easy to understand.
- Questionnaire is not too long.
- Right person fills in the questionnaire, so it should be sent to a specific person within the company.

Two specific problems arising from use of mail questionnaire

Bias from refusal. In case of unsatisfactory results from the mail questionnaire, the use of face-to-face interviews should ensure a sufficient spread of countries of origin for the sample of subsidiaries. Also a matter of concern; a determining factor for response rate is the interest the person has for the particular topic addressed.

Screen questions. The questionnaire covers a wide range of information from general knowledge about the host country environment to specific facts about the subsidiary's organisation and structure. For this particular problem, the pilot study should help to determine the person best suited to answer the questionnaire and avoid the need for several respondents.

Methodological issues

The choice of procedure corresponds to the total design method (Dillman, 1978, 1991), which shows that people respond to three incentives: the reward (being part of a carefully selected sample is considered as a reward), the cost (the time of answering the questionnaire is reduced by using a shorter questionnaire), and the trust (use of an official presentation).

A number of tools exist to be used in improving response rates for mail questionnaires. This study emphasised the use of white stationery, stamped return envelopes, shorter questionnaires, personalisation of respondents, anonymity and confidentiality; the social usefulness of the study and importance of respondents to study success was stressed, a copy of the survey results was offered and a repeated follow-up programme was used.

Practical issues

- *Personalisation.* Each questionnaire was sent to a particular person in the company, this required that the researcher knew the name of the respondent before sending the questionnaires. The first pilot survey showed the importance of addressing the questionnaire to the respondent him or herself. Therefore the strategy chosen was to try and talk directly to the manager on the phone before the survey, not only to get to identify his or her name, but further to warn him or her that he or she was going to receive a questionnaire.
- *Content.* Each envelope should contain a self-addressed, prepaid return envelope, so as to minimise the input of the respondent.
- *Official letter.* The layout of the letter looks professional and official sponsors are mentioned. We used white stationery with a professional heading for the introduction letter. This letter was explaining which institutions were involved in the survey, which government financed it, what type of companies were concerned, what the exact topic, aim and expected results of the survey were, and finally the letter insisted on the usefulness of the respondents' contribution.
- *Official sponsors.* To further emphasise the 'official' nature of the survey, the source of the funding for this field research was mentioned on the first page of the questionnaire. Additionally all institutions hosting this research were mentioned: the Bradford University School of Management, the Université de Paris-I, Panthéon-Sorbonne and the Universiti Sains Malaysia School of Management.
- *White stationery.* Dillman (1978) considers the use of white stationery as an important factor in ensuring response rate. As a result, the researcher looked for a good printing company once in Malaysia that would ensure both reliable service and good quality stationery. The cover of the questionnaire was light blue for greater recognition.

- *Questionnaire length and layout.* Despite warnings in the literature, encouraging the researcher to produce a short questionnaire, the final version of our questionnaire was quite long. To compensate for the length, the layout was designed to make the questionnaire user-friendly and to look both professional and pleasant. A double-column system was used, so that as a whole, the questionnaire was thin and looked easy to answer. The real time estimated to complete the questionnaire was one hour.
- *Gift.* A bookmark with the logo of the university was purchased and enclosed with each letter. This option was chosen to follow the reward concept defined by Dillman, whereby the researcher trusts that the respondent will respond and therefore sends a present before receiving the answer.
- *Access to results.* The respondents were offered a copy of the executive summary of the results. Thus, the respondent was filling in a questionnaire knowing he or she will be informed of results which could potentially be of use to him or her. Such an offer was repeated twice. The respondent was informed in the introduction letter that he or she could receive a copy of the results, if he or she filled in the prepared box on the last page of the questionnaire itself, placed there as an incentive for the respondent to first answer our questions.
- *Incentives.* Incentives to answer the questionnaire are a mixture of all the factors listed above. The overall layout of the questionnaire, its professional appearance and official background, the inclusion of a gift and the offer of a copy of the results are all important incentives to encourage the respondent to answer.

Establishment of company list

Sample delimitation and selection were defined above. Methodological considerations encouraged the researcher to focus on only one sector. Companies targeted within the electrical and electronics sector were firms with full-, majority- or minority-foreign ownership. Firms had to carry out either a manufacturing or an assembly type of activity.

The full list for the survey was then finally established and completed. The list of companies' part of the sample was compiled from three different sources:

- Penang Development Corporation list of companies (February 1996).
- List of companies in the electrical and electronics sector in Malaysia provided by the Malaysian Industrial Development Authority (December 1995).

- List published by the Japan External Trade Organisation (JETRO) of the Japanese companies operating in the Malaysian electrical and electronics sector (February 1996).

From these, a list of 320 companies was drawn up of foreign firms or firms with foreign partners. Two other directories were used to double-check this list:

- *Malaysia: Electronics and Electrical Industries* published by the Malaysian Industrial Development Authority and the *Business Times* in 1993.
- The *FMM Directory 1995*, the Federation of Malaysian Manufacturers' directory. This directory however does not aim to be complete, only the companies that take part in this Federation appear in the directory.

Pilot study

Fifty companies were chosen at random for the pilot sample. Thirty companies were contacted to obtain the appropriate contact name in the company, and 20 companies were sent the questionnaires addressed to purchasing/materials manager. Six questionnaires were returned after the first letter and a follow-up letter.

Some modifications were made to the questionnaire, some questions were reformulated, some abandoned and the questionnaire was compressed from ten to three double-sided pages, with two columns on each page.

Only one questionnaire was returned from the set sent without a specific name. The final questionnaire was sent to the companies that had not responded. Four additional questionnaires were returned; these were the questionnaires sent previously. Obviously, some of the respondents started answering and only finished answering when they received the second print of the questionnaire. As a result, it was decided that a second set of questionnaires would be sent to the whole of the sample.

Finalising company list

Our list of 320 companies was then completed by the list of names of the manager to whom the questionnaire had to be addressed. As mentioned earlier, personalisation has been identified in the literature as a major factor increasing response rate. This was confirmed by our pilot study and as a result it was decided to call every company. This strategy had multiple advantages in our research. First, the title of the person in charge of direct purchasing varies in each company. In most cases, the manager is simply called purchasing manager, or direct purchasing manager. But in other cases, as underlined in the responses to the questionnaire, the person in charge may have been the operations manager, or even the managing director him or herself (purchasing manager, materials manager, purchasing officer, production manager, managing director, or general manager, depending on the company). Hence, calling each company first was extremely useful to identify the official title of the person carrying out the direct purchasing for the subsidiary. The second advantage of calling was to know the name of the manager. This allowed for personalisation of the survey, and was a very important issue identified in the strategy. The third advantage of calling each firm first was to talk to the manager him or herself whenever possible. In numerous cases, the call was a means of identifying and introducing the research to the manager. The response was often negative, and questions such as 'How did you get our address and telephone number?' were frequent. This gave a good opportunity for the researcher to reassure the manager, emphasising that the address and telephone number had been provided from a list provided by the MIDA. The respondent was informed that he or she would receive a questionnaire shortly, and it would be left to his or her discretion to answer it or not. Last, an additional advantage of calling was to be able to confirm that the company was still in operation. In a number of cases, the telephone number was already outdated, and the message was that the company was not located there any more, or that it had stopped operation. This movement of companies is frequent in a developing country such as Malaysia, which is one of the reasons why company directories are often inaccurate in this region of the world.

Final methodology adopted: a step-by-step approach

The overall methodology was divided into four stages. Stage one related to designing the questionnaire and the methodology to be adopted for the research. Stage two consisted of the pilot study. Stage three included the establishment of the company list with contact names. The last stage took the researcher to Malaysia for the implementation of the survey with interviews, and the mail survey. This was the shortest stage in the process design. Indeed, compared to earlier stages, the operationalisation of the mail survey took place within only one month. This was primarily due to time constraints, as the researcher's stay in Malaysia could not be extended, and also simply due to the nature of the mail survey process.

Operationalisation of mail survey. As soon as the company list was complete, the researcher could proceed with the survey. This was planned to last no more than a month, as respondents were unlikely to send a questionnaire back if they had not done so within this time period.

Week 1 was the initial mailing to each company on the list. Each envelope contained a personalised letter of introduction, a present (paper clip/bookmark), the questionnaire and a stamped self-addressed return envelope. Each address was typed on a white sticker to improve the official layout of the envelope. The respondent was promised confidentiality of the results, and therefore if he or she did not mention the name of his or her company on the last page of the questionnaire, he or she was assured of anonymity. However, for the administration of the questionnaire, the researcher did keep track of the respondent to avoid further mailing in the case of response.

Week 2 was planned for the first follow-up letter. This letter was sent exactly one week after the first mailing to all companies. A marginally small number of questionnaires were returned after one week, and as a result the follow-up letter was sent to nearly all the firms.

Week 3 was the third step in the mail survey. The follow-up letter had been of little help, and only a few additional questionnaires were received. The companies were not called again during the follow-up, due to the extreme difficulty of this process. The second set of questionnaires was a preferred strategy; therefore, a third mailing to the companies who had not responded so far was made. It included a personalised letter, the questionnaire and the return envelope. Slight amendments had been made on the last page of the questionnaire so that the researcher could identify whether the questionnaire received was the one sent in week 1 or the one sent in week 3.

By the end of one month, 84 questionnaires had been received altogether. The mailing of the second questionnaire had been a successful strategy and many responses were received after week 3. It is interesting to note that the majority of the questionnaires received after the second follow-up were not the second set of questionnaires, but the first ones sent. This indicates clearly that the respondents had started to complete the questionnaire and only fully completed it after receiving a second copy. Twenty-four additional questionnaires were received after a period of six weeks, or two to three weeks after the second questionnaire had been sent.

List of Products Manufactured by Surveyed Companies

Product classification took place only after the respondents had answered the questionnaire for several reasons. It was the intention of the researcher to obtain a list of products manufactured/assembled so as to categorise them in a standard way. The process of categorising helps the researcher to be familiar with its answers, by working on a case-by-case basis on each questionnaire. This post-survey classification also avoids misclassification by the respondent. Misclassification is likely to occur when the respondent is unfamiliar about the term used in the classification chosen by the researcher, or when the company manufactures several products, which may belong to different product categories.

The classification method employed for this survey essentially follows the major product groups given by the four-digit International Standard Industrial Classification (ISIC) for the electrical and electronics products. Some further subdivision is used to differentiate the product according to the technological level. Other surveys were looked at in order to compare the various classifications possible. The classification finally established here is intended to provide the relevant information for this particular study.

Building product classification for survey

When comparing various publications, similar product categories are used, with changes only according to the overall objective of the study/publication. The UNCTC (1986) survey used very broad categories, as further details were not essential to the result of the survey. The OECD (1992) survey classification differs insofar as there is no distinction between types of components, and the emphasis lies on the industrial electronics. Finally, another example of classification provided is that of the Malaysian Ministry of Finance (1996). This publication aims at reporting the achievements of the electronics sector by type of products. Electrical products are included together with electronics products, albeit in an independent section.

While elaborating our own classification as presented in Table 5.2, several problems have been encountered. Because of a lack of very detailed classification sources, some simplifications have been included. We considered the following two issues:

- the difficulty of the product manufacturing (whether the product is standard or not);
- the place of the product in the production chain, from raw materials to primary products (components built for no predefined product) to secondary products (sub-assembly that can be used to make a product) to the tertiary products (marketed entity).

It was impossible to reach such a level of precision for lack of sufficient information on each product, and also because there exists no classification complying with our purpose.

Companies were required to name the four main products manufactured/assembled within the site. The first product in the list provided was chosen for coding, under the assumption that the first product mentioned was the most important one. We are aware of the limitation of such a process, but it is impossible to provide several classifications for each company; this would complicate the analysis of the results and would not bring any further understanding. When the term used for this first product was ambiguous, it was replaced by looking at the other types of products manufactured by the company. It was assumed a company will manufacture similar types of products (for instance 'switching system' could be an active component if diode/transistor, or an electromechanical component if timer; by looking at the other product manufactured by the firm, 'converters', we assume in this case that the switching system is an active component as well as the converter).

Table 5.2 Product classification adopted in this survey

Category	Products	Number of cases
Parts and components	1. AC active components[2]	27
	2. PC passive components[3]	48
	3. EM electro-mechanical components[4]	45
Finished parts for production purposes	4. PS power sources[5]	7
	5. PE production and other miscellaneous equipment[6]	5
Finished products	6. CO computers and peripherals[7]	16
	7. OIE other industrial equipment[8]	13
	8. CE consumer electronics[9]	42

Note: In many cases, subsidiaries manufacture/assemble more than one product.

Also, other sub-categories have not been included as they would have been too small and therefore irrelevant; the products were then included in the closest category (for instance, microwaves' semiconductors were included in the active components).

To achieve this classification, several directories were consulted. Each product indicated by the firm was checked individually in one of the directories given below:

- Technical index of the *Electrical and Electronics Directory*, 1996 issue;
- *Kompass Directory Index of Products*;
- *Singapore Electronics Manufacturers' Directory*, 1993, Economic Development Board;
- *Philips Reference Components' Guide*, 1989.

Four directories were used because none contained names of every single product that were given by the companies. The product list was changed several times after being first constructed with the help of these directories. The researcher consulted engineers familiar with specific electrical and electronic products. As a result, the classification took place in three steps:

(1) an overall list of categories was decided; (2) more specific directories were used to place each specific product in one of our categories; and (3) the list was double-checked by experts.

Recoding for further analysis

In order to carry out further analysis on only the subsidiaries that supply primary input within Malaysia, the list of products manufactured has been changed. The final category aims at giving a representation in terms of 'biggest' product category, and categories essential to the understanding of the topic under study. Thus, although very few firms manufacture/assemble active components, the products in this category are the most 'technology-intensive' products, and must therefore be considered. The categories 'passive components' 'electromechanical components' and 'consumer electronics' are the categories with the highest number of firms, and they were therefore kept unchanged. All other categories were placed under the heading 'other products'. Such a procedure means the researcher loses in terms of detail in the data, but this is essential for the accuracy of the statistical analysis, for which a minimum number of 'dummy' variables must be kept.

Mail Questionnaire Results

The objective of this section is to analyse the response rate (see Table 5.3) obtained during the mail survey and provide background information for respondent companies. Background information allows checking the validity of the sample by comparing it with real FDI flows in the MEE.

Method of data input

All questionnaires were first coded and then input was done into SPSS Word (Norusis, 1994), for ease of analysis. Some open-ended questions were first coded so as to ease data input.

Overview of results

Number of questionnaires received. A total of 108 questionnaires were received. However, it was found that some questionnaires were not usable within this research for a variety of reasons. The major reason was the inadequacy of the product manufactured. It can be argued that these cases would be added in the sample, but they are not part of the researcher's original delimitation, and were therefore excluded. Two questionnaires were returned incomplete and could not be processed, one firm had no purchasing activity and one was a sales' subsidiary only.

The response rate achieved was 33.75%. This response rate is considered high compared to other studies carried out in Malaysia, which showed substantially lower response rates. Among the 95 usable questionnaires, it is interesting to note that eight firms had no local direct purchasing. Such responses are crucial to this study; such firms are part of our sample, but still import all direct input from outside Malaysia.

Table 5.3 Mail questionnaire results

	Number of firms
Number of companies in survey list	320
Total number of questionnaires received	108
Response rate	33.75%
Number of usable questionnaires	95
Of which, number of firms with no local direct purchasing	8
Reason why other questionnaires not used:	
1. Locally owned firms[10]	5
2. No purchasing activity	1
3. Products outside the targeted area[11]	4
4. Incomplete questionnaires	2
5. Sales subsidiary with no manufacturing	1
Total unusable	13

Note: Three other companies (two Taiwan, one Japan) mentioned on phone that all inputs were imported from parent firm.

Responses by firms' characteristics. Firms' characteristics are analysed by looking at the country of origin, the type of product manufactured, the ownership pattern of the subsidiary and the time of establishment in Malaysia. These categories help to verify the adequacy of the results, we can compare the general description of our sample (see Table 5.4) with the actual FDI flows and stocks in Malaysia described in Chapter 4.

Country of origin. Japanese and firms from the NIEs form the main bulk of all foreign firms in the MEE, with, 52% and 28% respectively, of the total sample. Out of 95 firms, 48 were Japanese. Taiwan is the second largest investing country in the MEE with 22% of the total sample. The survey results follow the breakdown in foreign firms' origin in the MEE in 1996. Taiwan was the first overall investing country in Malaysia in the early 1990s.

Thus, the high number of Taiwanese firms reflects the situation in MEE in the early 1990s. Out of 27 firms from the NIEs, 21 were Taiwanese, 1 was South Korean, 1 was from Hong Kong and 4 were from Singapore. European and North American firms, respectively, accounted for less than one-tenth of the sample. Out of 9 European firms, 3 were from the UK, 2 were from France, 1 was from Germany, 1 was from Holland, 1 was from Ireland and 1 was from Switzerland. Eight firms originated from the USA, 1 originated from Australia and 1 originated from New Zealand.

Table 5.4 General description of subsidiaries by country of origin

Country of origin		Japan	NIEs	Europe	USA	Other	Total
Product classification							
Active components	Count	2	4	1	3	0	10
	Col %	4.1	14.8	11.1	37.5	0.0	10.5
Passive components	Count	13	8	2	1	0	24
	Col %	26.5	29.6	22.2	12.5	0.0	25.3
Electromechanical	Count	15	3	3	1	1	23
	Col %	30.6	11.1	33.3	12.5	50.0	24.2
Power sources	Count	1	1	0	0	0	2
	Col %	2.0	3.7	0.0	0.0	0.0	2.1
Production equipment	Count	1	3	1	0	0	5
	Col %	2.0	11.1	11.1	0.0	0.0	5.3
Computers and peripherals	Count	3	2	0	2	0	7
	Col %	6.1	7.4	0.0	25.0	0.0	7.4
Other industrial equipment	Count	2	1	1	1	0	5
	Col %	4.1	3.7	11.1	12.5	0.0	5.3
Consumer electronics	Count	12	5	1	0	1	19
	Col %	24.5	18.5	11.1	0.0	50.0	20.0
Total	Count	49	27	9	8	2	95
	Row %	51.6	28.4	9.5	8.4	2.1	100.0
Year of establishment							
1972-75	Count	6	0	1	2	0	9
	Layer %	6.3	0.0	1.1	2.1	0.0	9.5
1976-80	Count	4	0	1	0	0	5
	Layer %	4.2	0.0	1.1	0.0	0.0	5.3
1981-85	Count	2	0	3	0	0	5
	Layer %	2.1	0.0	3.2	0.0	0.0	5.3
1986-90	Count	30	19	3	3	2	57
	Layer %	31.6	20.0	3.2	3.2	2.1	60.0
1991-95	Count	7	8	1	3	0	19
	Layer %	7.4	8.4	1.1	3.2	0.0	20.0
Total	Count	49	27	9	8	2	95
	Layer %	51.6	28.4	9.5	8.4	2.1	100.0
Firm ownership							
Wholly owned	Count	39	19	5	8	2	73
	Col %	79.6	70.4	55.6	100.0	100.0	76.8
Majority foreign	Count	8	6	3	0	0	17
	Col %	16.3	22.2	33.3	0.0	0.0	17.9
Minority foreign	Count	2	2	1	0	0	5
	Col %	4.1	7.4	11.1	0.0	0.0	5.3
Total	Count	49	27	9	8	2	95
	Col %	100.0	100.0	100.0	100.0	100.0	100.0

In the case of four subsidiaries, the investment originated from another subsidiary of the parent firm located in a country different from the home country. Thus, one Japanese subsidiary located in Singapore is the parent of a subsidiary in the sample. One firm invested from Germany, but was originally from Holland. One firm invested from Australia, but was originally from Germany. Finally, one firm investing from Ireland was originally from New Zealand. The country of the parent firm as opposed to that of the investing subsidiary was computed. For ease of analysis, major groups of countries were taken into account, for example, Japan, the NIEs, Europe, the USA and 'others' (with Australia and New Zealand). Such groupings are common in all analyses of the activities of foreign firms in Asia. Yet Japan, Taiwan and the USA are the major investing countries.

Types of products manufactured. The survey sample shows a high specialisation in specific products of the MEE. Malaysia is specialised in the manufacture of parts and components. Product categories such as active components, passive components and electro-mechanical components total 60% of the sample. Finished products for production account for 7% of the firms; they include activities such as electrical equipment and other production equipment, which are essentially labour-intensive assembly-work activities.

Hence, two-thirds of the subsidiaries manufacture/assemble intermediate products, while finished products (computers and peripherals, other industrial equipment and consumer electronics) are manufactured/assembled by only 33% of the subsidiaries.[12]

The most frequently manufactured products are part of the passive components, electro-mechanical components and consumer electronics categories. These products are labour-intensive, standardised products. Hence, despite recent efforts for technological upgrading, Malaysia remains a manufacturing platform for labour-intensive, standardised products, most of which are intermediate goods. Technology-intensive products are included in the active components (which includes, among others, integrated circuits and semiconductors), and the computers and peripherals categories. US firms first started to invest in semiconductors in Malaysia in the early 1970s, with the introduction of free trade zones (FTZs). However, firms manufacturing the second type of products invested recently in Malaysia. These two sectors represent only 18% of foreign subsidiaries in the MEE, but this shows that Malaysia specialises in low-end as well as high-tech products.

Such a specialisation can be explained by different factors. Malaysia specialised in the manufacture of semiconductors as developing Asia became a rapidly growing market for semiconductor devices. Electronics firms in Asian NIEs have become world leaders in the export of a variety of consumer electronics products, as well as in certain types of computer peripherals and communication equipment, which incorporate semiconductor devices of various types in varying quantities. As the manufacturing of semiconductors is fairly complicated, once the investment has been made and the training provided, the firm prefers to keep the operation; thus as some labour-intensive investment firms tend to move to cheaper locations, the high-tech intensive ones would remain in the host country.

In the case of consumer electronics, firms (mainly from Japan) invested in individual countries to circumvent tariff and/or other trade barriers. A frequent condition for access to the domestic market of the Association of Southeast Asian Nations (ASEAN) was equity sharing with a local joint venture (JV) partner. This contrasts with the semiconductor sector, which is technology-intensive and for which government restrictions are minimal, and equity sharing is seldom required.[13]

For the category of 'industrial electronics', a large representation of Western Europe TNCs contrasts with the small role of the Japanese firms in this sector. This is traditionally an import-substituting investment (for instance, to sell telecommunication equipment). In this case, the participation of FDI is openly stipulated in public sector procurement contracts. Contracts often impose certain equity sharing and local sourcing requirements. One part of the sector is export-oriented (computers and peripherals, telephones, radios). In this case, there is intense price competition in the international market of the products. Malaysia has developed into one of the leading exporters of pocket pagers and mobile two-way radio units. This partly explains why European firms form more JVs. Finally, for the non-semiconductor components, both active and passive electronics components in ASEAN are highly labour-intensive production processes.

Japanese firms concentrate on the production of standardised products and the assembly activities of labour-intensive products. Thus passive components, electro-mechanical components or consumer electronics account for 82% of the Japanese subsidiaries. As 65.2% of the Japanese firms manufacture/assemble components and parts, Malaysia clearly represents an intermediate manufacturing platform for Japanese firms in the region. These figures also reflect that numerous Japanese small-and medium-sized firms have recently followed their big customers towards their host countries.

Some firms from Asian NIEs concentrate in more technology-intensive industries than Japanese firms; 14.8% of them manufacture active components, and 7.4% manufacture computers and peripherals. The bulk of these firms, however, manufacture standardised products. This trend can be explained by the fact that investment from the NIEs is a new phenomenon and only started in the late 1980s.

North American firms operate in higher-intensive sub-sectors; 63% are manufacturing active components or computers and peripherals. On the other hand, a majority of European subsidiaries assemble units or manufacture standardised products.

Firms' structure and network. Table 5.4 shows that four-fifths of the firms established a new production unit when they invested in Malaysia, while only 20% of the firms took over an existing company. There is no difference whether the mode of establishment was through 100% ownership or a JV. The most widely used ownership form is the wholly foreign-owned subsidiary, with 77% of all the firms in the sample. Majority ownership form ranks second, with 18% of the firms. European firms in the sample resort more to the JV form than other firms, with 45% of them under this ownership agreement. The NIEs rank second, with 30% of the firms under JV agreements. Japan favours the wholly owned form of investment, with 80% of the subsidiaries under this ownership form.

Time of establishment. Table 5.4 shows that there are two distinct waves of inward investment in Malaysia. The first one took place in the early 1970s; this corresponds to the time when the government actively developed FTZs in the country in an attempt to attract foreign companies within its national boundaries. The second and most substantial wave occurred in the late 1980s. These two waves were described in Chapter 4 by studying inward investment in all sectors in Malaysia. Sixty per cent of the firms invested in the MEE between 1986 and 1990 and another 20% in the first half of 1990. This second wave resulted, among other factors, from the New Economic Law enacted in 1986 by the government in order to promote both national and foreign investment within the country.

In the 1970s and early 1980s, foreign investment in Malaysia originated either from Japan, the USA or Europe. TNCs from the NIEs started to invest in Malaysia in the late 1980s, investing substantially in the 1990s.

Local purchasing. Subsidiaries were asked whether or not they did carry out some direct purchasing in Malaysia, or whether they imported all their direct supply, keeping only indirect supply locally. This is of crucial importance to this study, because firms with no local supply will not be included in further statistical analysis on local transfers. Foreign firms with no local supply have no relationship with local direct suppliers. Table 5.5 indicates that slightly less than 10% of the firms have no direct purchase at all. Nearly one-fifth of firms from the NIEs carry out no local direct purchasing.

Table 5.5 Foreign firms' local purchasing of direct inputs

	None		Some		Total
	No.	Share %	No.	Share %	No.
Japanese firms	3	6.1	46	93.9	49
NIE firms of which:	5	18.5	22	81.5	27
Taiwan	5	23.8	16	76.2	21
South Korea	0	0.0	1	100.0	1
Hong Kong	0	0.0	1	100.0	1
Singapore	0	0.0	4	100.0	4
Other country firms of which:	1	5.3	18	94.7	19
Europe	1	11.1	8	88.9	9
USA	0	0.0	8	100.0	8
Australia	0	0.0	1	100.0	1
New Zealand	0	0.0	1	100.0	1
Total	9	9.5	86	90.5	95

Main reasons why questionnaire may not have been answered

At the beginning of this section, we identified some known factors likely to increase response rates in a mail survey. The researcher identified a number of reasons why the questionnaire may not have been answered during data collection, data input and data analysis. Such an analysis may help prevent further mistake in other surveys and are therefore included in this chapter. The various reasons identified are listed below.

- *The structure of the firm (size or typography).* In large companies, the purchasing department may be very big, with several managers for each section (quality, logistics, purchasing and so on). Therefore, it is often difficult to know exactly to whom the questionnaire should be sent. Also the 'purchasing' or 'materials' manager may not be the appropriate name and may not have the responsibilities of the targeted manager. Or else one person may not be able to answer all the questions. For a small company, there is only one person in charge; therefore it is easier, as he or she is more likely to have all the information required. In this research, particular attention was given to knowing the title and the name of the respondent; however, it is likely that when the manager was not available, a substitute name may have been provided to the researcher.
- *Type of the product.* In some companies, several purchasing departments exist, one per main product or per type of purchasing. The questionnaire may have been sent in some cases to the general manager, who may not

have had the data to answer for the overall purchasing pattern. Some very small firms, if very specialised or high-tech, may have a big purchasing department compared to the size of the firm.

- *Problem of name.* Although each company was called, sometimes the name provided was inaccurate.
- *Language.* In numerous Japanese or Taiwanese companies, we were told that the manager did not speak English and therefore would be unable to answer the questionnaire.
- *Supplies origins.* At least five companies said on the phone that 100% of the input came straight from the parent firm or country of origin, and therefore it was useless to send the questionnaire. The questionnaire was sent anyway in an attempt to include these companies in the sample. One firm sent a letter back, specifying the origin of the supplies. eight firms completed the first part of the questionnaire, which was concerned with the characteristics of the firm, and did not complete the second half of the questionnaire, as the questions were related to local purchasing. Such responses were useful to the researcher to try and identify which type of company did not carry out any local supply.
- *Company policy.* Sometimes, the manager was simply not allowed to answer the questionnaire and reveal any type of information, despite the confidentiality of the research. In one case, the manager was not allowed to give anything in writing, but agreed to an interview with the researcher. In other cases, we were told it was simply not the company policy. In one case, the questionnaire was faxed to the parent company for them to check the questions, but the manager claimed he had not been allowed to answer the questionnaire.
- *Lack of interest.* Some managers/companies were not interested or too busy to answer.
- *Job obligations.* Some managers were travelling at the time of the survey and therefore the staff would not answer the questionnaire. Indeed, most purchasing managers have to travel regularly.
- *Rate of turnover.* One manager told us she had only just taken on the position and did not feel she could answer. Another company had not yet found a replacement for the manager. In those cases, the questionnaire was still sent, hoping another subordinate would be able to answer it.
- *Length of questionnaire and wide range of areas covered.* Several respondents specified that the questionnaire was handed over to the business planning department to fill in the first part. Also they treat their suppliers more case by case, therefore many managers found it difficult to generalise.
- *Nature of research.* Some could simply not answer, as they did not keep this type of information. The problem will be that for each supplier and

product, the behaviour will differ. Answering the questionnaire may have been too tedious, with too much generalisation, or even impossible for some of the respondents.

Missing values

Each researcher should understand the processes leading to the missing data in order to select the appropriate course of action. A missing data process is:

- any systematic event external to the respondent (such as data entry errors or data collection problems);
- any action on the part of the respondent (such as refusal to answer) that leads to missing values (Hair et al., 1998)

Some questions must be addressed, such as 'Are those individuals who did not respond similar to those who did?', and 'If so, do these differences have any impact on the analysis, the results, or their interpretations?'

The effect of missing data can further be detrimental on the interpretation of the analysis.

Examining patterns of missing data. To decide whether a remedy for missing data can be applied, the researcher must first ascertain the degree of randomness present in the missing data. We analysed the occurrences of missing values for the specific questions on management and technological transfer provided to the suppliers. These missing values are counted only for the 86 cases that do carry out local purchasing, as the other companies have been disregarded for further analysis.

In the case of financial transfers, the study of the case studies and the literature indicates that such transfer is minimal. This is confirmed by a descriptive analysis of the cases that answered the questions. Further analysis of transfers will not include financial assistance for these reasons.

For the other categories of transfers, and in order to avoid the detrimental effect of missing data on the sample size, some steps have been taken. The analysis of the missing values shows a low percentage of missing values for most questions, except for one sub-question (20% missing values). This sub-question was therefore taken out of the analysis altogether to avoid bias. This question may show a high share of missing values because of the inability of the respondents to understand it correctly. If this assumption is correct, it is likely that other respondents who did answer the question did not understand it either, in which case the results would bias the overall quality of the analysis. Five sub-questions showed 10% missing values.

Table 5.6 Missing values per case and question for questions related to knowledge transfer (20 altogether)

Share of missing values in total answers (%)	Frequency per case	Frequency per question
None	56	0
[0.1-10[	16	17
[10-20[	5	2 (one question with 10.5%, one question with 14%)
[20-30[	4	1 (one question with 20.9%)
[30-40[	1	0
[40-50[	2 (case 3 with 45%, case 26 with 50%)	0
50 and over	2 (case 21 with 75%, case 47 with 70%)	0

Approach for dealing with missing data. A remedy must be applied after an assessment of the missing data processes, the appropriateness of the selected remedy or the consequences it will have. We choose to delete case(s) and/or variable(s). In this approach, we first determine the extent of missing data for each case and variable, and then delete the case(s) or variable(s) exceeding a specified level. We find that the missing data are concentrated in a small subset of cases and/or variables, with their exclusion substantially reducing the extent of the missing data.

Three cases had half of the answers missing or more, as shown in Table 5.6. These three companies were therefore taken out of the analysis: company A, showing 15 missing variables (Taiwanese firm in the consumer electronics sub-sector); company B, with 10 missing variables (Taiwanese firm in the consumer electronics sub-sector); and company C, with 14 missing variables (Japanese company in the consumer electronics sub-sector). Although case 3 had 45% missing values for questions 37 to 40, the case was kept, as there would be very little change in overall shares of missing values if the case was eliminated. With three companies left out, the share of missing values per question is improved.

In the case of two sub-questions, the answers to these questions are important for further analysis. A replacement procedure has therefore been designed. To circumvent problems with those missing values, they have been replaced by estimates. Several methods exist to replace missing values. The series mean method is not a preferred method. The mean does not reflect one of the categories to be answered (1, 2 or 3). The median for the entire series fits the type of answers best and has been used, and the estimates were computed. This method has been extended to all the other questions to eliminate the remaining missing values.

Once we have taken the sub-question with a high share of missing values out of the analysis, and the three cases with more than 50% missing variables within the 'transfer' category, we can set an indicator from the remaining variables and cases. Such an indicator will be computed as the share of answers 'frequently' in total answers, without taking the missing values into account. The details of this indicator and its computing are given in Chapter 7.

INTERVIEW PROCESS

A series of interviews were conducted among purchasing managers of various foreign firms in the MEE. In this section, the semi-structured interview approach is described, before specifying the data coding and analysis, and giving the background of the sample firms.

Semi-structured Interviews

The questionnaire constructed for the survey was also used for the interviews. However, the aim of the researcher was to carry out semi-structured interviews, through which the respondent would be free to emphasise the most important factors for his or her subsidiary. Thus, whereas the original questionnaire was constructed for the purpose of a mail survey, with lists of closed questions, during the interview the respondent was left free to discuss and answer each question.

> In semi-structured interviews, the researcher will have a list of themes and questions to be covered, although these may vary from interview to interview. This means that you may omit some questions in particular interviews, given the specific organisational context which is encountered in relation to the research topic. The order of the questions may also be varied depending on the flow of the conversation. On the other hand, additional questions may be required to explore your research question and objectives given the nature of events within particular organisations (Saunders et al., 2000: 243-4).

Data recording. In the process of semi-structured interviews, some questions may be omitted, or some may be added. The respondent should be free to provide his or her own answer to the question. For this reason, it was important for the researcher to ensure a complete note-taking. In this case, the use of a tape recorder was chosen, so as to record every word of the conversation. This enables the researcher to be freer during the interview and not to focus on note taking. Most respondents were open to the use of a tape-recorder. This may have created certain uneasiness at the beginning of the interviews, but the respondent would soon forget the tape-recorder and answer the questions.

Data transcribing. All the interviews were transcribed word for word. The operation of transcribing was most useful to the researcher, as it allowed a second analysis of the interviews. Throughout the process of transcribing, the researcher kept notes on important topics that were not identified at first during the interviews. Patterns in the answers also start appearing during the process of transcribing, which are useful for later analysis of the interviews' content.

Strength of qualitative data collected during interviews. The use of qualitative data has many strengths that have been studied in the literature (for a complete review, see Miles and Huberman, 1994), but for this particular research a few appear of particular importance. The first one is that events studied through interviews are naturally occurring, ordinary events in natural settings, so that we have a strong handle on what 'real life' is like. Moreover, the local understanding of the data; since the researcher conducted the face-to-face interviews within the subsidiary, she had the opportunity to be close to a specific situation, particularly by visiting the manufacturing floor of the company. An additional advantage of interviews is to be able to discuss each question in-depth, which provides richness and holism of the situation. The researcher can more easily grasp the complexity of situations. Through this data collection method, additional questions and variables may be added to the topic studied that were not previously identified in the theory and literature reviews. Finally, the use of qualitative data was particularly useful, since the researcher needed to supplement, validate, explain, illuminate and reinterpret quantitative data gathered from the same research design and questionnaire.

Content analysis. Interview results were analysed first on a case-by-case basis, then by constructing categories of answers across all respondents per topic.

Sampling Companies to be Interviewed

Sampling for the purpose of interviews varies from that of the mail questionnaire. Most of the time, qualitative samples tend to be purposive, rather than random. The qualitative researcher may have additional constraints, such as time and objectives. For the purpose of this research, the critical factor in sampling was to provide data from each of the main countries of origin. As a result, the researcher paid a lot of attention to obtain interviews with firms from Japan, the USA, Europe and the NIEs. This purposive sampling was to be able to distinguish between each group of firms, according to their country of origin. Other sampling, such as sampling by product categories, or by time of establishment would have been more difficult, for reasons stated in the section on product categories, or simply because information on companies is often very restricted, and the easiest means of identification is that of the country of origin.

The interviews were aimed at getting into the complexity of the reality; each company has a different organisational structure, and each purchasing behaviour varies according to the situation. Moreover, the department in the factory may not carry out indirect and direct supplies. Each of the above-mentioned complexities can only be studied in-depth by going to the company and interviewing the manager on the specificities of his or her own company. The mail questionnaire by nature has to be formalised and applicable to all; it does not allow for flexibility needed for a complete view on the topic. However, a larger number of answers allow the researcher to ensure the reliability of the results.

Result of Interview Process

Thirteen companies were interviewed. Considering the original sample of companies, which was of 320 foreign firms in the sector under study, the sample of firms interviewed represents 4% of the total population, which is a good representation for a qualitative research. Of the companies interviewed, two were from Japan, one was from Singapore, one was from Taiwan, two were from Europe and six were from the USA. Of the 11 companies that fitted our targeted sample, two were manufacturing/assembling active components, three passive components, two electro-mechanical components, two consumer electronics, one computers and peripherals and one production equipment. The researcher interviewed two more companies that did not fit our predetermined sample; the first was a computer company but its aim was to mend damaged computers, whereas the second manufactured semi-conductors, but the person interviewed was the indirect materials manager.

Obtaining interviews. All interviews took place within the subsidiary, except for one: company 11, as the manager was first met during the company's 'supplier day', and then again in the university.

Various characteristics of firms' behaviour were identified during the process of the field research. We mentioned earlier how the researcher obtained the interviews. The first option open to the researcher to obtain interviews was the use of the mail survey. On the last page of each questionnaire, the respondent was asked to tick a box if he or she would agree to meet the researcher for an interview.

Four firms answered the mail survey, and as a result agreed to meet the researcher for an interview. During these interviews, the respondent was given the opportunity to add further details to his or her answers to the questionnaire, and provide his or her own interpretation and ideas. One of these four firms was a firm from the UK; the others were Asian firms. This was the only way the researcher managed to obtain interviews with Asian firms, apart from one interview with a Taiwanese firm, which was arranged through contact with a local training company providing its services to the Taiwanese firm. Managers of Asian firms were very reluctant to accept interviews, even when personal contact was available. In the case of firms from the NIEs, language was the main reason provided for refusing an interview. For both Japanese firms and firms from the NIEs, the mail survey was an easier method to obtain data. One-third of the Japanese firms were particularly interested in receiving a copy of the executive results, and this incentive seemed to have been their main motivation for filling in the questionnaire.

One more firm answered through the mail survey, but following a different motivation. The manager was working for a Dutch company, and he found it easier to answer the survey by meeting the researcher for an interview rather than by answering the questionnaire.

Personal contacts were extremely useful in obtaining interviews. Such personal contacts were mainly executives who were met during university courses and who were all working for foreign subsidiaries in Penang. This explains the geographical bias of the companies visited. The two companies visited outside the Penang area were companies contacted through the mail survey. The concentration of the results from the Penang area is justified by the fact that this is the preferred location for foreign firms in the MEE.

An additional means of obtaining interviews was the participation of the researcher in the government-organised Vendor'Net Programme, which took place in Penang in March 1996. Several foreign firms were thus approached and agreed to meet the researcher for interviews. This Programme was extremely useful to the researcher to meet not only foreign subsidiaries, but also local suppliers to foreign subsidiaries. Only one firm accepted an interview after a telephone call. This was a North American firm located in the Penang FTZ.

Overall, European firms and firms from the USA were more willing to accept interviews with the researcher, and some preferred that option rather than filling in the questionnaire. However, not all firms were cooperative; in particular several attempts were made to obtain interviews with other US firms, but the researcher was told that the parent company did not wish for the questionnaire to be answered.

More interviews could have been carried out as the researcher became more acquainted with the business environment at the end of her stay in Penang. However, it was impossible to extend the length of this stay.

Access to additional information. In continuity with the above discussion on the differences in obtaining interviews according to the country of origin, the overall openness of the respondents was also of interest. The Singaporean firm was the only Asian firm to accept, providing written information on purchasing to the researcher. All the other Asian firms were very cautious not to provide any written material. The researcher was not offered a visit to the plant, and could not gain access to the manufacturing floor of the company.

In the case of the European or US firms, additional printed materials were made freely available for the researcher during the interview and a visit to the company was organised in each case. However, such results may be due to the particular firms visited, and the small number of interviews did not allow the researcher to generalise such comments.

Other Data Collected

The information gathered through the interviews and the mail survey is undoubtedly the key data provided in this research. However, a complete picture of the topic came with the participation in events and use of secondary data sources. The researcher was invited to take part in two major events related to the research topic. These were:

1. Seminar-cum-exhibition for SMIs (small-and medium-sized industries) in the subcontracting business 'Vendor'Net 96, on 28-29 March, 1996, at the PSDC (Penang). It was organised by the Penang Entrepreneur Development Committee and the Penang Industrial Transformation and SMI Development Committee, as part of the Vendor'Net Programme launched by the Ministry of Entrepreneur Development, with the participation of 19 TNCs and over 100 local firms. This new programme was discussed and described within the new soft policies adopted by the Malaysian government to encourage positive linkages in the economy in Chapter 4.
2. One US company's 'supplier awards', day was a truly insightful experience, during which the researcher witnessed the close relations established by the firm with its local and regional suppliers.

By taking part in these events, the researcher was able to obtain a clearer picture of the buyer-supplier relations in Malaysia.

TRIANGULATION METHOD

The use of multiple methods within a single research project is one of the potential research strategies open to a researcher. Such a use is described as one of convergent methodology, multi-method/multi-trait, convergent validation or, what has been called 'triangulation' (Jick, 1979). This relates to the notion that qualitative and quantitative methods should be viewed as complementary rather than as rival camps.

Reasons for Combining Qualitative and Quantitative Methods

Despite the fact that most research projects favour the use of only one paradigm (that is either the qualitative or the quantitative paradigm), several authors have tried to show the advantages of using a combination of both methods (Jick, 1979; Greene et al., 1989; Morse, 1991; Creswell, 1994; Saunders et al., 2000). The main characteristic about using this combined approach is that it is difficult to provide a standard procedure on how to go about it. Like the qualitative method, it is likely that each research process using triangulation will be different from the other. Complementing research paradigms and methods means more than coexistence; it underlines how one approach can inform and guide the other (Salomon, 1991). Several advantages exist for using the triangulation method. First, it allows researchers to be more confident of their results. This is the overall strength of the multi-method design. The disadvantages of one design are the advantages of the other. For example, while disadvantages of positivism are often the weak understanding of social processes and of the meaning people attach to social phenomena, advantages of phenomenology are specifically to facilitate understanding of how and why, and enable the researcher to reach a good understanding of the social processes. Thus, if used appropriately, a mix of the two methods may lead to stronger research results.

Triangulation can also lead to inventive methods, new ways of capturing a problem to balance with conventional data collection methods. In addition, different methods can be used for different purposes in a study. The example of the use of case study methods is often advised as it allows getting a feel for the key issues before embarking on a survey. Using this method, one is more confident of having grasped the most important issues. Saunders et al. (2000) provide an example through which a research student has combined survey and case study methods. In this case, the use of semi-structured group interviews was a valuable way of triangulating data collected by other means, such as a questionnaire. The four tools utilised for this research were in-depth interviews, discussion groups, a questionnaire survey and semi-structured group interviews. The supervisors found that such a methodological approach gave the student a much better grasp of the content of the data. It also did much to enhance the credibility of this research. Hence, organisational researchers can improve the accuracy of their judgements by collecting different kinds of data on the same phenomenon. The method will allow for a more complete, holistic and contextual portrayal of the units under study.

Other advantages of triangulation include the fact that it may also help to uncover the deviant or off-quadrant dimension of the phenomenon, the possibility of a synthesis or integration of theories. According to Jick (1979), a thread linking all of the benefits from triangulation is the important part played by qualitative methods in triangulation. Qualitative data and analysis are seen as the glue that cements the interpretation of results.

Limitations of Triangulation

The majority of books advise the use of only one research paradigm. We have identified two main arguments against the use of triangulation. First, the two paradigms flow from a different use of the literature and the theory. The setting-up of research questions or hypotheses is different, whether it is for a quantitative research design or a qualitative one. While the quantitative approach is traditional, positivist, experimental and empiricist, the qualitative approach is constructivist or naturalistic and interpretative. A second reason against the use of a mixed approach is that it would take too long for the researcher to use both approaches. Such an argument is mostly to be taken into account when writing a journal article. Despite these two limitations, we have decided that the advantages outweigh the limitations.

Methods of Triangulation

Jick (1979) differentiates the 'between-methods' and the 'within-methods'. The between-methods are the most popular use of triangulation. It is largely a vehicle for cross-validation, through which the researcher can observe the same phenomenon by multiple methods. Multiple and independent measures, if they reach the same conclusions, provide a more certain portrayal of the leadership phenomenon. The within-methods also use multiple techniques, but instead of using two research paradigms, the difference lies in the way data are collected and interpreted. For instance, answers to a question in a questionnaire could be analysed both through scaling and through indices. This method involves cross-checking for internal consistency or reliability, and the researcher can be more confident in the emergent theory. On the other hand, the between-method triangulation tests the degree of external validity.

Models of Combined Designs

Creswell (1994) describes three different models of combined designs. The two-phase design combines methods to understand better a concept being tested or explored. The two methods are conducted independently, and the two paradigms are clearly separate. The second model is the dominant/less dominant design, which integrates the paradigms at several phases of the research process. This could be a quantitative study based on testing a theory in an experiment with a small qualitative interview component in the data collection phase. Last, the third design is the mixed methodology design that combines qualitative and quantitative approaches in a single study. This includes mixed aspects of both paradigms over various methodological steps in the design. The overall design thus best mirrors the research process of working back and forth between inductive and deductive models of thinking in a research study.

Triangulation in this Research

Contrary to various multi-methods' research design, we have chosen to focus on the advantages of triangulation, and emphasise the means of observing the data from the same respondents but through different means. The objectives through this method are to obtain more in-depth information about the purchasing relationship of the foreign subsidiary, to uncover hidden dimensions through the qualitative analysis of semi-structured interviews, to be able to describe other aspects of the relationship, impossible to grasp through the quantitative analysis of questionnaires, and to be able to build our own methodology, best fitted to the particular topic under study.

In this sense, we follow more closely a between-methods' strategy, with a mixed methodology approach combined with the two-phase approach. The multi-methods' strategy allows us to gather information from various angles to reinforce external reliability of the data. The two-phase approach is used, insofar as the qualitative data obtained through the semi-structured interviews are first analysed independently, leading on to the quantitative analysis of the questionnaires in an independent section. Both approaches then converge into a mixed methodology approach and all the results are merged into an independent section where both paradigms are used to reinforce the final results of this research.

CONCLUSION

Every attempt has been made in this chapter to provide the reader with all the detailed procedures followed in this research. Additional methodological procedures are provided throughout the following chapters, especially in terms of the statistical analysis operationalisation procedure.

The next three chapters follow the methodology presented in this chapter. The interview results are first presented in Chapter 6, providing the in-depth knowledge required to understand the complexity of studying backward linkages. The mail survey results give way to a series of logistic and linear regression models in Chapter 7. These aim at formalising the relationships between factors studied and backward linkages. In the final chapter of this book, the triangulation method leads to the convergence of both sets of results.

NOTES

1. Response rates vary greatly with the research method employed, and they differ significantly across countries. Harzing (2000) shows that response rates vary from country to country in cross-national surveys, with the highest rate in Denmark (42.1%) and the lowest in Hong Kong (7.1%).
2. Active components: discrete components, integrated circuits.

   > The semiconductor industry is the major sub-sector of the electronic components' industry producing active components, that is components which modify and control electrical signals by amplification, switching action or modulation of the signal in a circuit. [....] Active solid state components are based on material which has semiconducting properties (germanium, silicon, gallium arsenide) which is used in fabrication to produce a semiconductor device (OECD, 1985: 8).

   Included in this category are the discrete components (individual devices such as transistors, rectifiers and diodes) and integrated circuits. These could be subdivided differently according to the technology used in the fabrication process or according to their degree of integration on a chip. However, we are keeping these products under 'active components' as the companies surveyed were not asked to give further specification on their products.
3. Passive components/electro-mechanical components: the passive components do not change the energy of electrical signals. They form a slow-growing part of the electronic components industry. In most instances, the manufacture of passive components requires less knowledge and technology than that of active components.
4. Electro-mechanical components are classified as 'mature products' and require less knowledge than the 'active' and 'passive' components' categories.
5. Power sources are classified as 'mature products' and only require standard technological knowledge.

6. 'Miscellaneous equipment' includes a series of mature, low-tech-intensive products.
7. 'Computer and peripherals' covers products like personal computer, notebook PCs, motherboard cards and other computer-related sub-assemblies (rigid disk drives, tape drives and so on).
8. 'Other industrial equipment' refers mainly to Malaysia in the assembling of all other finished products related to industrial equipment.
9. 'Consumer electronics' covers products such as TVs, hi-fi systems, video cassette recorders, compact disc players and portable audio equipment. These products are mainly assembled in Malaysia.
10. Note that one firm was originally a subsidiary of a Hong Kong firm before becoming fully local; another firm was set up with the help of a foreign company.
11. Some companies were supporting industries for the electrical and electronics sector. The products concerned were hand-labellers, bearings for air-conditioning valves, various plating items and mechanical bells.
12. The overall variety of the firms' activities is not shown, since only the main products manufactured/assembled were input in the product category, thus eluding the frequent diversity in subsidiaries' production.
13. For an in-depth discussion on equity requirements in Malaysia, refer to Chapter 4, or to Giroud (1997). This paper analyses the ownership patterns of subsidiaries in the MEE.

6. Research Findings: Case-by-case Analysis

INTRODUCTION

The method to be followed in this book was described in Chapter 5. A series of interviews were conducted, so as to allow the researcher to obtain in-depth information about the specific nature of buyer-supplier relations in the Malaysian electrical and electronics sector (MEE). The results presented give a full record of what happens between a foreign subsidiary and its local suppliers, looking at particular types of firms with their strategy and network, and particular purchasing situations. Contrary to the mail survey questionnaire, which contained a series of closed-ended questions, the use of semi-structured interviews allowed the researcher to obtain a full description of the difference in transfers according to the type of product purchased, and the type of local suppliers. An in-depth description is essential to complement the results of a statistical analysis to be carried out from the mail questionnaires. This chapter concentrates on each specific relationship created by the foreign subsidiary with each of its local suppliers. First, a general description of each case study is presented. The analysis of the supply networks and strategies of each subsidiary follows, before a discussion of transfers occurring through backward linkages' creation.

GENERAL DESCRIPTION OF 11 CASE STUDIES

Despite an effort in selecting companies that would be comparable, some differences in terms of size, products manufactured and length of time spent in Malaysia are unavoidable, insofar as two companies are unlikely to be the same. However, all the companies are part of the same economic sector; they are all subsidiaries of international groups located in the same country. Out of the 11 companies, 2 are Japanese, 1 is Singaporean, 1 is Taiwanese, 3 are Europeans and 4 are from the USA (see Table 6.1).

Table 6.1a Summary of case studies

COMPANY	Company 1	Company 2	Company 3	Company 4	Company 5
Nationality	Japan	Japan	Singapore	Taiwan	UK
Year of establishment	1989	1973	1992	1989	1976
Reason for establishment	1. low-cost location 2. use of machinery	1. low-cost location 2. good infrastructure	1. low-cost location 2. favourable government policies	1. low-cost location 2. close to customer	1. large potential domestic market
Activity	Manufacturing	Manufacturing	Subcontractor	Manufacturing	Manufacturing
Products	Passive components	Electro-mech. components	Production equipment	Computer and peripherals	Electro-mech. components
Parent firm	Identical production	Identical production	Different range of products	1. R&D 2. upstream production	Higher value-added products
Ownership pattern	Wholly owned	Wholly owned	Wholly owned	Wholly owned	Majority owned
Type of investment	Greenfield	Greenfield	Greenfield	Greenfield	Greenfield
Local equipment *	20%	10%	10%	None	20%
Rest of equipment	All from home country	All from home country	All from countries other than home	All from home country	Majority from home country
Employment	3,500	1,245	1,200	1,200	390
Annual sales**	327,50	140	-	-	75
Export share in production	90%	97%	100%	100%	60%
Internal supplies	40%	10%	90%	30%	20%

Notes:
* This corresponds to the percentage (in US$) of machinery and equipment that the firm bought locally when establishing the subsidiary.
** Annual sales figures are in million Malaysian Ringgits.

Table 6.1b Summary of case studies

COMPANY	Company 6	Company 7	Company 8	Company 9	Company 10	Company 11
Nationality	Netherlands	Netherlands	USA	USA	USA	USA
Year of establishment	1988	1979	1992	1972	1989	1974
Reason for establishment	1. low-cost 2. good infrastructure 3. favourable policies	1. low-cost location 2. good infrastructure	1. low-cost location 2. use of machinery 3. good infrastructure	1. low-cost location 2. good infrastructure	1. low-cost 2. good infrastructure 3. favourable policies	1. low-cost location 2. strategic location in Asia 3. favourable policies
Activity	Assembling	Manufacturing	Assembling	Assembling and manufacturing	Assembling and manufacturing	Assembling and manufacturing
Products	Consumer electronics	Consumer electronics	Active components	Active components	Passive components	Passive components
Parent firm	Identical and different products	Different range of products	Other products, home parts used in subsidiary	Independent products	1. components for subsidiary 2. upstream prod.	Identical and different products
Ownership pattern	Wholly owned	Wholly owned	Wholly owned	Wholly owned	Wholly owned	Wholly owned
Type of investment	Take-over of existing company	Greenfield	Take-over of existing company	Greenfield	Take-over of existing company	Greenfield
Local equipment *	All	None	60%	10%	-	10%
Rest of equipment	None	Majority from countries other than home	All from home country	-	Additional equipment from home country	Majority from countries other than home
Employment	700	1,743	3,300	4,500	4,500	2,400
Annual sales**	-	270	-	-	-	350
Export share in production	98%	100%	100%	100%	-	95%
Internal supplies	None	15%	40%	10%	20%	30%

The time of establishment varies widely, from 1972 to 1992. Apart from the recent wave of inward investment after 1986, this shows that foreign companies in the sector have long-term investment in Malaysia. Very interestingly, no companies invested in Malaysia in the early 1980s when government policies were less flexible. A new wave of investment started in the second half of the 1980s onwards, after the government launched an 'attractive' policy. The companies that invested in the early 1970s are not currently all specialised in less technology-intensive products. Out of the five firms concerned, two are manufacturing electro-mechanical components, one is manufacturing passive components, one is producing consumer electronics products and one is manufacturing active components. The product range has changed over time, certainly for the company manufacturing active components, which in part shows that Malaysia has become the centre for active component and semiconductors' manufacturing.

When asked what were the main reasons for the establishment of the subsidiary in Malaysia, all companies answered 'low-cost location' as the first reason, then the good infrastructure and the favourable governmental policies towards foreign investment. These three motives are the most important ones. Most companies are located within the free trade zones (FTZs) that provide favourable business conditions.[1] Other motives for investment were the strategic location of Malaysia in the Asian region, the large potential market both domestic and regional and finally being close to the customers.

Ten out of the eleven companies are fully owned; keeping the subsidiary entirely under the parent firm control is the preferred option. Joint ventures (JVs) are scarcely used in this economic sector, often in an attempt not to share the technological advantage of the firm. Similarly, eight companies proceeded to a green field investment. The only cases of take-overs have been firms that were manufacturing similar types of products to the buying firm. This is less likely to happen for the high-technology firms in a country such as Malaysia where indigenous capabilities in this sector are scarce and are still arising from foreign firms.[2]

The main activities of these firms are either manufacturing or assembly work. No firms carried out other types of activities. This was the basis of the survey strategy, since more backward linkages are likely to exist for these activities. The Singaporean and one of the US subsidiaries are subcontractors for other multinational firms.

Very little machinery and equipment were bought locally when establishing the plants. In two cases, no equipment was bought locally at all, whereas the maximum was 20% (in US dollars). Out of the three take-overs, one still imported 40% of the machinery and equipment from the home country, while a second one imported additional equipment from the home country. In the remaining cases, the imported machinery originated only from the home country for the two Japanese subsidiaries and for the Taiwanese subsidiary. For one of the Japanese firms, the plant was fully equipped with the former equipment of the parent firm, because the parent firm was upgrading its production to higher value-added goods. The Singaporean subsidiary is the only firm for which the totality of the equipment was imported from countries other than the home country. This can be explained by the small size of this economy and specialisation.

All the firms are large-scale firms. Only two firms have fewer than 1,200 employees with 390 and 700 employees respectively and four have more than 3,000 employees. This may show the importance of low-cost labour in Malaysia. Although currently the wage rates are higher than in other countries in the Association of Southeast Asian Nations (ASEAN) region, this factor combined with other motives still makes Malaysia an attractive investment location. Certainly, the type of activities carried out by the subsidiaries remains labour-intensive.[3]

Five firms provided their annual sales' results. Analysing the total sales/total employment ratio, it is 93,714 and 112,449 Malaysian ringgits (RM) respectively, per employee for the Japanese companies, RM 192,307 and RM 154,905 for two of the European firms, and RM 145,833 for one of the US firms. This may show that Japanese firms in the sample are less productive, but considering the small number of firms, no generalisation is possible; moreover, the type of product manufactured is important for a full understanding.

These companies are all export-oriented. This is in part explained by the fact that export-orientation is one of the government requirements to obtain favourable treatment and it was one of the requirements to be allowed 100% ownership structure until 1998. Only one company exports less than 90% of its production. In this case, the company seems to have settled in Malaysia because of a large potential domestic market. However, it still exports 60% of its production.[4]

Internal supplies[5] are more substantial for companies that have only recently settled in Malaysia. They represent 90% of total supplies for the Singaporean company settled in 1992, 40% for the US company also settled in 1992 and 40% for the Japanese company settled in 1989. Otherwise, no specific tendency appears.

DESCRIPTION OF SUPPLY NETWORKS AND STRATEGIES

This section is a descriptive record of the supply networks and strategies for each of the foreign affiliates interviewed. The details of the networks and strategies of each firm help in understanding the background to the knowledge transfer.

Asian Companies

Four Asian companies were interviewed, two Japanese, one Singaporean and one Taiwanese. Asian firms' supply networks and strategies appear below.

Company 1 (Japanese)

Supply network. The company has two subsidiaries in Malaysia, both located in the same industrial estate, and a sales office in Singapore. Both subsidiaries in Malaysia have the same activities and produce the same goods. The purchasing activity is centralised in one department for both production plants. Decision making in terms of purchasing is limited, as most items are still bought from Japan. For the other inputs, the Japanese office specifies the product type, the supplier and the part number. The subsidiary then provides the product specifications to the supplier designated by the parent company. The company has not yet been certified with the International Organization for Standardization (ISO)

There are 80 external suppliers in Malaysia. Fifty are Japanese-owned and provide inputs such as electrical items, semiconductors and mechanical parts (18 have followed the company to Malaysia since 1989; one of them settled in the same industrial estate as the subsidiary, they provide mechanical and electrical items). thirty are Malaysian-owned, the major suppliers provide plastic injection goods. All the other suppliers provide secondary products such as carton boxes and packaging materials. The majority of the suppliers operate under a subcontracting relationship with the subsidiary (45-50 suppliers). The others only receive purchase orders, but are regular suppliers.

The stamping parts also used are standardised items that could be purchased locally, but they are provided internally by the parent company. Imported supplies include mechanical items, electrical items and printed circuit boards (PCBs). Most supplies come directly from Japan; the others are bought by the office in Singapore, Indonesia and Thailand, and then forwarded to the Malaysian plants.

Supply strategy. The main criteria for supplier selection are the quality, the service and the cost. The supplier's evaluation is first carried out by a purchasing team from the parent firm in Japan. In order to facilitate communication with this team, four Malaysian-owned suppliers hire Japanese personnel to facilitate dealings. The quality assurance department then carries out an audit of the supplier.

The suppliers are regularly followed up, and an average quality limit is evaluated. Once accepted, the supplier becomes part of the approved vendor list. Regular suppliers visit the subsidiary at least once a month, and they get a report on delivery and forecast.

New products originate in Japan. It is difficult for the subsidiary to adapt to new products, and it still has not been ISO certified. The just in time (JIT) system is only used for floppy disk drives, since the production storage area is quite big, but this will allow cutting down on space. However, major problems in Malaysia occur because of the shortage of supplies. Suppliers are unable to fulfil the needs and requirements of the subsidiary. The subsidiary practises short lead time. Short quantities are bought frequently, and the lack of flexibility to new orders is the major problem of local companies.

The main strategy is to have more than one supplier per product, but only one supplier supplies one item. Each item is part of the same product. The overall design of the purchasing strategy is done in Japan, which occasionally causes problems when translated into English for wider usage.

Prices in Japan are closely monitored, and for every cut in price in Japan, a similar cut is expected in Malaysia, plus a further negotiation is carried out to obtain lower prices.

Company 2 (Japanese)

Supply network. The firm has four other manufacturing subsidiaries in the region; one in Thailand, one in Indonesia and two have recently been established in China. There is also one sales' subsidiary in Singapore. There are no internal supplies among these subsidiaries because they all manufacture different types of relays. The subsidiary is mostly independent in its purchase strategy, since it started production a long time ago. It is ISO certified.

It has 54 local suppliers, of which Japanese (three Japanese suppliers followed the company and started operations in Malaysia) and other foreign-owned subsidiaries provide copper wires, contacts and plastic moulding parts, while locally owned suppliers only provide packaging materials. Six suppliers have subcontracting agreements. They provide assembly operations, mainly for coils, and assembling work on the components. Other contracts are purchase orders, but suppliers are long-term close suppliers. Imported supplies include raw materials and chemicals that cannot be found in Malaysia.

Imported external supplies include raw materials and chemicals from Japan and other countries in Asia. Those supplies cannot be found in Malaysia. Internal supplies are composed of metal items. The design technology for these products is too sophisticated for local suppliers, who cannot meet the specifications. For instance, they are not capable of manufacturing items as small and flexible as the standards required. These piece parts are only required in small volumes by the subsidiary, therefore they are bought by the parent firm and sent over.

Supply strategy. The subsidiary uses the QCD system (quality, cost and delivery) in order to carry out supplier selection. Suppliers are audited once at the beginning of the relationship. The audit covers the quality, the way the company is organised, the supplier facility, the management capability and the logistic issues.

After selection, suppliers are evaluated every month. A few categories are taken into account, a score is obtained by rating each category and the results are provided to the suppliers. The best, worst and mean scores are displayed (the identity of each supplier is not shown) on a graph, so that each supplier can compare him or herself to the rest of the group. The categories include quality/process (existing deviation or innovation, and so on), delivery performance, management response (existing problems, disputes, level of competitiveness and supportiveness of the suppliers) and cost (intended price/charge reduction, unnecessary charges elimination and so on). The subsidiary expects a yearly reduction in prices, from 10% to 20% on target, but it is not compulsory. Sometimes suppliers reach unexpected decreases of up to 50%. A continuing evaluation is also, carried out. Every month, new sources that are more cost-effective are identified. This leads to a renegotiation of the price with the accepted suppliers. If these negotiations are not successful, a change in suppliers is considered.

Some suppliers are chosen once a year for audit covering the process, quality control and innovations in the production. The supplier does not know in advance when the audit will take place. The subsidiary expects all suppliers to reach and obtain the ISO within two years of starting business with the firm. It gives no specific advice, as the supplier is expected to resort to a consultancy service, but on the second audit, the subsidiary follows the ISO standards step by step. This helps the supplier in knowing the requirements needed. Sometimes, the suppliers improve quickly and exceed the requirements fixed by the subsidiary.

Personal contacts exist up to the top management level. These relationships help when key information has to be shared amongst employees or change in strategy. Quality and purchasing strategies are shared with the top management staff of the suppliers.

The firm prefers to resort to several suppliers to avoid supply-dependence. However, some suppliers have the patent rights for one item, or the particular technological competence, in which case it is not possible to have more than one supplier. One example is the liquid fast-setting glue, which takes only five seconds' setting time. One specialised supplier only manufactures this glue; the subsidiary is obliged to have a single source.

Supplier reward system. Only one award is given yearly to the supplier reaching the highest score on the rating system (delivery, quality, cost, management's response). The yearly final assessment concerns only the 10 best suppliers in terms of volume and frequency of orders from the subsidiary. The award is allocated by the subsidiary, independently from the parent firm strategy. The results are shown on the 'performance appraisal board', so that the suppliers concerned adopt a competitive strategy to try to reach the first position. On a second board, they see the monthly results for delivery and quality. However, the cost is never displayed. The cost depends very much on the situation and what occurs.

Company 3 (Singapore)

Supply network. The company has several assembling plants throughout Asia, in China, Indonesia and Thailand. The Malaysian subsidiary carries out assembly work, mounting and testing of electronics components for multinational firms. As a subcontractor, this subsidiary is not independent in its purchase sourcing. Customers provide an approved vendor list (AVL) of suppliers with the exact parts wanted. The subsidiary is ISO certified.

Ninety per cent of total procurement in volume is internal supplies. The purchasing office for the group is in the headquarters in Singapore. Internal supplies are integrated circuits, connectors, PCBs, resistors and capacitors, none of which is manufactured by the group. The 10% external purchase achieved by the subsidiary does not represent important goods. The plant buys indirect materials locally,[6] and has some subcontracting relationship with local companies, which provide sub-assembly work. Most suppliers have regular six-month contracts. About 10 local suppliers have a subcontracting relationship in assembly and testing.

Supply strategy. Price, quality and lead time are the most important factors considered in the selection process. Lead time is the most critical factor because of the nature of the output. In a high-tech, fast-changing environment, the market has to be captured up-front before a competitor overtakes the product. Fast changes are the reason why the subsidiary keeps small inventories. Lead time and flexibility to changes therefore represent the two major requirements. For subcontractors, the subsidiary selects companies by quotations, taking three factors into account: price, service and lead time.

On a follow-up basis, regular reports are set up, either every quarter or half-yearly. Price, quality and service are looked at. The price factor is important. Contracts are only renewed every six months, because of frequent price changes. Depending on the market situation, the company may request the suppliers to lower their price. The quality is checked by trial and error. Random inspection is carried out to evaluate the reliability of subcontractors. The handling of the products, storage, safety and the workers' environment are examined. From the service point of view, suppliers need to be ready and flexible for the subsidiary.

The personal relationship is considered critical by the manager. After dealing with each other for a long time, the two parties get to understand each other. Personal contacts help in creating trust between the firms. They also allow each company to know about possible problems arising for one of them.

Company 4 (Taiwanese)

Supply network. The firm has two other manufacturing subsidiaries in Malaysia. There is no exchange of goods between them. The subsidiary is ISO certified.

There are 20 local suppliers, either Taiwanese or locally owned. They provide plastic parts (major components bought locally) and metal parts. There are six subcontractors doing sub-assembly of keyboards. All the other suppliers receive purchase orders.

The main input is the key tops, imported from only one supplier in Taiwan. Other inputs such as plastic panels and isolation films are imported from Taiwan, mainland China and Singapore. Internal supplies are mainly composed of plastic items, which are actually purchased by the parent firm for the whole of the group. The parent firm decides most of the sourcing strategy for materials' resources. It will tend to choose a supplier based in Taiwan that can supply both the parent firm and the Malaysian plant. The reason for this policy is that the product is changing fast as new models appear.

Supply strategy. Suppliers are selected by looking essentially at the quality, the delivery and the service factors, but the quality is the major factor and the cheapest price for the best quality will be chosen.

Vendors' evaluation is based on the quality control reports. These reports are passed on to the subsidiary who organises regular meetings at least once a month with its major suppliers to touch upon critical areas. Other topics may be discussed, such as the rejection rate (the in-house inspection rate for key-tops requires that if 2 out of 125 pieces are rejected, then the whole lot is rejected), the delivery and the means of improvement.

The quality is the most important issue; therefore, if a supplier has a difficulty, problem-solving will involve the subsidiary as far as possible. When a problem is faced, the subsidiary provides some criteria for the supplier to fulfil within a time limit in which case, the subsidiary intervenes directly in the management system of the supplier.

In terms of price, the subsidiary expects a decrease every three months, or an indicated 5% decrease quarterly. This is for secondary products. For major inputs like plastic and metal products, the price depends solely on the market price.

Bimonthly audits occur mainly for major suppliers, such as plastic parts' suppliers. Audits take place at random. They are conducted by a team composed of a purchaser, a member of the material control department, the quality control department and an engineer or one person from the research and development (R&D) department. The team is composed at least of a purchaser and a person from the material control department. The quality system, the method of control, the management and also the production are closely looked at.

Supplier reward system. Initial plans have been carried out to start a reward system, but they have not yet been implemented.

European Companies

Three European companies were interviewed; one British and two Dutch. European firms' supply networks and strategies are tackled in this section.

Company 5 (British)

Supply network. The firm has two manufacturing subsidiaries in Malaysia; one in the electrical sector and the other in the unrelated metal industry. There are no relationships between the two subsidiaries. While production in Malaysia remains in standardised products, the British parent firm manufactures higher-end products. The subsidiary is ISO certified. The subsidiary is mostly independent in its purchasing strategy, apart from press parts' components, which the parent firm insists on buying to use as well. Inputs such as ceramics castings could be found in Malaysia at competitive prices but are purchased internally for the same reasons.

There are 35 local suppliers. Both foreign-owned and locally owned firms supply plastic moulding, fasteners and screws and electrical cables. One foreign-owned firm supplies the more complicated press-stamping parts. No local suppliers could provide the latter when the subsidiary started purchasing them, but currently a few local suppliers could provide them. The majority of suppliers have a subcontracting relationship in processing or assembly work.

Imported supplies include plastic moulding powder and raw materials not available in Malaysia, such as copper, brass and steel imported from the UK, France, Australia, Japan and Korea.

Supply strategy. Quality and pricing are the most important criteria, followed by delivery and service. The service factor includes a positive flexible response to enquiry by the subsidiary.

Generally, it takes three to four months from the time the subsidiary enquires about a possible supplier for approval. After approval, the supplier is retained on a provisional basis, and its performance is constantly monitored at the beginning. It takes six to eight months before the supplier is considered a vendor.

Audits are carried out on the quality side. The audit covers the company process, its quality management system and its internal quality assurance/process. The team taking the final decision comprises the engineering, purchasing, finance and production managers; there are normally four people involved.

The subsidiary does not intervene directly in the suppliers' operations. However, regular feedback, quality reports and regular meetings inform the suppliers on the level of their services. The subsidiary follows the ISO 9002 quality system on quality and encourages the suppliers to apply for the certification.

There are no compulsory price cuts, but a decrease is expected to occur over a long period of time. Long-term relationships are preferred; most of the suppliers have worked with the subsidiary for more than five years. Once they are part of the AVL, it is rare for the subsidiary to change. Two suppliers have however been erased from the AVL; one for a decrease in the quality level and the other for over pricing compared to current market prices. In these cases, the advice given by the ISO 9000 was followed; after being told several times through regular reports of the problem, the suppliers were formally told to take corrective actions. They were given a time limit to make improvements, and were finally taken off the AVL.

Regular audits are carried out for suppliers in Malaysia, they occur once to twice a year. The suppliers are given advance notice. The checklist is similar to the one used for the first audit, before the supplier is accepted as a vendor.

The major problems with local suppliers are that their standard of design and process is not high enough in most cases for high-tech products, and that the quality tends to be low. Some suppliers still have difficulties in meeting the quality requirements.

Personal contacts are considered crucial in the Asian context by the manager. It implies that the suppliers know the subsidiary's requirements, priorities and strategy better, and therefore are more likely to provide a better service. Trust depends on these personal relations. Part of the relationship is 'intuition'.

Company 6 (Dutch)

Supply network. The subsidiary assembles a different range of products but in the same line as the parent firm. Sourcing is determined by the plant, except for certain key components, which match both the production of the parent firm and that of the subsidiary. The subsidiary is ISO certified.

There are 100 local suppliers; 40 are foreign-owned and 60 locally-owned. Key electronics components such as integrated circuits, transistors and power transformers are purchased from foreign-owned firms. Standard electronic components are bought from locally owned suppliers; they include injection moulding, mechanical fixing parts and indirect materials. The subsidiary has 15 subcontractors, which represents about 30% of the parts. Two of them assemble a complete product for the subsidiary (small radios). The subsidiary provides them with the material and the subcontractors provide the labour. The other suppliers receive purchase orders.

Some key electrical and electronic components are purchased internally, or are imported from Singapore, Indonesia, Thailand, Japan, Taiwan and Korea. Microprocessors are purchased from Japan.

Supply strategy. Various departments take care of the purchasing of the company. The sourcing department is responsible for initial buying and sourcing. The logistic department is responsible for production planning and control, materials planning and control, stores and warehouses and shipping of goods. The materials and control section is responsible for establishing purchase orders and buying and bringing in inputs. It is also responsible for the quality of the purchase. The sourcing department looks for new potential suppliers. Before a new supplier is accepted, a matrix is produced with the quotation, the delivery time, the purchase terms, the performance records, the cost and the quality trends. The supplier chosen for a new part can be a firm which already supplies the subsidiary. With this information, the PAT (part approval team) which has at least one member from each of the logistic, sourcing, quality assurance, R&D, and operations departments, will meet to decide jointly after deliberation. Every supplier has to go through this process to be an approved vendor.

Some indicators are used to evaluate the performance of the suppliers. They are in terms of delivery and quality. The suppliers' evaluation must reach a certain level, with a margin of more or less than 5% around the average determined by the subsidiary. Results are shared with the suppliers every month. A complete rating is done every year. Suppliers have already been erased from the list of vendors for reasons of quality defect.

The prices are reviewed every year by the sourcing department. Considering the products manufactured, since prices go down continuously, although material costs are stagnating, the subsidiary tries to get its suppliers to meet the cost requirements. Suppliers are expected to try and get ISO certification. Currently, about 30 of the suppliers have obtained the certification. These mostly supply key components.

Plastic is a sensitive area for this company; several suppliers per product is a preferred strategy to avoid supply-dependence and because one supplier would not have the capacity to handle the whole of the orders. The subsidiary has about 10 plastic suppliers.

Although there are few personal contacts between the suppliers' employees and the subsidiary, it is considered important to know your partners, insofar as knowing each other helps in being aware of respective responsibilities.

Supplier reward system. No specific award is given to suppliers; however, when they have performed well, they receive a 'congratulatory' letter.

Company 7 (Dutch)

Supply network. The firm has a manufacturing/assembly work subsidiary in Singapore, one in Hong Kong, one in Taiwan and one in China. There are also two other production subsidiaries in Malaysia. The subsidiary is ISO certified. It is not completely independent from the rest of the company in its purchase sourcing. For certain key modules, there is a regional 'competence team' (group of purchasing managers from subsidiaries in the region). Products and processes are decided in Singapore. Considering the short life cycle of the products, decisions have to be taken fast. Only about 30% of the purchases are decided by the subsidiary.

There are 80 local suppliers. Japanese-owned suppliers provide electrical components, other foreign-owned provide key modules, whereas locally owned suppliers provide less complicated parts such as metal/plastic parts and indirect materials. The company has only one subcontractor, but 10 partnership agreements in Malaysia. Two companies are Taiwanese-owned, and the other eight are Malaysian-owned. Close relationships in terms of process are kept with these companies.

About 15% of total inputs are purchased internally from other subsidiaries in Malaysia, Singapore, Taiwan, Hong Kong and Japan. Imported goods comprise electronic components from Hong Kong and Taiwan.

Supply strategy. Three departments in the company are responsible for the purchasing activities: the purchasing group, the materials' quality assurance group and the international/planning group. Depending on the type of product purchased, fixed specifications are laid down. In the specific case of key modules, the competence team takes the final decision. For all products, aspects looked at are quality, delivery reliability, lead time, accuracy of lead time, flexibility to new orders, response to the company's needs, questions and problems, integral cost and price. At the beginning, testing of incoming goods is carried out. Once there are no rejects and no more incoming inspections, the supplier becomes a 'non-inspected supplier'. The integral cost includes a variety of aspects, not only the price of the product. Costs such as transport cost, insurance and defaults in quality are all part of the integral cost.

After a supplier has been approved, a rating system determines its performance. This starts from the specifications, drawings and information given at the beginning to the suppliers. The rating criteria include quality, delivery liability, lead time, response and price. The evaluation is carried out on a monthly basis. For quality, the company surveys achieved developments and improvements. The delivery liability depends on the performance of the supplier. The response is a 'feeling' and a 'thinking' parameter. Suppliers are expected to be responsive to a problem and to keep up the relationship with the subsidiary. The competence team decides on the price in consultation with other subsidiaries if it is a shared component; otherwise, the subsidiary establishes the rating on price performance if it is a stand-alone product.

Each individual is part of an overall rating, which comprises the levels green, yellow and red. 'Green' means that the supplier is good, 'yellow' indicates that the supplier needs to carry out some changes, and is therefore given warnings and 'red' is the critical level. The supplier needs to improve; if not, it may not be kept as a vendor.

Results for the overall rating are shared with suppliers; if there is a problem, it will be exposed straight away, before it gets worse. Nearly all the problems are related to quality. For most suppliers, if the quality is good, all the other services tend to be good as well.

The high turnover in companies leads to real problems in terms of delivery. This means that problems are identified not in technology, but mainly in manpower and human resource management. The manager considers that local suppliers are slow in going over to automation, mostly because in this area the products change rapidly and an investment may not be profitable for a long time, but this is at the heart of most of the problems encountered with local suppliers.

Personal contacts are kept with the 10 suppliers that have a 'partnership' agreement with the subsidiary. Relationships with other suppliers are less important. However, personal links are considered to bring better service, better price-negotiating. The manager stated that 'in this part of the world, relationship is crucial'.

Supplier reward system. Every year, a 'supplier day' is organised. During this event, the subsidiary talks about its activity to its suppliers, about what is being achieved with them and what the future plans are. Suppliers get the opportunity to share their own ideas and points of view. Because the rating system is a fairly new practice in the subsidiary, suppliers are awarded only a 'certificate of appreciation'. Soon, the 'green' suppliers will be rewarded specifically.

The idea of the new system is to introduce competition among suppliers. The certificate was previously given to all suppliers, whereas with the ranking system, only the best suppliers are awarded. This ranking system is being standardised throughout the group world-wide.

North American Companies

Our sample comprises four firms from the USA. US firms' supply networks and strategy is described below.

Company 8 (USA)

Supply network. The company was taken over from a firm producing similar parts and components. After the take-over, a lot of improvements in the process were made and the number of employees increased. New equipment was imported from the USA. The subsidiary is ISO certified.

The company has 26 local suppliers. 6 suppliers provide direct materials, 3 Singaporean firms provide flexible circuit board (FCB) assembly work, and 3 locally owned firms provide actuators. Twenty suppliers are locally owned and they provide indirect materials, such as shipping materials, trays, boxes and packing materials. The subsidiary does not keep contracts; it tries to work towards a system whereby new orders are issued that cover a certain period of time. This is to gain time and cost in terms of administration work. Out of the six direct local suppliers, there are two subcontractors.

The company imports the majority of its direct inputs. Integrated circuits (ICs) are imported from two vendors in the USA, flexible circuit board assembling (FCBA) is done and raw flex come from Singapore and Japan, while actuators are purchased mostly from abroad throughout Asia.

Regarding the decision on the choice of suppliers, some suppliers are appointed by the customers (this however sometimes poses problems to the subsidiary because suppliers may listen to the subsidiary's customer and not to the subsidiary itself); others are appointed by the subsidiary. The subsidiary has recently put an effort into sharing its own needs with the customer, in order to choose jointly the appropriate vendors for both parties.

Supply strategy. In selecting a supplier, the subsidiary pays attention to quality, delivery and price. Because the subsidiary is a subcontractor, it is its customers that identify the suppliers. Currently however, the subsidiary intervenes in the final qualification of the supplier.

Several departments in the company intervene in the purchase decisions; the supply department is divided into the direct and indirect materials' sections. The quality assurance department is responsible for the incoming quality inspection, the human resource department takes care of the development and training programmes, the operations department oversees engineers for equipment evaluation and the logistic department looks after the production control.

Recently the subsidiary launched a special programme called the 'Supplier Management Programme', the supplier development and partnership programme. Engineers help to develop the suppliers, carry out a surveillance and help them with problems and issues. The company has a consultant who helps in developing the suppliers. The partnership programme implies that both parties know what the other wants, needs and how the subsidiary does business.

1. Reports are produced monthly and quarterly, following the monthly and quarterly reviews. A 'supplier business review' comes out, in which suppliers are told their performance on quality, delivery and pricing. The performance review is kept on the supplier scorecard. Sixty-five per cent of the mark is allocated to quality, depending on the defective parts per million, the lots acceptance rate, the vendor corrective action report request and the audits performed at the suppliers. Thirty-five per cent of the mark is allocated to delivery and service: schedule, service, material support and pricing. The last factor on the scorecard is the 'major issues' taken into consideration, such as the way the supplier's business is going and its responsiveness to the subsidiary's business changes and needs. These scorecards allow for comparison of suppliers' performances, and for suppliers to see how they stand compared to the others.
2. The subsidiary helps the suppliers with their 'quality system': manpower and employment are included. They are advised on the way to set up monitoring and gates.
3. Regular line audit and surveillance are organised, the subsidiary goes through the line, checks the lists and carries out inspection. Sometimes, awareness programmes on quality and technology issues are organised,

and the strategy is defined, together with the suppliers, through frequent meetings (usually twice a week). Audits are conducted on a random basis at least once a quarter.

4. The next step is to apply a 'customer source inspection' at the suppliers.

The main problem of the subsidiary is that the customers choose the suppliers; this is one of the reasons why the supplier development programme was created, and the subsidiary must deal with those vendors, even if specific problems arise. Two suppliers have been erased from the vendor list in the past because of quality, delivery and price defects. Despite the reports' results, warnings and advice, no improvements were achieved after three quarters, and the decision was taken to erase these vendors from the list. One of the two suppliers was locally owned. The other supplier was owned by a North American firm.

Suppliers are encouraged to work together on problem-solving. An example is two local suppliers doing similar sub-assembly work. Even though they are competitors, they are required to work together and cooperate. They both comment on their problems during weekly forums, and problem-solving techniques are used to try solving some common issues.

Some suppliers are met daily, depending on the sensitivity of the issues, either individually or in groups.

Some personal contacts exist, especially with local suppliers. It is considered important to know the suppliers' employees personally.

Supplier reward system. Every year, suppliers receive awards. The number and theme vary from year to year, and according to the situation. During the last annual supplier business review, five awards were given to five companies out of the fourteen major suppliers in the region. All suppliers are invited, top management people are very supportive and attended the presentation. As no supplier deserved the best supplier award, it was not given. However, following the scorecards' system, five suppliers were selected for: 'best on-time delivery' award, 'least fibre' award, 'most improved quality' award, 'most responsive' award and 'supplier partnership appreciation' award.

Company 9 (USA)

Supply network. The subsidiary manufactures active components. There are three subsidiaries located in Malaysia, one sales subsidiary and one other manufacturing subsidiary located in different states in the country. The company has three sites for the manufacture of components, one in Singapore, one in the USA and one in Malaysia. The parent firm has numerous subsidiaries for finished products' assembly work and production, and for other purposes. A materials council is in charge of the components' production. It meets twice a year in different places. Data and all types of information are shared. Yearly meetings are organised for the components' production, for the company world wide, when all the purchasing managers from all the different branches meet up (approximately 100 persons). The subsidiary is ISO certified, and one of its major short-term objectives is to reach the six sigma standard, together with the suppliers.

The subsidiary has 14 direct suppliers in Malaysia, 8 are locally owned, 6 are foreign-owned. Two provide lead frames, 4 provide plastic parts, 6 provide metal parts and support equipment, such as jig fixtures and die-set moulds. Finally, the subsidiary has two subcontractors doing sub-assembly work. There are five major suppliers for indirect inputs. The majority of the other suppliers have regular fixed contracts; these suppliers are considered as close partners and work closely with the company. A few informal contracts exist for the secondary products.

Imported supplies include PCBs from Thailand, lead frames from Singapore, precision plastic parts from Singapore and electronics components from the USA and Japan. Overall, 40% of total external supplies come from Southeast Asia. The subsidiary has subcontractors in Thailand and in the Philippines. Internal supplies are composed of computer measurement equipment from companies' sites in the USA and Japan, and various other components shipped from sites in Taiwan, Hong Kong, Korea, and small proportion from Europe.

The main reasons for supplying parts from abroad include the level of technology existing abroad (the USA and Japan) and the quality of the parts from these places. High-tech parts are not available in the region. One of the local suppliers currently shows a problem on the technical and management sides; it is possible that it will be erased from the vendor list.

Supply strategy. When selecting a supplier, the subsidiary uses a list of criteria gathered under the term 'TQRDC', which stands for technology, quality, reliability, responsiveness, delivery and cost. There is also a section on the environment. A team is sent to audit and evaluate the suppliers. It looks at the factory and the system, which leads to an assessment. The team is composed of engineers and buyers. The process is launched when there is a rise in production, or when a new product is to be manufactured. In the second case, the subsidiary will need new inputs. In all cases, priority tends to be given to existing suppliers.

Prices are evaluated either by following the market trend, by using the zero-base pricing (bottom-up pricing involves the evaluation of a component) or by consulting engineers. Regular visits to the suppliers' plant, or of the suppliers to the subsidiary, favour the creation of personal contacts. In this way, a closer partnership and relationship can arise.

Supplier reward system. Yearly awards are given to the best suppliers: the total performance award.

Company 10 (USA)

Supply network. The plant is a take-over of an existing company, which previously manufactured similar types of goods. A large part of the current equipment was replaced by imports from the USA. As a subcontracting company, the majority of the parts and components have to be purchased from the vendor list provided by the customers. The choice of the suppliers is not given to the subsidiary. Independence of the subsidiary in terms of supply only occurs for indirect supplies. The few local suppliers are explained partly by a certain dependence on the parent company and partly by customers that require the company to purchase among chosen suppliers. The subsidiary is ISO certified.

There are five direct materials' suppliers in Malaysia. two are foreign-owned; they originate from the USA and from Singapore and supply head gimbals. Three are Malaysian-owned, and carry out sub-assembly work for the company. Altogether, four suppliers do subcontracting work: assembly work of head gimbals. The last one also does assembly work for the subsidiary, but it is only linked through short-term flexible contracts. Imported external supplies originate from Singapore, Korea, Japan and the USA. Internal supplies are composed of electronics components from the US parent firm.

Supply strategy. The subsidiary is mainly concerned with the selection of its subcontractors. The scarcity of local external supply arises, on the one hand, from the unavailability of components locally, and on the other hand, on the necessity to comply with the customers' vendors. Several aspects are taken into consideration in the choice of subcontractors, such as, in order of importance, the quality of the product, the cost and the services, including the delivery time.

There is a very close relationship between the subsidiary and a handful of suppliers. Four suppliers assemble components for the company. Even when the relationship is close, some difficulties still occur that are overcome by a close follow-up, weekly visits and monthly audits.

Because four of the subcontractors are problematic, every shipment received is inspected. These inspections are carried out as well as monthly audits, during which the incoming quality assurance team visits the suppliers. Follow-up reports are then reported by the 'quarterly business review sessions' based on review results, and on feedback given to the suppliers.

Suppliers are told their performance over the three previous months, the problems are reviewed and they are given advice on how to improve their delivery, their quality and therefore the pricing. Personal contacts are considered crucial, as they allow the subsidiary to know exactly how to deal with the subcontractors.

Company 11 (USA)

Supply network. The company has a wide network of subsidiaries throughout the Asian region, for all types of activities. The subsidiary is mostly independent in its purchasing; however, 30% of the inputs are internal supplies. All the purchasing strategy is determined by the parent firm and used by the subsidiary. The subsidiary is ISO certified, but one of its main objectives is to reach the six sigma standard, together with the suppliers, and for this purpose is working closely with them.

There are 50 direct suppliers in Malaysia, some are foreign-owned, some are JVs and some are locally owned. Foreign-owned suppliers provide the company with semiconductors. Either foreign or local firms provide connectors and switches, while only locally owned firms provide the subsidiary with metal, plastic and rubber parts. Only about 10% of the suppliers have a subcontracting agreement with the subsidiary. Most suppliers have a stable relationship, with renewed contracts. For indirect materials, there are a few informal contracts. Imported materials are composed of various components from Taiwan, Japan, Singapore and the USA. Thirty per cent of all supplies are purchased internally. They include components of various types.

Supply strategy. After a detailed benchmarking of the existing suppliers on the market, the subsidiary establishes its choices. For new components, priority is often given to existing suppliers.

There are four key aspects to a good buyer-supplier relationship for this company: trust, alliances on technologies for key products, human resources matters with training provided, and the way to grow in the future. A great deal of attention is given to quality issues through the 'quality system review'. Responsiveness and delivery are also major issues.

Key suppliers are considered as partners. The company has designed key elements for a good partnership. These elements include having a common ground, communication, two-way problem-solving and continuous improvement. To have a common ground, suppliers should make efforts and give support; there is a need to agree on the key initiatives to grow. The subsidiary tells its suppliers about its expectations and intentions for better communication, and it expects its suppliers to provide it with their own ideas on the subsidiary. The two-way problem-solving involves training;

techniques are shared with suppliers, whereas suppliers bring ideas, for instance on design. Suppliers are expected to bring quality, improvement, teamwork, expertise in their field and growth to a world class. The suppliers undergo regular follow-ups, which include on-time delivery, inventory performance, quality rating and cost improvement.

Supplier reward system. Seven-yearly awards are given to the best suppliers for quality and customer satisfaction, best delivery and new product support. Several teams present various themes during the partnership for growth supplier day; the most recent research conducted was on the topic of 'Total Customer Satisfaction'. Each presentation follows a set format, which includes the identification of the problem that occurred, how the supplier set up the objectives, the benchmarking and looking for possible solutions, taking the appropriate and chosen action, followed by procedures and follow-up of the improvement, the evaluation of the results for both suppliers and the subsidiary, and future further improvements. Some suppliers present their teams. Each presentation is marked according to the teamwork and presentation (introduction, analysis techniques, remedies, results, institutionalisation and presentation), and the objectives (quality and delivery, cycle time/lead time, responsiveness, flexibility and institutionalisation of results).

ANALYSIS OF SUPPLY NETWORKS AND STRATEGIES

The previous section provided a case-by-case description of firms' supply networks and strategy. In this section, an integrative approach to the analysis of the latter is taken by combining information into preset categories. We first look at trends in local purchase before analysing contractual relations and supplying relations created by foreign firms with local suppliers.

Trends in Local Purchase

To understand fully buyer-supplier relations, we need to analyse the patterns of local purchasing carried out by foreign firms, and the way they choose suppliers.

Patterns of local purchasing

Two crucial points must be raised from Tables 6.2 to 6.4. The first is related to the place of Malaysian-owned firms in the supply patterns of foreign affiliates, and the second to the type of inputs purchased from these latter.

Table 6.2a Supply network and strategy of Asian foreign affiliates

Country of origin	Company 1: Japan	Company 2: Japan
Subsidiaries (excluding the firm)	3 in Malaysia (similar products) 1 sales office in Singapore	1 in Thailand, 1 in Indonesia, 2 in China and 1 sales office in Singapore
Number of local suppliers	80	54
Origin of suppliers	50 Japanese (18 followed the company from Japan); 30 Malaysian	Japanese and foreign-owned Malaysian suppliers only for secondary items
Type of inputs purchased in Malaysia	Japanese suppliers: semiconductors, electrical and mechanical components Malaysian suppliers: plastic injections, secondary items	Copper wires, contacts, plastic moulding parts
Contract	45 to 50 subcontractors the others through purchase orders	6 under subcontracting agreements
Imported inputs' origin	Japan, ASEAN	Japan, Asia
Type	Electrical and mechanical components, PCBs	Raw materials and chemicals
Share of internal supply	40%	10%
Type of internal inputs	Stamping parts, mechanical items	Complex metal items
Selection criteria	Quality, service, cost	Quality, cost, delivery
Quality checks	Visits and reports	Regular evaluations and audits, ISO expected after 2 years
New product development	In Japan	-
Supplier strategy	1 supplier per item purchased (but several suppliers per product)	Several suppliers
Pricing	Changes expected to follow Japanese prices	-
Supplier reward system	None	Rewards to suppliers reaching highest score in terms of delivery/quality/cost/management's response

Table 6.2b Supply network and strategy of Asian foreign affiliates

Country of origin	Company 3: Singapore	Company 4: Taiwan
Subsidiaries (excluding the firm)	1 in China, 1 in Indonesia, 1 in Thailand	2 in Malaysia
Number of local suppliers	10	20
Origin of suppliers	Malaysian	Foreign-owned Malaysian
Type of inputs purchased in Malaysia	-	Plastic and metal parts
Contract	10 subcontracting for assembling and testing	6 subcontractors (sub-assembling)
Imported inputs' origin	-	Taiwan, China, Singapore
Type	-	Key tops, isolation film, etc.
Share of internal supply	90 % centralised purchasing	30%
Type of internal inputs	ICs, PCBs, resistors, etc.	Plastic items
Selection criteria	Price, quality, delivery (service)	Quality, delivery, service, price
Quality checks	Regular reports	Regular reports, audits
New product development	-	-
Supplier strategy	-	-
Pricing	Key factor in the relationship, regular price cuts are expected	Regular price cuts expected
Supplier reward system	None	Planned for near future

Table 6.3 Supply network and strategy of European foreign affiliates

Country of origin	Company 5: Europe	Company 6: Europe	Company 7: Europe
Subsidiaries (excluding the firm)	1 in Malaysia	-	2 in Malaysia, 1 in Singapore, 1 in Hong Kong, 1 in Taiwan, 1 in China
Number of local suppliers	35	100	80
Origin of suppliers	Foreign-owned Malaysian	40 Foreign-owned 60 Malaysian	Foreign-owned Malaysian
Type of inputs purchased in Malaysia	Plastic mould, fasteners and screws, electrical components, cables	Foreign-owned: electronic components Malaysian: standard electronic components	Japanese: electrical components Other foreign: key modules Malaysian: metal/plastic parts
Contract	Majority of subcontracts in processing and assembling	15 subcontractors 2 assemble complete product	1 subcontractor 10 partnership agreements
Imported inputs' origin	UK, France, Australia, Japan and Korea	Singapore, Indonesia, Thailand, Japan, Taiwan, Korea	'Regional Competence Group' Hong Kong, Taiwan
Type	Plastic moulding powder, raw materials	Key electrical and electronic components, microprocessors	Electronics components
Share of internal supply	20%	Negligible	15%
Type of internal inputs	Press parts, ceramic cast	-	-
Selection criteria	Quality, pricing, delivery, service	Delivery, quality	Quality, delivery reliability, lead time, flexibility, management responsiveness, integral cost, price
Quality checks	Regular audits, reports and feedback	Monthly reports	Performance rating system in place
New product development	---		-
Supplier strategy	---	Several suppliers per product	-
Pricing	Par with market price expected	Prices reviewed yearly	-
Supplier reward system	None	'Congratulatory' letter	Yearly supplier day

Table 6.4a Supply network and strategy of US foreign affiliates

Country of origin	Company 8: USA	Company 9: USA
Subsidiaries (excluding the firm)	-	2 other subsidiaries in Malaysia, one sales subsidiary and another manufacturing site
Number of local suppliers	6 for direct inputs, 20 for indirect materials	14 direct suppliers Many indirect suppliers, 5 major ones
Origin of suppliers	3 Singaporean suppliers, 23 Malaysian	6 Foreign-owned 8 Malaysian
Type of inputs purchased in Malaysia	Singaporean: assembling of flexible circuit boards (FCBs) Malaysian: 3 provide actuators, 20 provide secondary materials	Metal parts and support equipment (jig, fixtures, die-set moulds) 2 suppliers carry out sub-assembling Indirect materials
Contract	Subcontracting, purchase orders	2 subcontracting relationships, fixed regular contracts
Imported inputs' origin	USA, Singapore, Japan, various other Asian countries	Sub-assembling in Thailand and Philippines, PCBs, lead frames, precision plastic parts, electronics components
Type	ICs (from 2 US vendors), FCBA, raw flex from Singapore, actuators	PCBs (Thailand), lead frames (Singapore), precision plastic parts (Singapore), electronic components (USA and Japan)
Share of internal supply	40%	10 %
Type of internal inputs	-	Computer measurement equipment the (USA, Japan), various components (Taiwan, Hong Kong, Korea, a few from Europe)
Selection criteria	Suppliers chosen by customers. Main issues: quality, delivery and price	TQRDC (technology, quality, reliability, responsiveness, delivery and cost)
Quality checks	Monthly reviews, scorecards of performance, audits. Close relations within supplier development programme	Audits
New product development	-	-
Supplier strategy	-	-
Pricing	-	Regular evaluations
Supplier reward system	Yearly supplier awards	Yearly supplier awards

Table 6.4b Supply network and strategy of US foreign affiliates

Country of origin	Company 10: USA	Company 11: USA
Subsidiaries (excluding the firm)	-	-
Number of local suppliers	5	50
Origin of suppliers	1 from USA, 1 from Singapore 3 Malaysian subcontractors	Foreign-owned, joint ventures, locally owned
Type of inputs purchased in Malaysia	Foreign-owned: head gimbals Both: sub-assembling	Foreign-owned: semiconductors Both types: connectors, switches Locally-owned: metal, plastic and rubber parts, indirect materials
Contract	4 subcontractors, 1 short-term flexible contract	5 subcontracting agreements, most other suppliers have renewable contracts, informal contracts for indirect materials
Imported inputs' origin	Singapore, Korea, Japan and the USA	Taiwan, Japan, Singapore and the USA
Type	-	Various components
Share of internal supply	20%	30%
Type of internal inputs	Electronic components from US parent firm	Various components
Selection criteria	Quality, cost, service, delivery time	Trust, alliances on technology, human relations, future growth and relation
Quality checks	Inspections, audits, visits to the subcontractors. Quarterly business review sessions	Quality review system, regular follow-ups
New product development	-	-
Supplier strategy	-	-
Pricing	Reviews every 3 months	-
Supplier reward system	-	Yearly award system

Table 6.2 indicates how few direct inputs are purchased from Malaysian-owned suppliers. Japanese firms purchase primarily from Malaysian-based Japanese affiliates, while firms from the newly industrialising economies (NIEs) still rely heavily on imports of materials and have a limited number of subcontractors in Malaysia. Furthermore, when Malaysian companies are contracted, it is often for low-tech products and components. Key parts are either purchased from foreign-owned suppliers or imported.

As shown in Tables 6.3 and 6.4, although the European and North American affiliates interviewed resort less to internal supplies, and use more locally owned suppliers, they tend to purchase low-tech products from Malaysian-owned suppliers.

Specific problems of local firms

The subsidiaries interviewed were asked to evaluate the local firms and, if any, explain the specific problems of those firms. The following discussion on local firms' performance is the result of those answers. Since each problem may be of particular concern for each firm, we have kept the answers in a table format, to be able to discuss the answer by category of foreign firms, and by category of products manufactured.

As shown in Table 6.5, the main problems of locally owned firms are the lack of flexibility (Companies 1 and 9), human resource management problems (Company 7), the lack of technical expertise (Companies 5 and 8), the difficulty to meet quality requirements (Company 5) and the disadvantage of being a new-comer in a fast-changing high-tech industry (Companies 3 and 8).

An indicator of local companies' performance is the type of products purchased. In Company 5, which started operations in 1976, press-stamping parts' suppliers are all foreign-owned based in Malaysia. These parts are complicated ones and when the subsidiary was benchmarking local suppliers, there was none locally owned. In the mid-1990s, a few locally owned companies developed in the sector, but the subsidiary had already established long-term relationships with other firms. The plastic industry is well developed locally, but for most subsidiaries the quality requirements are too high to purchase only from Malaysian companies. Moreover, in the case of Company 9, local supplies are not sufficient to cover the needs.

Hence, locally owned firms suffer from lack of experience and expertise; they also have a disadvantage, as many TNCs have already established long-term relationships with foreign-owned suppliers; and finally, many do not produce the required quantity to attract new customers.

Table 6.5 Subsidiaries' perspective of problems of local firms

Firm	Origin	Year est.	Product		Problems of local suppliers for subsidiary
1	Japan	1989	PC	1.	Shortage of supplies, suppliers do not fulfil the needs of the subsidiary
				2.	The lack of flexibility to new orders poses problems with the short lead-time strategy of the subsidiary
3	Sing.	1992	PE	3.	Locally owned firms have difficulties in succeeding in the electronics sector because of the nature of the products. They are technology-intensive, and go through constant upgrading
5	UK	1976	EM	4.	Low standard of design and process
				5.	Difficulty to meet quality requirements
7	Dutch	1979	CE	6.	Technically, local suppliers are not less efficient than foreign suppliers for the types of inputs purchased
				7.	Human resource is a major problem arising in Malaysia due to the high turnover rate. Employees change too often in local firms; they are less efficient and this leads to quality problems and further to a slowdown in the delivery*
				8.	Going for automation may be a solution to these problems;
				9.	A good indicator is the quality offered by the supplier; if the level of quality is good, it is rare that the supplier has other types of problems such as delivery or service problems. Quality is the key to the relationship
8	USA	1992	AC	10.	Lack of technical expertise, especially in the electronics sector where firms started long ago and have an ownership advantage over new players. For this reason, locally owned suppliers are less sophisticated than foreign-owned suppliers
9	USA	1972	AC	11.	Part of the plastic parts' sourcing is carried out in Malaysia; however, the needs are too high only to be covered by local sources

Notes:

Year est.: year of establishment
PC: passive components
PE: production equipment
EM: electro-mechanical
CE: consumer electronics
AC: active components
* Very interestingly, labour is a key factor in the efficiency of firms; on the other hand, manpower takes a negligible part in the whole set price of consumer electronics products.

Choice of suppliers

Additional factors intervene in the supplier selection process, which might act against the use of locally owned firms. Four can be distinguished: influence of parent firm, influence of customers, future improvement of local firms and government-induced relations.

Influence of parent firm. Firms that have only recently started their operations in Malaysia are still very dependent on their parent firms in terms of sourcing strategy (Companies 1, 3, 4 and 8). Firms of Asian origin seem to be more dependent in their overall strategy, and not only in terms of internal supplies. The geographical, management and time factors explain this dependency on the parent firms. Asian firms can import cheaply from within the region. Firms from the NIEs are at an early stage in their internationalisation process, and adaptation to the local host economy is still immature. Also, management decisions taken by the parent firms are still restricting the affiliates' behaviour. A drawback of this dependency for the host country is that parent companies very rarely benchmark companies located in Malaysia; they give priority to home country sourcing. This trend appears in the Japanese and the Taiwanese cases, particularly for Companies 1 and 4.

A similar dependence on the parent company appears for Western companies. Companies 5 and 6 are independent in their purchase decisions for standard products purchasing. However, they are not for a few major components that can be used in both the parents' and the subsidiary's productions.

The life cycle of the product often discourages firms from resorting to local purchase; as the product undergoes frequent changes and upgrading, the companies need to be able to rely on their suppliers. Companies 6 and 7 in the consumer electronics category expressed their concerns about finding reliable suppliers.

Influence of customers. Some companies' customers, who prefer to carry out the pre-selection themselves, provide the companies with manufacturing parts and components for particular suppliers. This is particularly the case for subcontracting companies. Companies 3, 8 and 10 thus have limited independence in their choices of suppliers. These practices lead to specific types of relationship problems, and these subsidiaries need to adapt and provide particular types of support to these pre-selected suppliers. Company 8 specifically wished to establish 'virtual vertical integration'.

Future improvement of local firms. Company 5 considers that local suppliers have shown signs of improvement on the technical and management sides. There are fewer problems related to schedules and quality control. However, the efforts are not yet sufficient. A good estimation is the parts' rejection rate, which remains too high. Company 9 is still pessimistic about future supplying trends towards locally owned firms:

> The subsidiary will buy more locally, we have an advantage in supplying locally. However we focus on working and developing with the existing supplying companies. More Japanese suppliers are moving overseas to Southeast Asia, and they will provide better choice to supply locally.

Government-induced relations. Government-induced relations with local suppliers have proved very beneficial, but for a very small number of suppliers. Two programmes have influenced foreign subsidiaries towards creating a specific link with local suppliers: the 'Umbrella Subcontracting Scheme' and the Vendor'Net Programme.[7] Company 1 launched a supplier in the plastic industry through the 'Umbrella Subcontracting Scheme'. Company 5 benefits from exemptions on independent duty for the needed imports of inputs by purchasing a minimum of 30% of total inputs locally. Company 6 signed a memorandum of understanding with two local vendors through the Vendor'Net Programme, and also benefits from the GSP Programme (Generalized System of Preferences) by purchasing 50% of total supplies within the ASEAN region. Company 11 signed a memorandum of understanding as well.

These specific relationships result from the government action to try to get foreign subsidiaries to generate positive spillovers. However, these apply to a limited number of cases. Only a few foreign firms take part, and when they do, it only benefits a handful of local vendors. From the multinational firm's perspective, these programmes are not very attractive. Here are comments provided by Company 9:

> The incentives or attractiveness is not so high for the TNCs, it is more important for the SMEs. In fact, it would be better logistically to have your suppliers next door. You develop strong relationships; this is what we find with the suppliers we have developed over the years. We are really close, they know exactly what our company wants, where it lies, and what it does not want. Those are the benefits that come. (...) The government programme was officially launched 3 years ago. Could be too early to judge. A bridge between SMI and TNC is a good start.

Contractual Relations

The types of contracts carried out by transnational corporations (TNCs) appear in Tables 6.2-6.4. There are major types of contractual relations applied in Malaysia; the subcontracting relationship, the regular contracts through which close relationships are created and purchase orders, which lead to no long-term commitment on either side. The subcontracting relationship leads to closer relationships with the partner firm. Yet suppliers are long-term suppliers, and although the purchase orders do not lead to any obligations on either part, once a supplier is selected by the TNC and is part of the PVL (preferred vendor list), a stable relationship is established.

The Asian firms interviewed favour a 'traditional' difference in suppliers, between close relations maintained with subcontractors versus other suppliers. Exceptions are found. For instance, one supplier in the plastic injection field was set up when Company 1 was established in Malaysia in 1989. The supplier was fully Malaysian-owned, it was given all the technical capabilities, material was provided and training was carried out by a Japanese team.

Western affiliates and affiliates from the USA have a panel of contractual relationships. They maintain close relations, not only with their subcontractors, but also with a number of carefully selected 'partners'. Although Company 6 does not have any system of partnerships, suppliers are considered as partners, and personal relationships are emphasised. Companies 7, 8, 9, 10 and 11 all have some strategies related to their relations with their suppliers. Company 10, in fact, has four subcontractors and one close supplier; therefore links are created mostly with subcontractors, three of which are Malaysian-owned. Close relations imply regular follow-ups, weekly visits and monthly audits. Company 9 developed a 'Supplier Development Programme', by which the subsidiary shares some management elsewhere and technical knowledge with the suppliers. Since the company wants to adopt the six sigma model, suppliers are trained in this management method. All 14 suppliers are considered to be close partners; 8 are locally owned and provide plastic and metal parts, support equipment and are subcontractors. Company 11 also considers suppliers as close partners, exchanging ideas, and having a two-way problem-solving system, combined with regular follow-ups on time delivery, inventory performance, quality rating and cost improvements. Many of these suppliers are Malaysian-owned; they supply metal, plastic and rubber parts, but also connectors and switches, and mostly have regular contracts with the subsidiary.

Company 7 has 10 formalised partnership agreements in Malaysia, 2 with Taiwanese-owned firms and 8 with Malaysian-owned firms, which supply metal and plastic parts. Such partnerships imply close relations in terms of production process. Personal contacts are kept with these suppliers; this is believed to bring better services and better price negotiations.

In the case of Company 8, the Supplier Management Programme is only slightly more than one year old. It involved the creation of strong links between the two partners. The subsidiary's engineers and a consultant are responsible for the development of their suppliers; they provide frequent supplier business reviews and carry out regular audits and assessment on quality. This programme was created because the customers choose suppliers, and the subsidiary has to deal with them, whether or not it considers the quality adequate. These partnerships exist with three Singaporean-owned firms, and three Malaysian firms, which supply actuators.

Undoubtedly, subcontractors keep a favoured place in the buyer-supplier transfers, and still benefit from closer links with the buyer. However, the discussion above shows that other suppliers should not be disregarded from the analysis of transfers. Recent supplier development schemes involve greater transfers between foreign affiliates and their local suppliers.

Supplying Relations

The trends in local purchase and contractual relations have been specified above. Supplying relations can now be analysed, paying particular attention to the importance of price, regular follow-ups and supplier awards systems.

Importance of price

Asian companies insist more on the price of the items they purchase. For Company 1, cost is a major selection and evaluation criterion, prices are closely followed in Japan, and for every cut in price there, a similar cut is expected in Malaysia, with further negotiations to reach lower prices. Company 2 expects a yearly reduction in price of 10-20%, although this is not compulsory. Company 4 specifies a 5% decrease in price is expected quarterly. European firms also consider cost and price reduction as important criteria; prices are reviewed and reductions are expected, but they are less systematic and may occur over a longer period of time. Although managers acknowledge the importance of price, many companies stated that price is a requirement less important than quality in this part of the world. Still, most of the companies' regular supplier reviews include pricing. Such aspects may at times be difficult for the suppliers, which may feel squeezed by anticipated price cuts. Company 1 expects similar price variations as those occurring from suppliers based in Japan.

Regular follow-ups

A major contribution provided by TNCs results from their regular evaluations and feedback of their local suppliers. All the TNCs have regular evaluations or reports based on the quality control, delivery reliability, lead time, flexibility to new orders, response to the TNC needs and problem-solving. These evaluations are shared with the suppliers during meetings or in reports. Company 4 declares that the quality is crucial, and therefore problem-solving on the matter takes place as far as possible between the buyer and the supplier. These practices may be heavy on the supplier's management, which is to share decisions with its customer. Companies 2 and 7 have a detailed rating system, which is recorded all year long and leads to the 'best supplier' award at the end of the year. Company 2 puts its 10 best suppliers into competition to reach the first position. Company 7 follows a different process; best suppliers are given a 'certificate of appreciation' at the end of the year. The 'supplier day' is seen as an opportunity for both parties to talk about each other's activities, achievements and future plans. A similar system to that of Company 2 is envisaged for Companies 6 and 7. Company 8 proceeds through a very detailed follow-up system, which consists of several stages, whereby the supplier is evaluated during a monthly supplier business review on quality, delivery and pricing. The supplier's evaluation is then reviewed on its scorecard, and it can compare results with those of other suppliers.

Pricing policies and quality assurance follow-up are common practices among TNC affiliates with their local suppliers in Malaysia. It is likely that these behaviours are heavy on the suppliers' management systems and operations. Strict quality controls, however, undoubtedly contribute to the enhancing of the local suppliers' performance in terms of quality, cost and delivery.

Supplier awards systems

Out of the 11 cases, Companies 2, 7, 8, 9 and 11 have a reward system based in most cases on the regular evaluations during the year. Company 4 already has plans to give rewards, while Company 6 gives congratulatory letters. All these practices receive positive answers from the suppliers, who try their hardest to run for the award. Not only does working for a foreign subsidiary provide a good reputation for the supplier, it also enables easier access to bank loans. Company 2 suppliers are making their best effort to reach the first place. The company makes sure of the competition between suppliers by keeping respective results on a board that each of the best suppliers has access to. For Company 8, rewards for suppliers that have helped in solving specific problems give incentives, which may influence other suppliers to increase their service to obtain the reward.

TRANSFER GENERATED BETWEEN SUBSIDIARIES AND SUPPLIERS

Background information for foreign firms' supply relations helps delineate actual transfers between firms, and to understand which firms are concerned.

There are often few local suppliers, and those are sometimes foreign-owned. Sample firms provided information about transfers to indigenous suppliers. Overall transfers taking place are described first; we then discuss the factors likely to affect transfers and other comments made by respondents.

Overall Transfers

There are four types of transfers mentioned by respondents: financial transfers, technology transfer, training-related transfer and management-related knowledge transfer; our own generic term 'knowledge transfer' would include all these categories. Knowledge transfer from foreign subsidiaries by country/region of origin appears in Tables 6.6-6.8.

Financial transfers

Financial transfers between foreign firms and their local suppliers are practically non-existent. Eight firms out of eleven never provide any kind of financial support to their suppliers. In the case of the firms that have provided some financial support, they were limited in time and scope, and are not regular company practices.

As indicated in Table 6.7, payment before delivery sometimes occurs but under special circumstances. Company 6 uses this practice only with suppliers that provide a good service or have good potential, and Company 7 has in the past used this method, but only for its closest suppliers. Another type of financial transfer occurred through the purchase of machinery and equipment, but again, this transfer is rare. Company 5 bought machines, which were lent to some subcontractors. Finally, one firm on only one occasion lent money to a supplier. The financial department of Company 7 decided on interest to be applied to the loan, and once the goods were delivered, the loan was deducted from the invoice. Such a loan could be considered to have been a payment before delivery, except interest was charged to the supplier. Financial transfers are rare and are not found to bring any benefits to the local Malaysian suppliers.

Knowledge and technology transfer

Various types of knowledge transfers are described in Tables 6.6-6.8. They can be divided into two different types. One type is related to the input itself and the technical help that may be generated by the foreign subsidiary. The other type of technological transfer is related to the production process itself.

Table 6.6 Knowledge transfer by Asian firms

	Technical transfers	Training transfers	Management transfers, additional comments
Company 1	Support provided through training.	Foreign experts: sometimes experts from Japan come mainly to provide technical help and advice, mostly when new moulds will be used. In this case, experts will visit suppliers as well.	1. Advice on management development, software is given 2. Means of improving delivery provided
Company 2	1. Appropriate equipment provided to subcontractors that do not have it. 2. In the case of coils, only a few good subcontractors exist. Therefore, if a subcontractor is selected, it is provided with the necessary design, machinery and equipment.	1. Production team of subsidiary gives occasional training, and provides work guidance for suppliers needing it 2. In the case of coils, production team sent to supplier's plant for about one month to train employees	Subcontractors mainly benefit from support
Company 3	None	None	Subsidiary provides no support to its suppliers. This results from heavy dependence on customers' decisions, and also on the fact that subsidiary was only set up recently in Malaysia
Company 4	1. For plastic parts, subsidiary provides method of production. R&D department of company designs these parts 2. Company experts go to supplier to set up production process	Training only takes place on technical matters. Technical staff may go to suppliers for on-the-job training only	Subcontractors benefit from support, notably for technical matters. Subsidiary advises on specific machinery to purchase. Some modifications of parts occur directly within subsidiary. Or else, subcontractors receive help in production requirements

Table 6.7a Knowledge transfer by European firms

	Financial transfers	Technical transfers
Company 5	Although no financial support provided, machines bought for some subcontractors. Machines still belong to subsidiary which only 'lends' them to suppliers	1 Initial specifications provided to suppliers 2. Original design sometimes changed to better fit suppliers' production facilities 3. For some subcontractors, subsidiary has provided help in layout and organisation of production facilities
Company 6	Subsidiary occasionally pays before delivery, but only in case of suppliers that provide good service or have good potential	1. In case of a mould breakdown, when supplier does not have repair technical capability, subsidiary's tool engineers intervene 2. For moulding, plastic parts and mechanical parts, engineers or quality control employees commonly sent to help and advise supplier 3. For plastic or metal tooling parts, early Vendor Participation Programme aims at getting suppliers acquainted from the beginning with technical requirements necessary, so they have time to prepare. Subsidiary provides drawing 4. No support provided; for components; fixed specifications given
Company 7	1. In only one case, subsidiary lent some money to supplier. Financial department decided on interest. Once goods delivered, loan deducted from invoice 2. For closest suppliers, payment before delivery has occurred	Rare

Table 6.7b Knowledge transfer by European firms

	Training transfers	Management transfers, additional comments
Company 5	1. Irregular training for suppliers 2. For two subcontractors working exclusively for subsidiary, training on production process provided. Main areas covered: quality control and process yield. Training only on-the-job	None
Company 6	1. Practical on-the-job and theoretical training on quality and technical matters for suppliers' managers, quality assurance inspectors, supervisors, technicians and engineers 2. Subcontractors manufacturing/assembling complete product trained on product testing, points to look for, ways to read equipment 3. Training provided more frequently for subcontractors. However, may occur in case of problems for other suppliers	1. Suppliers considered as partners: 'There is a relation with the supplier, it is two-way. The modern concept is not only financial, we prefer to consider each other as business partners, rather than buyer/customer'. 2. Subsidiary advises AQL and process control charts to suppliers. Quality control employees of subsidiary trained two companies created by subsidiary in these concepts
Company 7	1. No training for normal suppliers 2. Experts sent to subcontractors; employees trained at subsidiary 3. Operation training for plastic suppliers that carry out sub-assembling for subsidiary	1. Subsidiary supplier development team helps to upgrade suppliers. Incoming quality teams say when something is wrong. In this case, team with engineers and quality people is sent to suppliers, and tries to upgrade their process 2. Suppliers advised to use AQL quality system (batch size, failure rate, sample size), because they are rated on this system. Subsidiary even teaches system to the suppliers that do not know it 3. For close regular suppliers, subsidiary tries to help them to obtain ISO certification

Table 6.8a Knowledge transfer by North American firms

	Technical transfers	Training transfers
Company 8	Suppliers shown at subsidiary how parts are used, and what parts should be	1 Most common training is related to management. Fundamentals like manufacturing process techniques, quality management, certain steps and problem-solving techniques taught to the suppliers 2. Training can take place at suppliers' site
Company 9	Particular area where support takes place regards 'process control requirements'	1. Training takes place either at supplier or at subsidiary. Suppliers trained when necessary on assembly process, and way subsidiary proceeds 2. Management techniques taught, such as TQM system 3. Engineers go to suppliers when designing their equipment or doing try-out (up-front stage during R&D stage very important)
Company 10	1. Subsidiary provides specifications and sometimes method of manufacture 2. Subcontractors shown how to assemble, and subsidiary may help in settling manufacturing operations 3. In one case, subsidiary has helped in establishing production plant as a whole, supplied machinery, tools and inputs	1. Engineers often visit suppliers and advise them. 2. Suppliers' employees are trained on use of machines and technology used 3. On-the-job training most frequent method used. 4. At beginning of relationship, suppliers often visit plant and sometimes receive classroom training
Company 11	Manufacturing specifications of products often provided, and knowledge shared to try improving quality and reliability of products	1. Regular training provided to suppliers, both on technical and management aspects 2. Off-the-job training consists in showing or sharing recent improvements in technology, as well as teaching new management techniques used by the subsidiary 3. Suppliers' employees sometimes invited to subsidiary for on-the-job training on various manufacturing processes 4. Managers or engineers of suppliers benefit more from this training than operation staff

Table 6.8b Knowledge transfer by North American firms

	Management transfers, additional comments
Company 8	1. Same level of support provided to subcontractors and other suppliers 2. Specific management methods taught on a theoretical and practical basis: TQM, Soshin methods 3. Statistical process control adopted by suppliers on demand of subsidiary
Company 9	1. Supplier Development Programme by which subsidiary shares some management and technical knowledge with suppliers 2. Six sigma standard is objective, and suppliers are taught about it and best means of reaching best level possible
Company 10	Subcontractors required to follow subsidiary's quality management system: statistical process control and finished goods' inspection
Company 11	1. Quality system not imposed on suppliers, but subsidiary has influence through recommendations occurring during quality reviews 2. As subsidiary is trying to use six sigma system, suppliers are trained on system in order to comply with its requirements.

It is common for a buyer to provide the product specifications to its suppliers, Companies 6 and 10 provide these specifications, while companies 2, 5 and 6 provide the original design and drawings of the inputs. Most commonly, the items purchased are designed by the R&D department of the foreign affiliate (or by the parent company), and suppliers are expected to follow instructions provided to them. Company 5 often has to change the design of the input to fit the suppliers' production possibilities. Specifications are not provided for components in the case of Company 6, mostly because these are standard plastic and metal tooling parts.

Subsidiaries frequently also provide information on the production process itself (manufacturing specifications) to its suppliers. Machinery and equipment may be provided to the subcontractors, as in the case of Company 2. The method of production is given to the suppliers of plastic parts by Company 4, and provided to their suppliers by Companies 10 and 11. Company 4 experts often set up the production process of the suppliers, and Companies 5 and 10 often help in the layout and organisation of the production facilities. Similarly, Company 9 provides support on the process control requirements.

In specific cases, the subsidiaries may intervene at various stages in the manufacturing process. In the case of a mould breakdown and if the supplier cannot mend it, Company 6 tool engineers intervene. In one case, company 10 has helped in establishing the production plant as a whole for a supplier. It supplied the machinery, the tools and the inputs.

Overall, technical help is often related to training by the subsidiary. When companies provide the production process to their suppliers, this further implies that the suppliers' employees are trained in how to use these processes. Companies 1, 4, 8 and 10 clearly state that technical support occurs through training. For Company 8, the suppliers are shown on site how the parts are used and what the parts should be. Company 10 shows the subcontractors how to assemble the products.

Technical support depends on the product purchased. Apart from the specific relationships that exist with the subcontractors, the suppliers that most frequently benefit from transfers supply plastic and metal parts (Companies 4 and 6). In Company 6, for moulding, plastic parts and mechanical parts, it is common for the subsidiary to send engineers or quality control employees to help and advise the supplier. For tooling parts, whether plastic or metal, the early 'Vendor Participation Programme' aims at getting the suppliers acquainted from the beginning with the technical requirements necessary, which gives them time to prepare. Fixed specifications are provided for components but with no further support. Such a trend indicates that locally owned suppliers substantially benefit from transfer of technology and knowledge, since they provide plastic and metal parts to most of the subsidiaries in the sample.

Training-related transfers

Training is the most frequent means of knowledge transfer between a foreign subsidiary and its suppliers. All the firms provide some kind of training, except for Company 3. Training is a major means of technical support, but also plays a major role in management-related transfer of technology and knowledge (Company 11). For instance, the most common training for Company 8 is provided to management. Company 9 trains on Total Quality Management (TQM) techniques. As Company 11 states, managers and engineers of the suppliers benefit more from training than operation staff.

Training takes place either on-the-job, when employees are shown how to proceed on the work itself, or off-the-job through courses in-house or in specialised institutions. On-the-job training is the most common type of training provided by subsidiaries to their local suppliers; Companies 2, 4, 5, 6, 7, 8, 9, 10 and 11 all provide some kind of on-the-job training. Practical on-the-job and theoretical training are provided on quality and technical matters to suppliers' managers, quality assurance inspectors, supervisors, technicians and engineers. This training is aimed at helping them meet the subsidiary's requirements. Training is more often provided for subcontractors (Companies 5, 6 and 7); however, it also occurs in case of problems for other suppliers, and for suppliers under partnership agreements.

Training can be related to management methods (quality control and process yields shown by Companies 4 and 5, product testing and problem spotting by Company 6, training in new management techniques by Company 10), or to the production process (occasional training by the production team of Company 2, engineers of Company 9 show design of new equipment and try out, engineers' visits for Company 10's suppliers, or Japanese experts' visits to suppliers of Company 1, training to plastic suppliers who provide sub-assembly work to Company 7, training on new improvements in technology by Company 10 and training at the subsidiary for manufacturing processes by Company 11).

Off-the-job training is closely related to theoretical training. Company 6 teaches suppliers' managers on quality assurance. Similarly, fundamentals of the manufacturing process techniques, quality management and problem-solving techniques are taught by Company 8 to its suppliers. Company 10 provides classroom training at the beginning of a relationship with a supplier on machines and technology, while Company 9 emphasises the assessment processes and the way the subsidiary proceeds.

Training is frequent at the subsidiaries (Companies 10 and 11), but also occurs at the suppliers (Company 8 and 10), or in both places (Company 9).

As in the case of technology transfer, training can sometimes be provided to suppliers of specific inputs. The production team of Company 2 provides monthly training at the suppliers in the case of coils, and plastic parts' sub-assemblers receive special training by Company 7.

Management-related knowledge transfer

Management systems can either be advised by the subsidiary or imposed on its supplier. Company 1 gives advice on management development and software, and on the means to improve delivery. Companies 6 and 7 advise their suppliers to use the approved quality system (it includes batch size, failure rate and sample size), and process control charts (Company 7). As a quality management system, Companies 8 and 10 impose the use of statistical process control on their suppliers, and Company 10 advises on finished goods' inspection. Company 8 teaches management methods such as the TQM system, while Companies 9 and 11 both try to apply the six sigma system and therefore train their suppliers on its basis and requirements. Hence, foreign affiliates do influence local suppliers' management styles and methods. In some cases, the suppliers may find the relation difficult.

Factors Influencing Transfers from Foreign Affiliates to Local Suppliers

The structure for this section results from a careful analysis of foreign firms' comments on transfers to local suppliers, within the framework devised in Chapters 2 and 3. Firm-related, purchase-related, investment-related factors and the time factor are considered.

Transfers and firm-related factors

There are five firm-related factors: country of origin, purchasing strategy related to product manufactured, types of firms and product manufactured, management strategy and size of firms.

Country of origin. It is difficult to highlight a clear distinction of transfer occurrence according to the country/region of origin of the affiliate, mostly because all companies vary in size and product manufacture, and in terms of other factors likely to influence purchasing behaviour. There seems to be a tendency, however, in North American and European firms to create stronger links with a wider variety of suppliers, and not solely with subcontractors. Training and management transfers also seem to take place more often in Western firms than in Asian firms. Such comments, however, must be taken with caution, due to the small number of firms taken into account in this qualitative analysis.

Major difference in purchasing according to product manufactured. High-tech companies have a closer relationship with their suppliers to ensure the high quality of the products delivered. Companies 8, 9 and 11 manufacture/assemble active components and passive components, and have specific supplier development programmes, in which the aim is to work closely together with their suppliers to obtain the best quality of product and services possible for the subsidiaries' and the suppliers' benefit. Relationships with the suppliers are not so close for affiliates that manufacture/assemble consumer electronics' products. These affiliates transfer less technology and knowledge to their suppliers.

In all cases, the reputation of the subsidiaries is a key to explain the type of relations they maintain with their suppliers. Defective inputs lead to defective products. As big firms maintain their sales on their good reputation, they try to avoid input problems.

Type of firms and product manufactured. Two cases are subcontracting firms: Company 3 of Singaporean origin, and Company 10 of US origin. These firms are under the strict requirements of their customers, and have less flexibility in their purchase sourcing. Company 3, faces specific problems related to having a particular supplier imposed, which forces the subsidiary to find specific ways to help the suppliers to improve their service. Although not a subcontractor, but because of the specificity of its products, Company 8 also faces similar problems:

> We are in a difficult situation: our customers appoint many of our suppliers. Sometimes they will listen to our customers instead of us. But we cut the purchasing orders and we pay them, so it is a balancing act.

The position of the product in the life cycle, the level of standardisation reached and the level of technology included in the product determine to a large extent the type of relations with external suppliers. The level of technology included in the product manufactured by the subsidiary is important, insofar as low-tech products (such as electro-mechanical components) or assembly work (consumer electronics), lead to more local purchase.

Management strategy. Two of the companies interviewed are paying particular attention to suppliers' management staff training (Companies 9 and 11). They are both aiming at reaching a 6σ performance standard, and therefore are training the suppliers' management staff in this management style. Previous methods, such as training on the TQM system, were also provided. Company 8, also a US company, includes management training in its training programmes to suppliers. For these firms, overall performance is closely related to that of their suppliers:

> You have a plan that covers all your business fundamentals: not only your activity against the plan, but also monitoring and effectiveness of your plans (Company 9).

Size of firms. The size and organisation of the firm are very important in the relations with the suppliers: whether there are several departments responsible for dealing with the suppliers, like the Company 6 system, or whether the company is small and every aspect is dealt with within the same department.

Transfers and purchase-related factors

The discussion on contractual relationships has highlighted the increasing importance of regular contracts and the creation of close links between foreign firms and their suppliers in Malaysia. Tables 6.6-6.8 indicate that knowledge transfer occurs mostly towards subcontractors in the case of Companies 2, 4, 5 and 10. However, close partners also benefit from transfer of technology and knowledge in the other companies. Although local purchasing is limited in scale, the companies' analysis has shown that some Malaysian-owned firms benefit from these close links, and therefore receive knowledge transfer from the foreign subsidiaries. Such transfers occur for subcontractors, but also suppliers, essentially supplying plastic, metal and rubber parts, and electrical components or support equipment. The items purchased are not technology-intensive, but the suppliers benefit highly from the production process and management skills of their foreign partners.

Transfers and investment-related factors: government policies

Most of the companies interviewed are not given any supply requirements by the Malaysian government. To benefit from the GSP (General Supply Preferences) Programme, Company 6 is required to acquire at least 50% of total purchase within Malaysia and/or the ASEAN region. Moreover, specific inputs such as transistors all have to be purchased locally. These requirements apply for each product manufactured.

Other types of government actions seem more beneficial to the creation of local linkages between multinational firms and local firms. Company 1 has promoted a local supplier in the plastic injection area as part of the Umbrella Subcontracting Scheme. The supplier was fully Malaysian-owned; it was given all the technical capabilities, materials and training by a Japanese team. This supplier was created to provide the subsidiary with specific plastic injection. The set-up lasted for two and a half years. Currently, some

Japanese competitors have settled in Malaysia, and about 60% of the purchases for plastic injection have been diverted to these companies.

Companies 6, 9 and 11 are taking part in the Vendor'Net Programme. Companies 9 and 11 are present during this yearly event, but have not yet signed any contract with a local vendor through this scheme. Company 6 signed a memorandum of understanding with two local vendors; one vendor is now assembling consumer electronics final products for the firm and the other vendor is assembling PCBs. Those vendors have been entirely developed by the subsidiary and were given machinery, technical support, equipment, engineers' help and training. The local authorities recommend local vendors to TNCs, which then choose potential vendors.

The most efficient programmes in terms of beneficial linkages between TNCs and local suppliers are the various actions such as the Umbrella Marketing Scheme or the Vendor'Net Programme through which the government puts the best local vendors in contact with the TNC subsidiaries. Although a very limited number of local firms are concerned, the result is more beneficial for the local vendor.

Time factor

The time spent in the host country influences transfers for two reasons. Subsidiaries increase their knowledge of the local business environment, and they slowly gain more independence from their parent firms.

Other Comments

Respondents provided additional comments related to buyer-supplier transfers that are worth mentioning in the context of these results. Comments on personal relationships and evaluation systems were given. Finally, additional information was gathered by the researcher, when interviewing companies with no local direct purchasing.

Personal relations

Personal relationships only arise after the suppliers have been selected. However, close relationships between buyer-supplier can be the result of personal contact between managers. Its advantages are well described by Company 5, which puts forward concepts such as trust and 'intuition' in keeping the supplier, and in the long-term evolution of the relationship. Both Companies 5 and 7 mentioned that: In this part of the world, relationship is crucial.

Evaluation system

All companies have a specific evaluation system (follow-up system), to keep constant records of the suppliers' performance and to make sure problems are discovered before becoming too important. By so doing, subsidiaries are in a way managing their local suppliers.

The system of scorecards of Company 8 is no different from the ranking system of Company 2, or even from the traffic light system of Company 7. The good point is to get the supplier always to improve, and to know how to do so. On the other hand, it could also be very heavy on the supplier's management.

The major difference once the evaluation is carried out is whether the subsidiary helps the supplier to improve or not. Whereas Companies 2 and 7 do not provide any support, Company 8 works together with the firm to find a solution to the problem.

The level of support provided may well only depend on the level of technology required in the final product. Companies 2 and 7 are looking for customised items, and do not provide support to their suppliers, whereas Company 8, which is looking at a very high quality and technology standard, is ready to intervene in the supplier's activity to help and improve the results.

The first important step for the subsidiaries is that the suppliers should be aware of their performance, and should know which are their possible weak points.

Other types of linkages (Companies 12 and 13)

Two additional interviews were conducted with firms from a different sample than that predetermined by the researcher. Company 12 shows that in the case of a service activity such as repairs, virtually no long-term relationship exists between the subsidiary and its suppliers. In this case, the only likely positive impact of the subsidiary is the occasional contracts with local firms.

Company 13 shows further the complexity of the topic. In this case, the subsidiary is a large group, with several departments responsible for the purchasing activity. Part of the activity of the indirect materials' department can be found in some of the cases studied in this research in the case of subsidiaries that do not differentiate so clearly between direct and indirect materials purchasing. For this high-tech company, it is noticeable that some secondary products, such as chips' packaging, may contain a great deal of technology and know-how, and therefore may lead to more relations between the supplier and the subsidiary than in the case of metal parts' purchasing for another subsidiary.

CONCLUSION

The continuous transfers from the TNC take place through reports and evaluations of the suppliers. Both the TNC and its suppliers work together in case of problems in the quality/delivery of the input. TNCs resort the most frequently to knowledge transfer when they need to solve a specific problem. These transfers include technical or operational help, or training provided to the suppliers. Technical help varies according to the type of input requested. It occurs more frequently for subcontractors. In terms of cost, European firms consider price reductions as important criteria of the suppliers' performance, but they are less systematic and may occur over a longer period of time than for the Asian firms. Government policies have proved successful when getting TNCs to work closely with a limited number of locally owned suppliers, through projects such as the Umbrella Marketing Scheme or the Vendor'Net Programme. Through these programmes, TNCs have transferred a substantial amount of technology and know-how to their local vendors.

The Japanese are distinctive in their reliance on home country transplants. The relocation of Japanese suppliers to Malaysia or other Southeast Asian countries further benefits TNCs from other parts of the world, as they also rely on the Japanese networks for their own supplies. We can perceive that NIE firms/countries are geographically closer to Malaysia, so use of imported suppliers is not unusual. As these firms move up the technology ladder, they could evolve towards the US/European supply pattern. Results in this chapter indicate a twofold pattern/network strategy between Asian firms, on the one hand, with heavy reliance on either imported goods within the firm, or on relocated home country suppliers, and the Western companies, on the other hand, with closer supplying links in the host economy.

The last section analysed the knowledge transfer carried out by firms in our sample. Product and process technology transfer and transfer through training were identified as the key transfers existing. Some factors influencing buyer-supplier transfers were identified and discussed. This qualitative analysis grasped the complexity of backward linkages between foreign firms and local suppliers. To complement the information gathered through interviews, and to confirm relationships identified, a statistical analysis on a greater number of responses obtained by means of a mail questionnaire is conducted in Chapter 7.

NOTES

1. All these firms, except for two, were located in the Penang FTZ. The other two were located in the existing industrial estates (one in Butterworth, and one in Kuala Lumpur).
2. This indicates the small importance of ownership in explaining transfers

between affiliates and suppliers. The preferred option is the wholly owned subsidiary. A large share of the foreign affiliates were set up through greenfield investment that means fewer previous contacts with local suppliers.

3. Large-scale employment does not always equate to labour-intensity. The size of subsidiaries could be the result of scale economies, government incentives, size of ultimate markets and so on. Labour-intensity for these companies was the key initial attractive factor, but patterns show that higher capital intensity sets in with age of subsidiary. Within the Penang area, a move towards higher capital-intensity is further linked to agglomeration effect. High-tech companies attract other high-tech companies.
4. Giroud (1997) investigates reasons why foreign investors may choose the JV format, although they prefer wholly owned format. JVs are found less export-oriented and less profitable.
5. Internal supply refers to all items purchased from within the firm, rather than through outsourcing.
6. The distinction between direct and indirect supply was made in Chapter 5. Direct supplies include parts and components that require some established knowledge from local suppliers. Indirect supplies would include boxes, pencils or items that would not directly enter into the production process. Supply of indirect material does not lead to buyer-supplier transfers, and for this reason, firms that only purchase these are not included for further analysis in this research.
7. Term defined in Chapter 4. This programme is one of several soft policies adopted by the Malaysian government. It encourages large firms to provide assistance to local vendors.

7. Factors Explaining Existence and Degree of Knowledge Transfer

INTRODUCTION

In the previous chapter, the analysis of 11 interviews resulted in a description of knowledge transfer between foreign subsidiaries in the Malaysian electrical and electronics sector (MEE). In this chapter, the same relationships are under study, but the results presented were gathered by means of a mail questionnaire. The methodology employed and the analytical process were fully described in Chapter 5. The empirical research was conducted in 1996 among a total of over 320 companies in the sector. One hundred and eight questionnaires were returned, and 95 were found usable for further analysis. Out of the 95 firms in the sample, some did not carry out any local direct purchasing, and consequently, they are not included in further analysis of foreign affiliates' relationships with their local suppliers. The descriptive analysis of the sample in Chapter 5 includes firms with no local direct purchase.[1]

This chapter is divided into three major sub-sections. First, a descriptive analysis of the transfer by types, modes and country of origin of the affiliates enables a grasp of the phenomenon under study. This leads to the operationalisation section and the building-up of several transfer indicators that will be used in the statistical analysis stage. We compute six indicators of transfer; three indicators show the existence of transfer and three show the degree of transfer. The three categories of transfer include an 'overall' transfer indicator, and two specific indicators; the first is related to the transfer of product and process technology transfer, and the second represents the transfer occurring through training. The in-depth statistical analysis stage starts with a series of logistic regression analyses. The logistic regression method is adopted because it allows for computation of the odds ratio,[2] thus enabling the researcher to differentiate firms that operate transfers from those with no transfer. We determine all the factors that influence the existence of transfer between foreign affiliates and their local suppliers. The linear regression analysis stage follows the logistic regression analysis. We choose to run linear regression analysis, because this method allows identification of factors that significantly explain the degree of transfer. We then highlight all the factors that explain the degree of transfer between partners.

KNOWLEDGE TRANSFER TO LOCAL SUPPLIERS: DESCRIPTIVE ANALYSIS

Before embarking on any statistical analysis, it is essential to provide a description of the variables to be studied. The descriptive analysis of the different types of transfer between foreign firms and local suppliers appears in this section. We then provide a scale reliability analysis, in order to explain how composite variables of transfers were computed, and to prove the reliability of each variable.

Descriptive Analysis of the Different Types of Transfer

In Chapter 3, two key types of transfers were identified, product and process technology transfer, and transfers through training. These types of transfers[3] are analysed below. Although financial transfers were identified as infrequent and not substantial, these transfers were included in the questionnaire and their analysis is provided.

Financial transfers

Table 7.1, shows that few financial transfers exist between foreign subsidiaries and their local suppliers in the MEE. Slightly more than 10% of the firms contribute to the risk capital of their suppliers and/or pre-finance tools and materials for these. Other types of financial transfer are almost non existent.

Product technology transfer

Product technology transfer is frequently provided to the local suppliers by the foreign firms' subsidiaries. In Table 7.1, 84% of the firms frequently provide the product specification to their suppliers, and slightly over 70% of them provide physical/technical specifications about the products. Such exchange of information is less frequent in the case of the Japanese and NIE firms. Forty-two per cent of newly industrialising economics (NIE) firms and 71% of Japanese firms provide physical/technical specifications, against 100% in the case of the European and US firms.

The specific method of manufacture is normally less likely to be provided, except for the subcontractors. Yet, 75% of the North American firms share these with their suppliers. Such figures are low for the other groups of firms.

Joint design and joint operation settlement are also primarily shared with the subcontractors, or in the case of partnership, between the foreign firm and its suppliers. European and North American firms generate these flows more frequently than the Asian firms, with 63% and 50% respectively, of the former jointly designing the input with their suppliers. This trend is confirmed for the joint settlement of the production operations.

Table 7.1 Transfers between foreign firms' subsidiaries and local suppliers (Percentage of firms with frequent transfer for each item)

	Japan	NIEs	Europe	USA	Other	Total
Financial transfer						
Risk capital contribution (76 firms)	9.5	23.5	14.3	0.0	0.0	11.8
Loans (73 firms)	2.4	0.0	0.0	0.0	0.0	1.4
Pre-finance of tools and materials (76 firms)	11.6	0.0	0.0	37.5	0.0	10.5
Payment before delivery (80 firms)	4.7	15.5	0.0	0.0	0.0	5.0
Product technology transfer						
Product specification (84 firms)	86.7	90.5	87.5	100.0	100.0	84.3
Physical/technical specifications (79 firms)	71.4	42.1	100.0	100.0	100.0	70.9
Method of manufacture (81 firms)	26.7	22.2	0.0	75.0	0.0	27.2
Input joint design (77 firms)	11.6	23.5	62.5	50.0	0.0	23.4
Joint operation settlement (68 firms)	16.2	6.7	42.9	42.9	0.0	19.1
Process technology transfer						
Information on plant establishment (81 firms)	4.7	15.0	12.5	25.0	0.0	9.9
Supply of machinery (81 firms)	20.5	5.0	12.5	42.9	0.0	17.3
Supply of tools (83 firms)	31.8	14.3	12.5	62.5	0.0	27.7
Information on input procurement (83 firms)	27.3	20.0	37.5	62.5	50.0	30.5
Information on supply management (81 firms)	11.4	5.3	25.0	50.0	0.0	14.8
Information on technical management (82 firms)	34.1	15.0	50.0	75.0	0.0	34.1

Note: Figures shown are the percentage of firms resorting to transfer 'frequently'. Firms that provide these flows only 'rarely' or that 'never' provide any do not appear in the table.

Process technology transfer

One-third of the firms on average share information on input procurement and/or on technical management with their suppliers, while one-quarter of them supply tools and materials to their suppliers. Process technology transfer occurs more frequently within the European and US firms for almost all types of process technology transfer in Table 7.1. The predominance of the US firms can be explained by their concentration in high-tech sub-sectors, where closer cooperation is often found.

Transfer through training and types of training

Table 7.2 shows that the major type of training provided by subsidiaries to their suppliers is concerned with the technology used in the manufacturing process of the inputs purchased. Training of management staff is the least frequent type of training, and is mostly delivered by North American firms.

Table 7.2 Transfers through training between foreign firms' subsidiaries and local suppliers (percentage)

	Japan	NIEs	Europe	USA	Other	Total
Types of training provided by subsidiaries to suppliers						
Management training (83 firms)	6.7	0.0	12.5	42.9	0.0	8.4
Operation staff training (83 firms)	15.6	0.0	25.0	42.9	0.0	14.5
Training on technology used (85 firms)	26.7	18.2	37.5	75.0	0.0	29.4
Modes of training provided						
Visits to subsidiary (84 firms)	24.4	31.8	87.5	85.7	50.0	38.1
Off-the-job training: seminars at subsidiary (78 firms)	11.6	5.6	0.0	50.0	0.0	12.8
Off-the-job training: seminars at suppliers (79 firms)	4.7	0.0	0.0	37.5	0.0	6.3
Off-the-job training: seminars in institutions (78 firms)	4.8	5.3	0.0	12.5	0.0	5.1
On-the-job training: at subsidiary (78 firms)	27.9	15.0	28.6	85.7	0.0	29.5
On-the-job training: at suppliers (74 firms)	29.3	5.3	14.3	100.0	0.0	25.7

Note: Figures shown are the percentage of firms resorting to information flows 'frequently'. Firms that provide these flows only 'rarely' or that 'never' provide any do not appear in the table.

Transfer and means of training

More than one third of the subsidiaries receive visits from their suppliers within the subsidiary. European and US firms receive visits from their suppliers more than the Asian firms. Off-the-job training is the least common means of training, and is only substantially provided by the US firms. In the latter case, half of the firms organise seminars for the suppliers within the subsidiary itself. Table 7.2 also shows that more than one quarter of the firms provide on-the-job training for the suppliers' staff. Such a method is more often carried out within the subsidiary itself than at the supplier. While European firms emphasise on-the-job training within the subsidiary (28.6% of firms), the Japanese firms refer mostly to on-the-job training at the supplier (29.3% of firms). All the North American firms provide some kind of on-the-job training at the supplier, and 85% of them also provide this training within the subsidiary. On the other hand, NIE firms hardly ever carry out any training.

Scale Reliability Analysis: Building of Knowledge Transfer Index

Each question related to a specific type of transfer was described earlier, and was used for the computation of overall transfer indicators. In order to identify individual questions to be grouped together, reliability tests are run. With this information, the operationalisation of various transfer indicators can be performed. After describing reliability tests, we then discuss the operationalisation procedure and finally provide a descriptive analysis of each transfer indicator.

Reliability tests

Reliability is a statistical measure of how reproducible the survey instrument's data are. Internal consistency reliability is used to assess survey instruments and scales. It is applied not to single items but to groups of items that are thought to measure different aspects of the same concept. Internal consistency is an indicator of how well the different items measure the same issue. This is important, because a group of items that purports to measure one variable should indeed be clearly focused on that variable. Internal consistency is measured by calculating a statistic known as Cronbach's coefficient alpha.

Coefficient alpha measures internal consistency reliability among a group of items combined to form a single scale (Litwin, 1995: 24). It is a statistic that reflects the homogeneity of the scale. That is, it is a reflection of how well the different items complement each other in their measurement of different aspects of the same variable or quality.

We evaluate the reliability of the newly coded questions related to transfer below (1 frequently, and 0 rarely/never). Cronbach's alpha values are used; they range between 0 and 1.0, with higher values indicating higher reliability among the indicators. A level of 0.7 or more is generally accepted as representing good reliability.

Operationalisation

The list of questions related to knowledge transfer was analysed earlier. A descriptive assessment and evaluation of the various types or means of transfer between the foreign subsidiaries and local suppliers showed that little transfer occurs frequently. Firms that did not purchase direct components from local suppliers were not included in the statistical analysis. The types and modes of transfer are listed in Tables 7.1 and 7.2.[4] Foreign subsidiaries indicated whether each transfer took place frequently or not. Statistics in these tables show the share of firms that 'frequently' transfer each type and mode. A full analysis of the missing values and their treatment was conducted. Variables related to financial transfers and one variable related to the product technology transfer is not analysed further, because of a high share of missing values. Similarly, three companies did not answer satisfactorily, and are not included in the statistical analysis of the transfer phenomenon between foreign subsidiaries and their local suppliers.

Once the initial missing value treatment was carried out, transfer variables were tested against each other, applying the Cronbach alpha reliability test. The initial test included all the transfer variables,[5] except for those removed in the initial missing value treatment process.

Overall indicator of knowledge transfer. The initial Cronbach alpha reliability test highlighted the fact that some variables would not bring additional information if used for further statistical analysis. Such variables[6] were removed from the test, because they were not bringing additional information on transfer. The final combination of variables has an alpha of .8873, and a standardised item alpha of .8864, thus showing high reliability. It includes 12 types and modes of transfer[7] between the foreign subsidiary and its local direct suppliers.[8]

The reliability of our set of variables is high, with a high alpha. A replacement method is applied for the computation of the overall indicator of knowledge transfer indicator. The share of companies with frequent transfer was computed, including all 12 variables, excluding potential missing values. The indicator is a computation of the share of subsidiaries that frequently transfer specific technology, including product and process technology transfers, and types and means of training. This numerical indicator is referred to as 'Transfer'. The overall indicator 'Transfer' is used for statistical analysis using logistic regression models and multiple regression models. It is subdivided into two indicators: one related to process and product technology transfer, and one related to transfer through training.

Indicator of process and product technology transfer related to purchasing. The Cronbach alpha reliability test is run only with variables indicative of process and product technology. One variable is removed from the reliability test, because it does not bring additional information on process and product technology transfer.[9] The final combination includes five variables[10] and has an alpha of .7975, and a standardised item alpha of .8022. This alpha coefficient is reliable, but not as high as that of the overall indicator. Transfer of process and product technology occurs less frequently than training, thus the analysis of such transfer is more delicate and requires careful attention. Results shown by this process and product technology transfer indicator are less reliable.

A replacement method is applied for the computation of the indicator. The share of companies with frequent transfer was computed, including all five variables, excluding potential missing values. The indicator is a computation of the share of subsidiaries that frequently transfer product and process technology. This numerical indicator is referred to as 'Transtec' and is used for statistical analysis using logistic regression models and multiple regression models.

Indicator of transfer through training. The Cronbach alpha reliability test is run only with variables indicative of types and means of training. The final combination includes six variables[11] and has an alpha of .8121, and a standardised item alpha of .8123. This alpha coefficient shows a high reliability.

A replacement method is applied for the computation of the indicator. The share of companies with frequent transfer was computed, including all five variables, excluding potential missing values. The indicator is a computation of the share of subsidiaries that frequently transfer product and process technology. This numerical indicator is referred to as 'Transtra' and is used for statistical analysis using logistic regression models and multiple regression models.

Indicators used for logistic regressions. The indicators computed above were then transformed into dummy variables for the purpose of the logistic regression. Firms with no transfer were input as '0', and firms with some transfer were input as '1'. The rationale for the use of a dummy is that we wish to distinguish two types of firms: on the one hand, firms classified as '1' vary linearly in terms of transfer; on the other hand, firms classified as '0' transfer no technology or knowledge. Each dummy was later entered as the dependent variable in the binomial logistic regression analysis. The dummy variable is referred to as 'transit' for the overall transfer indicator, as 'tecdummy' for the process and product technology transfer indicator and as 'tradummy' for the training indicator.

Descriptive analysis of indicators

The descriptive analysis of the transfer indicators in Table 7.3 confirms the findings of the previous sections. Firms from the NIEs transfer the least technology. For instance their share of overall transfer is 11%. Firms from the NIEs show a higher share of process and product technology transfer, with 16% against 12% for transfer through training. A similar trend is noticeable for Japanese firms. On the other hand, firms in the category 'other firms' show a higher share of training transfer, with nearly half of such transfer occurring frequently, while the share is equal to 40% in the case of process and product technology transfer. Overall, only a quarter of the different types of transfer occur frequently.

Table 7.3 Descriptive analysis of transfer indicators

	Japan	NIEs	Other countries	Total
Frequency	45	20	18	83
Transfer * (%)				
Mean	23.88	11.41	43.30	25.01
Std Deviation	28.22	16.40	30.71	28.36
Minimum	0	0	0	0
Maximum	91.7	66.7	100	100
Firms with no transfer (Frequency/share)	17 (37.78%)	10 (50%)	1 (5.56%)	28 (33.73%)
Transtec ** (%)				
Mean	28.26	16.00	40.00	27.85
Std Deviation	36.66	26.44	34.30	34.56
Minimum	0	0	0	0
Maximum	100	100	100	100
Firms with no transfer (Frequency/share)	22 (48.89%)	12 (60%)	5 (27.78%)	39 (46.99%)
Transtra *** (%)				
Mean	23.15	11.66	48.89	25.97
Std Deviation	30.20	21.00	31.47	31.09
Minimum	0	0	0	0
Maximum	100	83.3	100	100
Firms with no training (Frequency/share)	23 (51.11%)	13 (65%)	2 (11.11%)	38 (45.78%)

Notes:

* Share of frequently transferred factors in total factors for each case.
** Share of frequently transferred factors in total factors related to process and product technology.
*** Share of frequently transferred factors in total factors related to types and modes of training. Shares calculated excluding missing values, which explains why shares vary according to indicator being calculated.

A descriptive analysis of the logistic regression indicators in Table 7.4 shows that overall, 66% of the firms do not transfer any technology to their local suppliers. Again, firms from the NIEs show a lower tendency to transfer some technology, with just half of the firms transferring some (the intensity of transfer is not included in these indicators). Firms from the other category have the highest share of transfer, with 94% of the firms transferring technology. Such trends are similar in the case of the 'tecdummy' and 'tradummy' indicators. Transfer from firms from the NIEs and Japan primarily takes place for process and product technology, rather than through training. Firms from other origin transfer technology primarily through training. Eighty-nine per cent of firms from other origin provide 'some training', while 72% provide process and product technology transfer.

Table 7.4 Descriptive analysis of logistic regression indicators

	Country/region of origin							
	Japan		NIEs		Other countries		Total	
	Freq.	Col %	Freq.	Col %	Freq.	Col %	Freq.	Col %
Transit: dummy variable of transfer of technology (transfer)								
None	17	37.8	10	50.0	1	5.6	28	33.7
Some	28	62.2	10	50.0	17	94.4	55	66.3
Tecdummy: dummy variable of transfer of product and process technology (transtec)								
None	22	48.9	12	60.0	5	27.8	39	47.0
Some	23	51.1	8	40.0	13	72.2	44	53.0
Tradummy: dummy variable of transfer through training (transtra)								
None	23	51.1	13	65.0	2	11.1	38	45.8
Some	22	48.9	7	35.0	16	88.9	45	54.2

Note: 'None' refers to firms with no transfer at all; 'Some' refers to firms that carry out some transfer, degree of transfer disregarded.

The reliability and descriptive analysis of the above indicators being achieved, further statistical analysis of the factors impacting on these indicators can be carried out.

DETERMINANT FACTORS OF EXISTENCE OF KNOWLEDGE TRANSFER

This section contains a series of logistic regression analyses, using the three indicators 'transit', 'tecdummy' and 'tradummy' as the dependent variables. This statistical analysis will help in the determination of the key factors explaining the existence of transfers between foreign affiliates and their local suppliers. First, we explain methods and measures used, and then we carry out the statistical analysis to determine factors explaining transfers between foreign subsidiaries and local suppliers. Logistic regression models are determined for all firms first, and then for Japanese subsidiaries only. This section ends with an overall conclusion for the models developed, and a discussion on each of the factors explaining the existence of knowledge transfer.

Methods and Measures

To carry out logistic regressions, we need to define methods adopted and the model variables.

Methods

The regression coefficients estimate the impact of the independent variables on the probability that the subsidiaries will transfer technology. The procedure that calculates the logistic coefficient compares the probability of firms transferring technology with the probability that firms will not. This ratio can be expressed as

$$\frac{P(transfer)}{P(notransfer)} = \ell^{\beta_0 + \beta_1 x_1 + \beta_2 x_2 + \beta_3 x_3 + \ldots \beta_n x_n}$$

where 'β_0' is the intercept, $\beta_1 \ldots \beta_n$ are the regression coefficients and $x_1 \ldots x_n$ the independent variables.

Two selection methods were used. First, the 'enter' method, to include all the model variables and evaluate their role in the transfer process. Then, the 'backward conditional' elimination method that removes variables that do not significantly improve the probability of the observed results. In both methods, all variables were entered in a block. In the reduced model, only the best predictor variables were introduced. The rationale behind this choice is to allow the sample of observations to meet the highest ratio possible for discriminant analysis (the minimum to validate the results is a 5:1 ratio), which means a stronger reliability of the results. Thus, the smaller the number of independent variables, the better for the validation of the results. We have a good sample size of 83 firms, which is four times the minimum of 20 observations required for such calculations (Hair et al., 1998: 282). Some of the independent variables are nominal variables and were treated through the deviation method for the analysis.[12] To evaluate the impact of each independent variable, we examine the sign and magnitude of the standardised weight assigned to each variable. Independent variables with larger weights contribute more to the discriminating power of the function. The two main problems of this method are that the weight may look insignificant if multicollinearity exists (as mentioned above, this is not the case among our variables), and discriminant weights are subject to considerable instability.

The Wald statistic provides the statistical significance for each estimated coefficient. It can be interpreted as t or F values. The partial correlation of each predictor variable with the dependent variable is also indicated. A positive sign for the coefficient means that a variable increases the probability of transfer. A negative sign indicates the reverse.

Model variables

Because no similar studies were done before, we base our assumptions on the factors likely to affect the existence of knowledge transfer on the literature related to the evaluation of such transfers. In Chapter 2, a theoretical framework was defined, showing the influence of several factors on knowledge transfer by foreign firms to local suppliers. There is no generally accepted list of variables explaining the existence or not of transfer. Prior research has reported different variables as contributing factors to the degree of transfer (see Chapter 3 for a comparison of such variables). Government policies and investment environment were analysed independently in Chapter 4 and further discussed in Chapter 6. The independent variables included here were defined from research questions in Chapter 5 as a result of discussion in earlier chapters. Preliminary independent variables to be tested include the length of time the subsidiary has been established in Malaysia, the type of product manufactured, the market orientation of the subsidiary, the size of the company, the share of intra-firm purchase, the type of contractual relationship and the purchasing strategy of the firm. Several models are computed. The higher the level of classification rate the better, but then the effect of independent variables on the dependent must be found significant. The model is shown below:

Transfer/no transfer = *f* (time, product, export, size, intra-firm, subcontract, purchasing strategy, origin)

Time is a metric variable and refers to the time of establishment in the country, and the number of years thus spent in the host country until the survey was conducted. Product is a nonmetric variable, composed of several categories for each product likely to be manufactured/assembled by the foreign firm. Export is a metric variable assessing the export orientation of the firm. It refers to the share of total output exported for each company. Size is a metric variable showing the number of employees in the subsidiary at the time when the data were collected. Intra-firm supplies is a metric variable; it is the share of total input purchased by the firm that come from other affiliates or group belonging to the same firm. Subcontract is a metric variable. We measure the share of subcontracting relationships in all the contractual agreements signed by the foreign firm in the host country. Purchasing strategy is a nonmetric variable; it is the purchasing strategy followed by the firm. We use a dummy variable to evaluate whether the foreign firm favours one single supplier per item purchased or several suppliers per item. Finally, origin is a nonmetric variable. It is composed of several categories for each country/region where the parent company is located.

To test the model, a logistic regression analysis is run with all variables entered in one step. Logistic regression attempts to build a regression model that best describes which variables explain the existence of transfer. The parameters of the model are estimated using the maximum likelihood method. The significance level of each variable is reported to determine which variables do and do not significantly distinguish between the 'transfer' and the 'no transfer'.

Factors Explaining Transfers between Foreign Subsidiaries and their Local Suppliers

The indicators being proved reliable and computed, further statistical analysis can proceed. A series of logistic regression models are shown: first, for the overall transfer indicator, 'transit' then, for the process and product technology transfer indicator, 'transtec' and finally, for the transfer through training indicator, 'transtra'.

Factors explaining overall transfers

First, the reliability and validity of variables for each group of companies; those with transfer, and those without transfers are introduced in this section. In order to identify the factors that significantly influence overall transfer from foreign firms to local suppliers, logistic regression models are then computed.

Reliability and validity. Table 7.5 containing the descriptive results includes a comparison between cases that do not transfer technology and those that do. Whereas such a comparison should be carried out on an identical sample size, it is rare to obtain such perfect identical sizes[13] from a survey. In this respect, the table loses some of its explanatory value. This table shows that companies that transfer technology have spent more time on average in Malaysia. They have a higher average export orientation share (83.9% against 65.2% for companies that do not transfer technology). They are bigger in size (but have a much higher standard deviation, which indicates disparities in the sample). They have a slightly higher share of internal supplies. They have a substantially higher share of subcontracting relationships; they tend to manufacture more technology-intensive products (namely, a higher share of active and passive components, as opposed to electro-mechanical components). Finally, in category 8(c) other, nearly all the companies transfer technology, against 60% of the Japanese firms and half of the firms originating from the NIEs.

Table 7.5 Descriptive results of overall 'transit' indicator

	No transfer (28 firms)			Transfer (55 firms)		
	Mean		S.D.	Mean		S.D.
1. Time spent in Malaysia (years)	8		4	10		6
2. Export orientation	65.2%		35.1	83.9%		21.2
3. Size (number of employees)	577		559	1,470		1,589
4. Share of internal supplies	32.7%		29.2	38.9%		27.8
5. Share of subcontracting relations	10.4%		18.6	29.2%		34.9
	Freq.	*Share %*		*Freq.*	*Share %*	
		Row	*Col.*		*Row*	*Col.*
6. Type of product (frequency)	28	33.7		55	66.3	
(a) Active components	0	0	0	9	100	16.4
(b) Passive components	7	33.3	25	14	66.7	25.5
(c) Electro-mechanical components	11	52.4	39.3	10	47.6	18.2
(d) Consumer electronics	7	43.7	25	9	56.2	16.4
(e) Other products	3	18.7	10.7	13	81.2	23.6
7. Purchasing strategy (frequency)	6	21.4		22	78.6	
8. Country of origin (frequency)	28	33.7		55	66.3	
(a) Japan	17	37.8	60.7	28	62.2	50.9
(b) NIEs	10	50	35.7	10	50	18.2
(c) Others	1	5.5	1.8	17	94.4	30.9

Correlation analysis was run to assess the degree of multicollinearity among the independent variables. While some of the correlations are significant, all are below +/-.40, and most are below .20. This confirms that multicollinearity between independent variables will not occur, thus it will not affect the coefficients and will not interfere with the interpretation of the results.

Logistic regression models for the overall indicator 'transit'. The reliability of variables being shown, logistic regression models can now be computed. In the first model, all independent variables are entered into the model. A second model is computed, keeping only the significant variables.

Full Model

As mentioned above, we base our assumptions on the factors likely to affect the existence of knowledge transfer on the literature related to the evaluation of such transfers. The independent variables are described above. We first enter all the previously defined variables into the logistic regression. The model converged after six iterations. The model has a high and highly significant model chi-square, a good overall model fit and a very good sensitivity (its ability to predict firms that transfer technology, 90.20% classified correctly). With an overall correct classification of 84.21%, the model is highly significant.[14]

Because we chose first to enter all the variables into the regression, the Wald coefficients indicate that not all the independent variables are important explanatory factors for the existence of knowledge transfer.

Table 7.6 Parameter estimates for 'transit' logistic regression model

Variable name	ß	Wald	*R*	Significance
1. Time spent in Malaysia	.0215	.0556	.0000	.8135
2. Share of export market	.0210	1.6727	.0000	.1959
3. Size: number of employees now	.0014	5.0197	.1693	.0251*
4. Share of internal inputs in total inputs	.0078	.2602	.0000	.6100
5. Share of subcontracting agreements in supply	.0105	.5026	.0000	.4784
6. Product manufactured		7.2427	.0000	.1236
(a) Active component	6.6449	.7496	.0000	.3866
(b) Passive component	1.5614	2.0312	.0172	.1541
(c) Electromechanical component	.5851	.2399	.0000	.6243
(d) Consumer electronics	-2.7877	4.5728	-.1563	.0325*
7. Single supplier	-2.4941	5.3082	-.1772	.0212*
8. Country/region of origin of subsidiary		5.1253	.1033	.0771+
(a) Japanese origin	-.9462	.8581	.0000	.3543
(b) Parent from NIEs	-2.2006	2.0199	-.0138	.1552

Notes:

Constant insignificant, removed from model.
Variables significant at *** for p<0.001, ** for p<0.01, * for p<0.05 and + for p<0.1
Categories below country/region are compared to the other products category.
Purchasing strategy indicates companies that favour single supplier per item purchased, or various suppliers per item.
Share of export is share of final products manufactured by subsidiaries that is exported.
Number of employees in subsidiary represents subsidiary size.

The model is more accurate in predicting cases where 'some transfer' takes place than otherwise; this is expected because of the larger size of the 'some transfer' group of subsidiaries. The model correctly predicts 90.20% of the transfer phenomenon and 72% of the cases where no transfer occurs. The appropriate tests for significance of individual variables in logistic regression are based on the parameter estimates. Table 7.6 contains the estimated beta coefficients (ß) for the independent variables, and the significance of the Wald statistics appears in the last column of the table. Three sub-categories are significant at the .05 level (size, consumer electronics for the type of product and purchasing strategy/single supplier), and one at the .1 level (country/region of origin). Note that the share of export market and types of product manufactured are close to the .1 significance level.

Reduced Model

The 'backward conditional' elimination method was used to determine the reduced model. It removes variables that do not significantly improve the probability of the observed results. All variables are first entered in a block, and then a model with the selected significant variables is run. In the reduced model presented below, only the best predictor variables are introduced in the model. Now, the sample of 83 observations, when split into analysis and holdout (validation) samples, not only meets the suggested 5:1 ratio for discriminant analysis, but it does so by providing a 16:1 ratio of observations to independent variables. Also, as suggested, the coefficients are volatile, and tend to vary substantially. A reduced model allows the identification of independent variables that may be reliable, but did not appear as such through the use of the 'enter' model.

The model converged after seven iterations. The model has a high and highly significant model chi-square, a good overall model fit and a very good sensitivity (its ability to predict firms that transfer technology, 90.38% classified correctly). With an overall correct classification of 85.90%, the model is highly significant.[15] Because we focus on the reduced model, all the independent variables included are significant, and explain why transfer may or may not occur between foreign subsidiaries and their local suppliers. A parallel analysis of the Wald coefficient indicates that all the chosen independent variables are important explanatory factors in the existence of technology transfer. The reduced model has an increased reliability, and the significance of each independent variable is improved.

The reduced model is similar to the previous model but is more accurate at predicting cases when 'some transfer' takes place than otherwise; this is expected because of the larger size of the 'some transfer' group of subsidiaries. The model correctly predicts 90.38% of the transfer phenomenon and 76.92% of the cases when no transfer occurs. The appropriate tests for significance of individual variables in logistic regression are based on the parameter estimates. Table 7.7 contains the estimated beta coefficients (ß) for the independent variables, and the significance of the Wald statistics appears in the last column of the table. One sub-category is significant at the .01 level (consumer electronics within the type of product category). Three categories are significant at the .05 level (size of the subsidiary, purchasing strategy and country/region of origin), and two at the .1 level (types of product manufactured, and share of export market). In the reduced model, significance of the share of export market and types of product manufactured is confirmed.

Table 7.7 Parameter estimates for reduced logistic regression model

Variable name	ß	Wald	*R*	Sig.
2. Share of export market	.0222	3.6726	.1244	.0553+
3. Size: number of employees now	.0015	6.5561	.2053	.0105*
6. Product manufactured		8.5530	.0715	.0733+
(a) Active component	4.9656	1.1932	.0000	.2747
(b) Passive component	1.1564	1.3139	.0000	.2517
(c) Electro-mechanical component	.9719	1.0774	.0000	.2993
(d) Consumer electronics	-3.1799	7.0073	-.2152	.0081**
7 Single supplier	-2.2985	5.6478	-.1837	.0175*
8. Country/region of origin of subsidiary		6.2721	.1450	.0435*
(a) Japanese origin	-.8002	.7603	.0000	.3832
(b) Parent from NIEs	-2.4538	3.5752	-.1207	.0586+

Notes:

Constant insignificant, removed from model.
Variables significant at *** for p<0.001, ** for p<0.01, * for p<0.05, and + for p<0.1.
Categories below country/region are compared to other products category.
Purchasing strategy indicates companies that favour single supplier per item purchased, or various suppliers per item
Share of export is share of final products manufactured by subsidiaries that is exported.
Number of employees in the subsidiary represents subsidiary size.

Direction of relationship among variables. The size of each sub-sample is unequal, with 55 firms with 'some transfer', and 28 with 'no transfer'. Thus, results must be taken with caution. The descriptive statistics showed some substantial differences between the two groups in terms of export orientation, size, subcontracting relationship, technology-intensity and country of origin. The full logistic model also confirms that the time spent in Malaysia and the share of internal inputs in total inputs are not important explanatory factors of the existence of transfer between foreign subsidiaries and their local suppliers. Although the descriptive analysis seems to indicate a substantial difference in the share of subcontracting between the 'some transfer' and the 'no transfer' groups, the factor is not significant in the logistic regression. All other factors were kept within a reduced logistic regression model.

The most significant independent variables in the explanation of the overall transfer are the size of subsidiary, the purchasing strategy and the country/region of origin. The positive sign indicates that the larger the subsidiary, in terms of the number of employees, the more likely it is to transfer some technology to its local suppliers. The negative sign for the purchasing strategy indicates that firms that favour a single supplier per input are less likely to transfer technology to their suppliers. Finally, within the Country/region of origin of the subsidiaries, Japanese is not a significant explanatory factor, whereas the negative sign for firms from the NIEs indicates that these firms tend to transfer less technology, and finally, 'other' firms tend to transfer technology.

Less significant factors in the explanation of 'some transfer' include the export orientation of the firm, and the type of product manufactured. A positive sign indicates that the higher the share of export in the total market share of the subsidiary, the more likely it is to transfer technology to its local suppliers. Within the types of product manufactured, the sub-category consumer electronics is significant at the .01 level, with a negative sign. The companies that manufacture consumer electronics products are less likely to transfer technology than other subsidiaries.

Logistic regression model for 'tecdummy': product and process technology

Before carrying out the logistic analysis of 'tecdummy', we provide a descriptive analysis of the indicator.

Descriptive analysis. In Table 7.8, the sample size for firms that make 'no transfer' and those that make 'some transfer' is nearly identical, which avoids bias in the reliability of the results. The differences in independent variables between the two groups seem substantial for a few independent variables. In the case of the size of subsidiary, firms that operate 'some transfer' have an average of 1,315 employees against 1,023 for those that do not. The category share of subcontracting relations shows firms that operate 'some transfer' have an average of 30.4% of subcontracting relationship in total supplying relations, against 14.1% for those that do 'no transfer'. Within the types of products manufactured, the sub-categories active components, electro-mechanical components, consumer electronics, and other products[16] all show significant differences in means according to the two groups of subsidiaries. The independent variable purchasing strategy also shows contrast, with 64% of the firms operating with single suppliers having 'some transfer'. Finally, within the variable country/region of origin, only the other countries' sub-category shows substantial differences according to the two groups.

Logistic regression models. As in the previous section, in the first model, all independent variables are entered into the model. A second model is computed, keeping only the significant variables.

Table 7.8 Descriptive results of 'tecdummy' indicator

	No transfer (39 firms)			Transfer (44 firms)		
	Mean		S.D.	Mean		S.D.
1. Time spent in Malaysia (years)	9		5	10		7
2 Export orientation	70.0		32.3	84.3		21.7
3. Size (number of employees)	1,023		1,610	1,315		1,202
4. Share of internal supplies	32.5		29.2	40.7		27.1
5. Share of subcontracting relations	14.1		23.2	30.4		35.8
	Freq.	*Share %*		*Freq.*	*Share %*	
		Row	*Col.*		*Row*	*Col.*
6. Type of product (frequency)	39	47		44	53	
(a) Active components				9	100	20.5
(b) Passive components	12	57.1	30.8	9	42.9	20.5
(c) Electro-mechanical components	13	61.9	33.3	8	38.1	18.2
(d) Consumer electronics	11	68.8	28.2	5	31.3	11.4
(e) Other products	3	18.8	7.7	13	81.3	29.5
7. Purchasing strategy (frequency)	10	35.7		18	64.3	
8. Country of origin (frequency)	39	47		44	53	
(a) Japan	22	48.9	56.4	23	51.1	52.3
(b) NIEs	12	60.0	30.8	8	40.0	18.2
(c) Others	5	27.8	12.8	13	72.2	29.5

Full model

The full model operates for 'tecdummy' as it did for 'transit'. We first enter all the previously defined variables into the logistic regression. The model converged after six iterations. The model has a high and significant model chi-square, a good overall model fit and a very good sensitivity (its ability to predict firms that transfer technology, 71.43% classified correctly). With an overall correct classification of 75%, the model is highly significant.[17]

Because we chose to first enter all the variables into the regression, the Wald coefficients indicate that not all the independent variables are important explanatory factors for the existence of technology transfer.

The model presented is more accurate at predicting cases when 'no transfer' takes place than otherwise; this is expected because of the larger size of the 'no transfer' group of subsidiaries. The descriptive analysis of our indicators showed that transfer through training occurs much more often than transfer of process and product technology. This explains why 'no transfer' has a better prediction in the 'tecdummy' logistic regression. The model correctly predicts 79.41% of the 'no transfer' and 71.43% of the 'some transfer' group. The appropriate tests for significance of individual variables in logistic regression are based on the parameter estimates. Table 7.9 contains the estimated beta coefficients (ß) for the independent variables, and the significance of the Wald statistics appears in the last column of the table. One category is significant at the .05 level (types of product manufactured). Each sub-category within the 'types of product' variable is significant, either at the .5 level or at the .1 level. Note that the 'time' spent in Malaysia and the 'purchasing' strategy are close to the .1 significance level.

Table 7.9 Parameter estimates for logistic regression model for 'tecdummy'

Variable name	ß	Wald	*R*	Sig.
1. Time spent in Malaysia	.1122	2.2547	.0492	.1332
2. Share of export market	.0134	1.0460	.0000	.3064
3. Size: number of employees now	.0003	.6827	.0000	.4087
4. Share of internal inputs in total inputs	.0049	.1191	.0000	.7300
5. Share of subcontracting agreements in supply	.0052	.2105	.0000	.6464
6. Product manufactured		12.3304	.2027	.0151*
(a) Active component	9.3856	3.0064	.0977	.0829+
(b) Passive component	-2.7025	3.4514	-.1174	.0632+
(c) Electro-mechanical component	-2.9205	4.7241	-.1608	.0297*
(d) Consumer electronics	-3.2515	4.3588	-.1496	.0368*
7. Single supplier	-.5644	2.5101	-.0696	.1131
8. Country/region of origin of subsidiary		.1398	.0000	.9325
(a) Japanese origin	.1539	.1081	.0000	.7424
(b) Parent from NIEs	-.1961	.0879	.0000	.7668

Notes:

Constant insignificant, and removed from model.
Variables significant at *** for p<0.001, ** for p<0.01, * for p<0.05 and + for p<0.1.
Categories below country/region are compared to the 'other products' category.
Purchasing strategy indicates companies that favour single supplier per item purchased, or various suppliers per item.
Share of export is share of final products manufactured by subsidiaries that is exported.
Number of employees in subsidiary represents subsidiary size.

<u>Reduced model</u>

The 'backward conditional' elimination method is used to determine the reduced model. In the reduced model presented below, only the best predictor variables are introduced.

The model converged after six iterations. The model has a high and significant model chi-square, a good overall model fit and a good sensitivity (its ability to predict firms that transfer technology, 70.45% classified correctly). With an overall correct classification of 74.07%, the model is highly significant.[18]

Table 7.10 Parameter estimates for reduced logistic regression model for 'tecdummy'

Variable name	ß	Wald	*R*	Sig.
1. Time spent in Malaysia	.0730	2.3507	.0559	.1252
5. Share of subcontracting agreements in supply	.0160	2.9144	.0902	.0878+
6. Product manufactured		12.6322	.2031	.0132*
(a) Active component	3.8782	4.0320	.1345	.0446*
(b) Passive component	-1.2297	3.3394	-.1092	.0676+
(c) Electro-mechanical component	-1.4687	5.2479	-.1701	.0220*
(d) Consumer electronics	-1.7834	6.6968	-.2045	.0097**

Notes:

Constant insignificant, removed from model.
Variables significant at *** for p<0.001, ** for p<0.01, * for p<0.05 and + for p<0.1.
Categories below country/region are compared to 'other products' category.
Purchasing strategy indicates companies that favour single supplier per item purchased, or various suppliers per item.
Share of export is share of final products manufactured by subsidiaries that is exported.
Number of employees in subsidiary represents subsidiary size.

The reduced model is similar to the full model but is more accurate at predicting cases when 'no transfer' takes place than otherwise. This is expected because of the larger size of the 'no transfer' group of subsidiaries. The model correctly predicts 78.38% of the 'no transfer' phenomenon, and 70.45% of the cases when transfer occurs. The appropriate tests for significance of individual variables in logistic regression are based on the parameter estimates. Table 7.10 contains the estimated beta coefficients (ß) for the independent variables, and the significance of the Wald statistics appears in the last column of the table. In the reduced model, the type of product manufactured is still significant in the explanation of the existence of process and product technology transfer. Importantly, the significance of purchasing strategy did not prove significant in a reduced model, although its significance was close to the .1 level in the full model. Time spent in Malaysia has a significance level of .1252, and is close to the .1 level. Most importantly, the use of the reduced model has highlighted the importance of share of subcontracting relationships in total supply relations as a significant variable. With a significance level of .0878, the share of subcontracting becomes one of two explanatory variables in the existence of process and product technology, just after type of products manufactured, with a significance level of .0132. The sub-category consumer electronics has a definite negative impact on the 'tecdummy' indicator; it is significant at the 0.01 level.

Direction of relationship among variables. Thirty-nine subsidiaries make 'some transfer' of process and product technology, and 44 do not. The descriptive statistics, in Table 7.8, showed some substantial differences between the two groups in terms of size, subcontracting relationship, type of product manufactured and purchasing strategy. The full logistic model confirms/refutes some of these findings. For instance, in the full logistic model, the share of subcontracting appears as an insignificant independent variable in the explanation of process and product technology transfer. However, the reduced model shows a slight importance of this factor at the .01 level. In the full model, the size of subsidiary, and the purchasing strategy are insignificant variables.

The most significant independent variable in the explanation of the 'tecdummy' transfer is the type of product manufactured. Within this category, each sub-category is significant. As in the case of 'transit', the sub-category consumer electronics is the most significant sub-category; showing a negative sign it indicates that subsidiaries manufacturing consumer electronics are less likely to operate 'some transfer'. Similarly, subsidiaries that manufacture electro-mechanical components are less likely to operate 'some transfer', and the logistic regression coefficient shows a negative sign. On the other hand, subsidiaries that manufacture active components are more likely to operate some transfer of process and product technology.

In the reduced model, the only other independent variable to be considered in explaining whether subsidiaries are likely to transfer some process and product technology is the share of subcontracting relationships in total supply relationships. The positive sign indicates that the higher the share of subcontracting relationship, the more likely the subsidiary is to transfer process and product technology to their local direct suppliers.

Logistic regression model for 'tradummy': product and process technology

This section begins with the descriptive analysis of 'tradummy', before logistic regression models are shown.

Descriptive analysis. Table 7.11 shows the sample size for firms that carry out 'no training' (38 firms) and those that operate transfer through 'some training' (45 firms) is nearly identical, which avoids bias in the reliability of the results. The difference in independent variables between the two groups is substantial for most independent variables. In the case of the size of subsidiary, firms with 'some training' have an average of 1,543 employees against 736 for those that do not. The category share of subcontracting relations shows firms that operate 'some training' have an average of 30.6% of subcontracting relationship in total supplying relations, against 13.8% for those that do 'no training'. Within the types of products manufactured, the sub-categories active components, consumer electronics and other products all show significant differences in means according to the two groups of subsidiaries. The independent variable purchasing strategy also shows contrast, with 67.9% of the firms operating with single suppliers having 'some transfer'. Finally, within the variable country/region of origin, subsidiaries originating from the NIEs and from the other countries' sub-categories show substantial differences according to the two groups. The correlation matrix showed low correlation levels among variables.

Table 7.11 Descriptive results of 'tradummy' indicator

	No training (38 firms)			Training (45 firms)		
	Mean		S.D.	Mean		S.D.
1. Time spent in Malaysia (years)	9		5	10		7
2. Export orientation	70.2		33.3	83.8		20.8
3. Size (number of employees)	736		666	1,543		1,720
4. Share of internal supplies	30.7		27.9	42.0		27.8
5. Share of subcontracting relations	13.8		24.8	30.6		34.7
	Freq.	*Share %*		*Freq.*	*Share %*	
		Row	*Col.*		*Row*	*Col.*
6. Type of product (frequency)	38	45.8		45	54.2	
(a) Active components				9.0	100	20.0
(b) Passive components	11	52.4	28.9	10	47.6	22.2
(c) Electro-mechanical components	12	57.1	31.6	9	42.9	20.0
(d) Consumer electronics	11	68.8	28.9	5	31.3	11.1
(e) Other products	4	25.0	10.5	12	75.0	26.7
7. Purchasing strategy (frequency)	9	32.1		19	67.9	
8. Country of origin (frequency)	38	45.8		45	54.2	
(a) Japan	23	51.1	60.5	22	48.9	48.9
(b) NIEs	13	65.0	34.2	7	35.0	15.6
(c) Others	2	11.1	5.3	16	88.9	35.6

Logistic regression models. In the full logistic regression model, all independent variables are considered. The second model considers the significant variables only.

<u>*Full model*</u>

The full model operates for 'tradummy' as it did for 'transit' and 'tecdummy'. We first enter all the previously defined variables into the logistic regression. The model converged after six iterations. The model has a high and highly significant model chi-square, a good overall model fit and a very good sensitivity (its ability to predict firms that transfer technology, 82.93% classified correctly). With an overall correct classification of 80.26%, the model is highly significant.[19]

Because we chose to first enter all the variables into the regression, the Wald coefficients indicate that not all the independent variables are important explanatory factors for the existence of technology transfer.

The model presented is more accurate at predicting cases when 'some transfer' takes place than otherwise. This is expected because of the larger size of the 'some transfer' group of subsidiaries; 54.2% of the firms transfer technology through training, while 45.8% do not. The model correctly predicts 82.93% of the 'some training' and 77.14% of the 'no training' group.

Table 7.12 Parameter estimates for logistic regression model for 'tradummy'

Variable name	ß	Wald	*R*	Sig.
1. Time spent in Malaysia	-.2102	4.2376	-.1457	.0395*
2. Share of export market	.0174	2.1386	.0363	.1436
3. Size: number of employees now	.0014	4.5956	.1570	.0321*
4. Share of internal inputs in total inputs	.0344	3.5163	.1200	.0608+
5. Share of subcontracting agreements in supply	-.0009	.0049	.0000	.9443
6. Product manufactured		10.5204	.1547	.0325*
(a) Active component	4.4686	2.9139	.0931	.0878+
(b) Passive component	-.6815	.4582	.0000	.4985
(c) Electro-mechanical component	-1.4301	2.4208	-.0632	.1197
(d) Consumer electronics	-3.3498	7.0080	-.2180	.0081**
7. Single supplier	-.8740	3.9561	-.1363	.0467*
8. Country/region of origin of subsidiary		9.0999	.2200	.0106*
(a) Japanese origin	-1.2573	3.8253	-.1316	.0505+
(b) Parent from NIEs	-2.3902	5.8102	-.1902	.0159*

Notes:

Constant insignificant, removed from model.
Variables significant at *** for $p<0.001$, ** for $p<0.01$, * for $p<0.05$ and + for $p<0.1$.
Categories below country/region are compared to other products category.
Purchasing strategy indicates companies that favour single supplier per item purchased, or various suppliers per item.
Share of export is share of final products manufactured by subsidiaries that is exported.
Number of employees in subsidiary represents subsidiary size.

The appropriate tests for significance of individual variables in logistic regression are based on the parameter estimates. Table 7.12 contains the estimated beta coefficients (ß) for the independent variables, and the significance of the Wald statistics appears in the last column of the table. One sub-category is significant at the .01 level (consumer electronics within the types of product manufactured).

Five categories are significant at the .05 level (time spent in Malaysia, size of subsidiary, types of product manufactured, purchasing strategy and country/region of origin). One category and two sub-categories are significant at the .1 level (share of internal supply, active components and Japanese origin). Note that country/region of origin is close to the .01 level, with a significance of .0106.

Reduced model

The 'backward conditional' elimination method is used to determine the reduced model. In the reduced model presented below, only the best predictor variables are introduced.

The model converged after seven iterations. The model has a high and highly significant model chi-square, a good overall model fit and a very good sensitivity (its ability to predict firms that do 'some training', 81.40% classified correctly). With an overall correct classification of 81.01%, the model is highly significant.[20]

The reduced model is similar to the full model but is more accurate at predicting cases when 'some training' takes place than otherwise, but by a small margin. The model correctly predicts 81.40% of the 'some training' phenomenon, and 80.56% of the cases when 'no training' occurs.

The appropriate tests for significance of individual variables in logistic regression are based on the parameter estimates. Table 7.13 contains the estimated beta coefficients (ß) for the independent variables, and the significance of the Wald statistics appears in the last column of the table. In the reduced model, the category country/region of origin has become the most important factor in the explanation of the training transfer. This variable is significant at the .01 level, and with a significance level of .0043, it is close to the .001 significance level. Three variables are significant at the .05 level (size of subsidiary, share of internal supply and types of product manufactured). The variable share of internal supply is close to the .01 significance level, together with the sub-category consumer electronics (with a negative sign). Two variables are significant at the .1 level (time spent in Malaysia and purchasing strategy). The category purchasing strategy is close to the .05 significance level.

Direction of relationship among variables. Forty-five subsidiaries operate transfer through training, and 38 do not. In Table 7.11, the descriptive statistics showed some substantial differences between the two groups in terms of size, subcontracting relationship, the type of product manufactured, the purchasing strategy and the region/country of origin. The full logistic model confirms these findings. Time spent in Malaysia, size, type of product manufactured, purchasing strategy and region/country of origin are significant independent variables. The logistic regression does not highlight share of subcontracting as a significant variable in the explanation of the existence of training.

Table 7.13 Parameter estimates for reduced logistic regression model for 'tradummy'

Variable name	ß	Wald	*R*	Sig.
1. Time spent in Malaysia	-.1321	2.9392	-.0926	.0865+
3. Size: number of employees now	.0013	5.3493	.1749	.0207*
4. Share of internal inputs in total inputs	.0432	6.5759	.2044	.0103*
6. Product manufactured		11.4177	.1767	.0222*
(a) Active component	3.6465	4.5027	.1512	.0338*
(b) Passive component	-1.0329	1.6254	.0000	.2023
(c) Electro-mechanical component	-1.3424	2.8559	-.0884	.0910+
(d) Consumer electronics	-2.4697	6.5656	-.2042	.0104*
7. Single supplier	-.7807	3.6099	-.1212	.0574+
8. Country/region of origin of subsidiary		10.8842	.2507	.0043**
(a) Japanese origin	-1.1240	3.5960	-.1207	.0579+
(b) Parent from NIEs	-2.3229	6.2247	-.1964	.0126*

Notes:

Constant insignificant, removed from model.
Variables significant at *** for p<0.001, ** for p<0.01, * for p<0.05 and + for p<0.1
Categories below country/region are compared to other products category.
Purchasing strategy indicates companies that favour single supplier per item purchased, or various suppliers per item.
Share of export is share of final products manufactured by subsidiaries that is exported.
Number of employees in subsidiary represents subsidiary size.

In the reduced model, the most significant independent variable in the explanation of the 'tradummy' transfer is the country/region of origin. It is the category 'other countries', within this category, that significantly influences the existence of training between a foreign subsidiary and its local suppliers. Both Japanese firms and firms from the NIEs show a negative sign, which indicates that these firms are less likely to operate 'some training'.

The logistic regression results confirm the descriptive statistics related to the type of product manufactured and the existence of training. The sub-categories active components, electro-mechanical components and consumer electronics are significant. With a negative sign, subsidiaries manufacturing electro-mechanical components and consumer electronics are less likely to operate 'some training', while a positive sign indicates that subsidiaries manufacturing 'active components' are more likely to 'transfer through training'.

There are other significant independent variables in the explanation of the 'transfer through training'. Hence, the larger the subsidiary in terms of number of employees, the more likely the existence of training. The larger the share of internal supplies in total supplies, the more likely it is that training exists.

Two more independent variables are of some significance in the reduced model. First, a negative sign indicates that firms that only established recently tend to 'transfer through training'. This result could indicate the initial effort provided by foreign subsidiaries in terms of training to their suppliers when they initially establish in Malaysia. Second, purchasing strategy shows a negative sign. Subsidiaries that favour only one supplier per item are less likely to 'transfer through training'.

Determinant Factors in the Case of Japanese Subsidiaries

The above detailed logistic analysis of the three transfer indicators involved all countries in the sample. Similar statistical analysis will now focus on Japanese firms only, because the sample of Japanese firms is large, and in stock value, Japanese firms were the major investors in MEE in 1996. A series of logistic regression models are calculated; first, for the overall transfer indicator, 'transit'; then for the process and product technology transfer indicator, 'transtec' and finally, for the transfer through training indicator, 'transtra'.

Factors explaining existence of overall transfers

The indicator 'Transit' must be proved reliable and valid before starting with the computation. A series of logistic regression models are then calculated.

Reliability and validity. Table 7.14 includes the descriptive results for cases that do not transfer technology and those that do. The sample sizes for the two groups are unequal, with 17 Japanese firms that do not transfer any technology against 28 firms that do. In this respect, Table 7.14 loses some of its explanatory value. This table shows that, companies that transfer technology have spent more time on average in Malaysia. They have a substantially higher average export orientation share (82.8% against 59.3% for companies that do not transfer technology). They are substantially larger in size (but have a much higher standard deviation, which indicates disparities in the sample). They have a higher share of internal supplies (although it is not substantially higher) and show a higher share of subcontracting relationships. Firms manufacturing active components, passive components and other products are more likely to operate transfers. Finally, from the firms that choose a single supplier per item, the majority transfer technology.

Correlation analysis was run to assess the degree of multicollinearity among the independent variables. The correlation matrix showed that most of the correlations are relatively low. While some of the correlations are significant, all are below +/-.40, and most are below .20. This confirms that multicollinearity between independent variables will not occur, thus it will not affect the coefficients and will not interfere with the interpretation of the results.

Table 7.14 Descriptive results of Japanese transfer indicator

	No transfer (17 firms)			Transfer (28 firms)		
	Mean	S.D.		Mean	S.D.	
1. Time spent in Malaysia (years)	9	5		11	7	
2. Export orientation	59.3	37.1		82.8	23.7	
3. Size (number of employees)	602	653		1,926	1,682	
4. Share of internal supplies	27.9	22.1		42.8	26.5	
5. Share of subcontracting relations	11.8	20.3		26.7	32.9	
	Freq.	*Share %*		*Freq.*	*Share %*	
		Row	*Col.*		*Row*	*Col.*
6. Type of product (frequency)	17	37.8		28	62.2	
(a) Active components	0			2	100	7.1
(b) Passive components	3	25.0	17.6	9	75.0	32.1
(c) Electro-mechanical components	8	57.1	47.1	6	42.9	21.4
(d) Consumer electronics	5	45.5	29.4	6	54.5	21.4
(e) Other products	1	16.7	5.9	5	83.3	17.9
7. Purchasing strategy (frequency)	5	27.8	29.4	13	72.2	48.1

Logistic regression model for Japanese firms. As carried out above, we base our assumptions on the factors likely to affect the existence of knowledge transfer on the literature related to the evaluation of such transfers. Again, the independent variables preliminary tested included the length of time the subsidiary has been established in Malaysia, the type of product manufactured, the market orientation of the subsidiary, the size of the company, the share of intra-firm purchase, the type of contractual relationship and the purchasing strategy of the firm.

Several models were computed that showed a higher level of classification rate, but the effect of independent variables on the dependent was then found insignificant. This is why we choose to present a reduced model with significant results for the selected independent variables. The probability of a Japanese subsidiary transferring technology to its local suppliers is explained in the reduced model by the variables type of product manufactured, share of export in total sales, number of employees and tendency to use a single supplier per range of input. We took a data-driven approach to final variable selection. The remaining variables highlighted in the literature were not found significant in the determination of 'some transfer' versus 'no transfer'. A correlation analysis was run among the selected variables to identify possible multicollinearity effect. All the correlations were low and insignificant.

Some Japanese subsidiaries do not transfer any technology to their local suppliers in Malaysia. Out of the 45 firms with local direct purchasing patterns, 17 firms (37%) mentioned that they did not transfer any technology, or such transfers were too small to be recorded. A logistic regression model enables identification of factors that explain why a subsidiary may engage in knowledge transfer or not. The two groups of subsidiaries in the model are first a group with subsidiaries that transfer technology, and second a group with subsidiaries that do not. The reduced model is presented; the model converged after two iterations.

The overall result showed that the model has a high and highly significant model chi-square, a good overall model fit, a very good sensitivity (its ability to predict firms that transfer technology, 88.89% classified correctly) and an excellent specificity (its ability to predict firms that do not transfer technology, 76.47% classified correctly). With an overall correct classification of 84.09%, the model is highly significant.[21]

Because we focus on the reduced model, almost all the independent variables included are significant, and explain why a Japanese subsidiary may or may not transfer technology to its local suppliers. A parallel analysis of the Wald coefficient indicates that all the chosen independent variables are important explanatory factors for the existence of knowledge transfer.

The model presented in Table 7.15 is more accurate in predicting cases when 'some transfer' takes place than otherwise; this is expected because of the larger size of the 'some transfer' group of subsidiaries. The model correctly predicts 88.89% of the transfer phenomenon and 76.47% of the cases when no transfer occurs. The appropriate tests for significance of individual variables in logistic regression are based on the parameter estimates. Table 7.15 contains the estimated beta coefficients (ß) for the independent variables, and the significance of the Wald statistics appears in the last column of the table. Three sub-categories are significant at the .05 level (active components, electro-mechanical components and consumer electronics for the product categories), and one at the .01 level (size of Japanese subsidiary). Note that the subsidiary's purchasing strategy and share of export are close to being significant at the .05 level; with p = .0505, and p = .0513, respectively; they are significant at the .1 level.

Table 7.15 Parameter estimates for logistic regression model

Variable name	ß	Wald	*R*	Significance
2. Share of export	.0583	3.8001	.1718	.0513+
3. Subsidiary size	.0029	6.7055	.2777	.0096**
6. Product category		7.0680	.0000	.1323
(a) Active components	29.330	4.6636	.2090	.0308*
(b) Passive components	1.9634	.9881	.0000	.3202
(c) Electro-mechanical components	3.2093	3.9448	.1786	.0470*
(d) Consumer electronics	-6.0524	6.2487	-.2639	.0124*
7 Purchasing	-3.7208	3.8243	-.1729	.0505+

Notes:

Constant insignificant, removed from model.
Variables significant at *** for p<0.001, ** for p<0.01, * for p<0.05 and + for p<0.1.
Categories below country/region are compared to other products category.
Purchasing strategy indicates companies that favour single supplier per item purchased, or various suppliers per item.
Share of export is share of final products manufactured by subsidiaries that is exported.
Number of employees in subsidiary represents subsidiary size.

Direction of relationship among variables. The most significant independent variable in the explanation of knowledge transfer is the size of subsidiary, in terms of number of employees. The positive sign indicates that the larger the subsidiary, the more likely it is to transfer technology to its local suppliers. Although the independent variable product category does not appear as significant, sub-categories within this variable are significant. Japanese subsidiaries that manufacture active components show a positive influence for 'some transfer' to take place, but this finding is based on two observations only; similarly, but to a lesser extent, for firms manufacturing electro-mechanical components. A strong negative relationship is established between consumer electronics and technology transfer. This indicates that Japanese subsidiaries manufacturing final products are mostly assembly facilities, which would in part explain why they are less likely to transfer technology to their local suppliers.

As mentioned, the share of export in total sales and the subsidiary's purchasing strategy were close to being significant at the .05 level. The share of exports in total sales shows a positive relationship with technology transfer.

This sign indicates that the higher the share of export in total sales, the more likely the subsidiary is to transfer technology to its local suppliers. Finally, the negative sign associated with the purchasing strategy indicates that the firms which tend to have a single supplier per item purchased tend to transfer less technology to their local suppliers.

Logistic regression model for 'tecdummy': process and product technology

Before carrying out the logistic analysis of 'tecdummy', we provide a descriptive analysis of the indicator.

Descriptive analysis. In Table 7.16, the sample size for firms that do 'no transfer' and those that do 'some transfer' is nearly identical. There is little difference in independent variables between the two groups of firms. The only independent variable that shows a substantial difference is share of internal supplies.

Table 7.16 Descriptive results of Japanese 'tecdummy' indicator

	No transfer (22 firms)			Transfer (23 firms)		
	Mean		S.D.	Mean		S.D.
1. Time spent in Malaysia (years)	10		6	10		6
2. Export orientation	65.2		35.3	82.3		24.7
3. Size (number of employees)	1,420		1,992	1,432		915
4. Share of internal supplies	27.6		20.6	46.3		27.1
5. Share of subcontracting relations	16.7		27.2	24.8		31.3
	Freq.	*Share %*		*Freq.*	*Share %*	
		Row	*Col.*		*Row*	*Col.*
6. Type of product (frequency)	22	48.9		23	51.1	
(a) Active components				2	100	8.7
(b) Passive components	5	41.7	22.7	7	58.3	30.4
(c) Electro-mechanical components	8	57.1	36.4	6	42.9	26.1
(d) Consumer electronics	8	72.7	36.4	3	27.3	13.0
(e) Other products	1	16.7	4.5	5	83.3	21.7
7. Purchasing strategy (frequency)	6	33.3	28.6	12	66.7	52.2

Logistic regression model. In previous sections, in the first model, all independent variables are entered into the model. A second model is computed, keeping only the significant variables.

<u>Reduced model</u>

The 'backward conditional' elimination method is used to determine the reduced model. In the reduced model presented below, only the best predictor variables are introduced in the model.

The model converged after six iterations. The model has a high chi-square significant at the .01 level, a good overall model fit and a good sensitivity (its ability to predict firms that transfer technology, 78.26% classified correctly). With an overall correct classification of 81.40%, the model is significant.[22]

Table 7.17 Parameter estimates for reduced logistic regression model for 'tecdummy'

Variable name	ß	Wald	*R*	Sig.
2. Export market share	.0405	4.0009	.1832	.0455*
3. Size of the firm	.0012	5.2327	.2329	.0222*
5. Share of subcontracting relationship	-.0484	4.4422	-.2024	.0351*
6. Product manufactured		8.2610	.0662	.0825+
(a) Active component	11.3372	3.2301	.1437	.0712+
(b) Passive component	-2.3152	2.1226	-.0454	.1251
(c) Electromechanical component	-3.5078	4.0328	-.1847	.0446*
(d) Consumer electronics	-6.3598	6.6077	-.2780	.0102*
7. Single supplier	-1.1541	4.2932	-.1961	.0383*

Notes:

Constant insignificant, removed from model.
Variables significant at *** for p<0.001, ** for p<0.01, * for p<0.05 and + for p<0.1.
Categories below country/region are compared to other products category.
Purchasing strategy indicates companies that favour single supplier per item purchased, or various suppliers per item.
Share of export is share of final products manufactured by subsidiaries that is exported.
Number of employees in subsidiary represents subsidiary size.

The reduced model is slightly more accurate in predicting cases when no process and product technology transfer occurs. The model correctly predicts 85.00% of the 'no transfer' phenomenon, and 78.26% of the cases when transfer occurs. The appropriate tests for significance of individual variables in logistic regression are based on the parameter estimates. Table 7.17 contains the estimated beta coefficient (ß) for the independent variables, and the significance of the Wald statistics appears in the last column of the table. In the reduced model, four independent variables are significant at the .05 level, namely export market share, size of the firm, share of subcontracting relationships and single supplier. Type of product manufactured is significant at the .1 level, but this does not reflect the significance of the sub-product category. Hence, electro-mechanical components and consumer electronics sub-categories are significant at the .05 level, and active components at the .1 level.

Direction of relationship among variables. Twenty-three subsidiaries transfer 'some process and product technology', and 22 do not. In Table 7.16, the descriptive statistics showed little differences between the two groups' independent variables, except for share of internal supplies. The reduced logistic model did not highlight this independent variable as significantly impacting on the existence of process and product technology transfer.

The most significant independent variables in the explanation of the 'tecdummy' transfer are export market share, size of the firm, share of subcontracting relationship and purchasing strategy. The signs on the *R* coefficients indicate that the more export-oriented the firm is, and the larger the subsidiary, the more likely it is that the process and product technology transfer will occur. The negative sign associated with the share of subcontracting relationship is unexpected. It indicates that the larger the share of subcontracting relationship, the less likely it is that the Japanese subsidiary will operate transfer.

The type of product manufactured category is significant at the .1 level. Two sub-categories are significant at the .05 level and show a negative sign: hence, subsidiaries manufacturing electro-mechanical components and consumer electronics are less likely to transfer process and product technology. On the other hand, firms manufacturing active components are more likely to transfer technology; however, this latter observation is based on two companies only.

Logistic regression model for 'Tradummy': transfer through training

Before carrying out the logistic analysis of 'tradummy' for Japanese firms, we provide a descriptive analysis of the indicator.

Descriptive analysis. Table 7.18 shows a descriptive analysis of independent variables according to whether Japanese subsidiaries transfer technology through training or not. Significant differences among the two groups of firms can be observed for the export orientation of the subsidiaries (firms with training export an average of 82% of their production against 66% for those that do not offer training), the size of the firms (1,994 employees on average for the 'training' group), share of internal supplies and share of subcontracting. Clear distinctions between the two groups of firms can also be noticed in specific product sub-categories, such as active components, electro-mechanical components, consumer electronics and other products. Slight differences can be noticed for purchasing strategy as well. The correlation matrix between each independent variable showed low and insignificant correlation levels.

Logistic regression model. Because the sample of Japanese firms is comprised of 45 firms, a full model would not provide highly reliable results. Instead, we present the reduced model that considers the significant variables only.

Reduced model

The 'backward conditional' elimination method is used to determine the reduced model. In the reduced model presented below, only the best predictor variables are introduced.

The model converged after six iterations. The model has a high and highly significant chi-square, a good overall model fit and a good sensitivity (its ability to predict firms that transfer technology, 90.91% classified correctly). With an overall correct classification of 86.67%, the model is significant.[23]

Table 7.18 Descriptive results of Japanese 'tradummy' indicator

	No training (23 firms)			Training (22 firms)		
	Mean	S.D.		Mean	S.D.	
1.Time spent in Malaysia (years)	10	6		10	7	
2.Export orientation	66.0	36.0		82.2	23.4	
3.Size (number of employees)	883	767		1,994	1,889	
4.Share of internal supplies	28.7	22.5		45.9	26.4	
5.Share of subcontracting relations	10.8	18.0		32.0	35.3	
		Share %			*Share %*	
	Freq.	*Row*	*Col.*	*Freq.*	*Row*	*Col.*
6.Type of product (frequency)	23	51.1		22	48.9	
(a) Active components	0			2	100	9.1
(b) Passive components	6	50	26.1	6	50	27.3
(c) Electro-mechanical components	9	64.3	39.1	5	35.7	22.7
(d) Consumer electronics	7	63.6	30.4	4	36.4	18.2
(e) Other products	1	16.7	4.3	5	83.3	22.7
7.Purchasing strategy (frequency)	8	44.4	34.8	10	55.6	47.6

The reduced model is slightly more accurate in predicting cases when 'some training' occurs. The model correctly predicts 90.91% of the 'some training' phenomenon, and 82.61% of the cases when training does not occur. The appropriate tests for significance of individual variables in logistic regression are based on the parameter estimates. Table 7.19 contains the estimated beta coefficient (ß) for the independent variables, and the significance of the Wald statistics appears in the last column of the table. In the reduced model, two independent variables are significant at the .05 level, namely size of subsidiary and type of products manufactured. Both categories are close to the .01 significance level. Within the types of product manufactured category, consumer electronics is significant at the .05 level, again close to .01 level of significance, while active components and passive components are significant at the .1 level. Finally, time spent in Malaysia and export market share are both significant at the .1 level.

Table 7.19 Parameter estimates for reduced logistic regression model for 'tradummy'

Variable name	ß	Wald	*R*	Significance
1. Time spent in Malaysia	-.2176	3.7078	-.1655	.0542+
2. Export market share	.0362	3.2653	.1424	.0708+
3. Size of the firm	.0025	5.7587	.2455	.0164*
6. Product manufactured		7.3856	.0000	.0116*
(a) Active component	15.3886	3.6266	.1615	.0567+
(b) Passive component	-3.9004	3.2635	-.1423	.0708+
(c) Electro-mechanical	-2.4832	2.0529	-.0291	.1519
component	-7.6610	5.6638	-.2423	.0173*
(d) Consumer electronics				

Notes:

Constant insignificant, removed from model.
Variables significant at *** for p<0.001, ** for p<0.01, * for p<0.05 and + for p<0.1.
Categories below country/region are compared to other products category.
Purchasing strategy indicates companies that favour single supplier per item purchased, or various suppliers per item.
Share of export is share of final products manufactured by subsidiaries that is exported.
Number of employees in subsidiary represents subsidiary size.

Direction of relationship among variables. Sample sizes were equal between Japanese subsidiaries offering 'some training' and those that did not. Some significant factors were highlighted through the reduced logistic regression model. Size of the firm and types of product are key explanatory factors for the existence of training by Japanese subsidiaries. The larger the Japanese subsidiary is in terms of number of employees, the more likely it is to provide training to its local direct suppliers. Japanese subsidiaries manufacturing consumer electronics are less likely to provide training, together with firms manufacturing passive components, while firms manufacturing active components are more likely to train their suppliers.

Although less significant, the longer the Japanese subsidiary has been established in Malaysia, the less likely it is to provide training. On the other hand, the more the subsidiary is oriented towards export, the more likely it is to provide 'some training'.

hand, the more the subsidiary is oriented towards export, the more likely it is to provide ‘some training’.

Summary of Findings on Factors in Existence of Knowledge Transfer

Overall discussion

The series of logistic regression analysis presented in the previous section is summarised in Table 7.20. The significance level and the sign of the relationship between the independent variable and the existence of transfer are indicated for each independent and dependent variables. The summary table enables comparison of results across all variables by first taking into account results when all firms are introduced in the models, regardless of their nationality of origin, and then by focusing on the results if Japanese subsidiaries only are analysed.

The independent variables preliminarily tested include the length of time the subsidiary has been established in Malaysia, the type of product manufactured, the market orientation of the subsidiary, the size of the company, the share of intra-firm purchase, the type of contractual relationship and the purchasing strategy of the firm. Several models, full or reduced, were computed.

The regression models can now help in describing which variables explain the existence of various transfers. Table 7.20 shows the results obtained from the logistic regression analysis. This table is the basis for the discussion on the determinant role of individual independent variables in the existence of transfer between foreign affiliates and their local suppliers.

Determinant Factors of Existence of Knowledge Transfer: Discussion

With the help of the summary, Table 7.20, we now discuss the role of each independent variable in explaining the existence of knowledge transfer through backward linkages.

Time spent in Malaysia

The length of time spent in Malaysia by the foreign subsidiary, that is how long it has been operating in Malaysia, was found insignificant in most logistic regressions. This variable seems to influence transfer through training, but with a low significance level. The ‘time’ variable shows a negative sign in the logistic regression for the existence of transfer through training.

Table 7.20 Summary of significant factors in explanation of existence of transfer between foreign subsidiaries and their local suppliers

	All firms (83 firms)			Japanese firms (45 firms)		
Variable name	Transit	Tecdum.	Tradum.	Transit	Tecdum.	Tradum.
1. Time spent in Malaysia	NS	NS	SIG (-)+	NI	NI	SIG (-)+
2. Share of export market	SIG (+)+	NS	NS	SIG (+)+	SIG (+)*	SIG (+)+
3. Size: number of employees now	SIG (+)*	NS	SIG (+)*	SIG (+)**	SIG (+)*	SIG (+)*
4. Share of internal inputs in total inputs	NS	NS	SIG (+)*	NI	NI	NI
5. Share of subcontracting agreements in supply	NS	SIG (+)+	NS	NI	SIG (-)*	NI
6. Product manufactured	SIG+	SIG*	SIG*	NS	SIG*	SIG*
(a) Active component	NS	SIG (+)*	SIG (+)*	SIG (+)*	SIG (+)+	SIG (+)+
(b) Passive component	NS	SIG (-)+	NS	NS	NS	SIG (-)+
(c) Electro-mechanical component	NS	SIG (-)*	SIG (-)+	SIG (-)*	SIG (-)*	NS
(d) Consumer electronics	SIG (-)**	SIG (-)**	SIG (-)*	SIG (-)*	SIG (-)*	SIG (-)*
7. Single supplier	SIG (-)*	NS	SIG (-)+	SIG (-)+	SIG (-)*	NI
8. Country/region of origin of subsidiary	SIG*	NS	SIG**	NA	NA	NA
(a) Japanese origin	NS	NS	NS	NA	NA	NA
(b) Parent from the NIEs	SIG (-)+	NS	NS	NA	NA	NA

Notes:
NI: not included in reduced model.
NS: not significant in full model.
NA: not applicable, country of origin cannot apply for Japanese sample only.
SIG: significant variable.
(+/-): sign of relationship.
Variables significant at *** for $p<0.001$, ** for $p<0.01$, * for $p<0.05$ and + for $p<0.1$.

The negative sign indicates that the younger the subsidiaries, the more likely they are to operate transfer to their local suppliers through training. This contradicts some previous findings that emphasise that stronger transfers occur after long-term relationships between the subsidiary and its suppliers have been created. On the other hand, the negative sign highlights the need that subsidiaries have to train their suppliers at the beginning of the relationship in order to ensure the quality of the inputs and related services. Thus, this confirms the new purchasing strategic orientation adopted by TNCs recently. Over the past few years, TNCs have put the emphasis on building strong lasting relationships with suppliers, and on treating key suppliers as business partners. The creation of strong relationships may influence the existence of training between the partners. In our sample, subsidiaries that were recently established in Malaysia are more likely to provide training for their suppliers, but overall, the 'time' factor is not a significant factor in the existence of overall knowledge transfer between foreign subsidiaries and their local suppliers.

Export orientation of the subsidiary

The market orientation of the subsidiary shows a low significance level in the explanation of the existence of transfer. However, this market orientation factor is significant in four regressions out of six, which indicates clearly that this factor is important. The market orientation is significant for the overall transfer indicator when all firms are considered. It seems particularly important in the case of Japanese subsidiaries, and especially so in the case of product and process technology transfer from Japanese affiliates to their local suppliers. The positive sign indicates that the higher the share of final output sold on the export market, the more likely it is that a subsidiary will transfer technology. It must be emphasised that the level of significance of this factor is low (at the .1 level) in most logistic regressions.

Size of subsidiary

This 'size' factor is a significant factor in five regressions out of six, showing a good level of significance (from .05 to .01 level). The size of the company is a significant explanatory factor in both 'overall' transfer indicators, and both 'training' indicators. The size of the company is very significant in the explanation of transfer for Japanese subsidiaries. The positive sign indicates that the larger the subsidiary is in terms of number of employees, the more likely it is to transfer technology to its local suppliers.

Share of internal inputs in total inputs

The share of internal inputs in total inputs shows the extent to which the subsidiary is dependent on its parent firm or other subsidiaries for its input supply. This factor indicates the degree of intra-purchasing that occurs within the transnational firm. The share of internal inputs does not significantly explain the existence of transfer. This factor is significant in one of the regressions only, with a .05 level of significance. The positive sign indicates that when all firms are taken into account, the higher the share of internal inputs, the more likely the subsidiary is to transfer knowledge through training. This factor is not significant for any of the logistic regressions related to Japanese firms.

Share of subcontracting agreements

This factor shows the extent to which the subsidiary relies on subcontractors for its direct input purchase. It is significant in two of the six regressions, with low significance levels (from .1 to .05). The share of subcontracting is significant in both logistic regressions related to transfer of product and process technology, with a higher level of significance in the case of Japanese subsidiaries only. The sign of the relationship differs; it is positive when all firms are taken into account, but negative in the case of Japanese firms. Thus, for all firms, the higher the share of subcontracting relations, the more likely the subsidiary is to transfer product and process technology. For Japanese firms, the higher the share of subcontracting relations, the less likely the subsidiary is to transfer such technology.

Product manufactured

The type of product manufactured is a key explanatory factor in the existence of transfer between partners. This factor is significant in five regressions, with a significance level of .05 in four of these. The level of significance and sign of the relationship differ according to the product sub-categories. Out of the five sub-categories, only four appear in the logistic regression table, because computation is operated against the fifth category, that is the other products category. Table 7.20 shows that negative signs for the sub-categories passive components, electro-mechanical components and consumer electronics indicate that firms manufacturing/assembling these products are less likely to transfer technology or to carry out training. The consumer electronics sub-category has high significance levels, showing that firms assembling these final products are unlikely to operate transfers to their local direct suppliers. On the other hand, the manufacturing of active components, which are more technology-intensive products in nature, is likely to explain the existence of transfers between foreign firms and their local suppliers. Thus, it appears from these results that the level of technology or the nature of the subsidiary's activity explains whether the firm will transfer technology or not. Assembling activities are less likely to lead to transfer, while the manufacturing of technology-intensive products is likely to do so.

Single supplier
The single supplier strategy shows whether the subsidiary favours a single supplier per item purchased or not. This factor is significant in four regressions, with low significance levels (two at .1 level and two at .05 level). In each case, a negative sign indicates that subsidiaries that do not favour a single supplier per input are more likely to transfer technology to their suppliers. This indicates that when a single supplier is chosen, it might be for its technological capabilities. A good supplier might require less transfer on the part of the subsidiary.

Country/region of origin
The origin factor is a key factor in the explanation of the existence of transfer. This factor does not apply when only Japanese firms are taken into account. It is significant in two out of three regressions (for all firms), with good significance levels (one at the .01 level and one at the .05 level). The three sub-categories are Japan, NIEs and other firms. The logistic regression analysis is run against the third category, and therefore the other firms' category does not appear in Table 7.20. A negative sign for the NIE firms indicates that NIE subsidiaries are less likely to transfer technology.

The above discussion has emphasised the significance and signs of each independent variable in the explanation of the existence of knowledge transfer. The following section looks at the impact of independent variables on the degree of knowledge transfer between foreign subsidiaries and their local direct suppliers.

DETERMINANT FACTORS OF DEGREE OF KNOWLEDGE TRANSFER

In this section, a series of linear regression analysis with the three indicators 'transfer', 'transtec' and 'transtra' as the dependent variables are computed. This leads to the determination of the factors that affect the degree of transfer between foreign affiliates and their local suppliers.

Methods and Measures

After indicating the most important explanatory factors in the existence of knowledge transfer, using multiple regression analysis, we try to ascertain which of the characteristics discussed in Chapter 3 has the highest explanatory power with regard to the degree of knowledge transfer.

The application of multiple regression analysis requires that various assumptions are satisfied: linearity, equality of variance and normality. It is often difficult to comply with all assumptions.

In our case, linearity is not a problem. In the bivariate analyses, we systematically checked relationships for non-linearity, but did not find any indications for this. Plotting standardised residuals against standardised predicted values for each of the regression analyses confirmed this observation completely. The same plots also indicated that there was no reason to suspect violation of the equality of variance assumption. Normal distribution may be the only concern for the knowledge transfer indicator. The application of several 'correction' actions did not lead to any improvement in the normality of the indicator. However, histograms of standardised residuals and cumulative normal probability plots showed that the deviation from normality was not very strong for most variables, and that the K-S Lilliefors normality analysis would be appropriate in our case.

Table 7.21 shows a series of regression models of the influence of various variables on the degree of knowledge transfer from subsidiaries to their local suppliers. Since we wanted to create parsimonious models, stepwise selection (a combination of forward selection and backward elimination) was used to identify the subset of variables that were good predictors of the dependent variable. In order to allow marginally significant variables (two-tailed p-value between 0.05 and 0.10) to be included in the model, the entry requirement was modified to 'probability of F-to-enter is less than or equal to 0.10'. The variable enters the regression analysis if the probability associated with the F-test is less than or equal to 0.10.

Numerous dummy variables were created; one for each country of origin subsidiary's parent firm, and for each type of product manufactured. To this end, values for each country of origin or type of product were recoded into 0 and 1, where 1 signified that the case originated from the country in question, and 0 that it did not. Since there were no specific theoretical reasons to choose a particular order of variables, all independent variables were entered in one block.

The statistics reported for regression models differ considerably between publications. Some publications do not mention adjusted R-squared, which is the most conservative measure and most suitable for comparing models between different data sets, since it also takes sample size and the number of predictor variables into account. Further, regression models with small sample sizes and a relatively large number of predictor variables can easily result in an 'over-fitting' of the variate to the sample.

Table 7.21 Linear multiple regression models

	All cases			Japanese firms		
MODELS: Stepwise method	Transfer	Transtec	Transtra	Transfer	Transtec	Transtra
Company-related factors						
Products manufactured						
Active components	.251*** (30.1)	-	-	-	-	-
Passive components	-	-	-	-	-	-
Electro-mechanical components	-	-	-	-	-	-
Consumer electronics	-	-	-	-	-	-
Other products	.231** (20.2)	-	.242** (22.509)	-	-	-
Country of origin						
Japan	-	-		NA	NA	NA
NIEs	-	-		NA	NA	NA
Other countries	.179* (14.7)	-	.204* (17.509)	NA	NA	NA
Company characteristics						
Size (nb. of employees)	.250** (5.314E-3)	-	.309*** (7.723E-3)	-	-	
Time of establishment	-	-	-	-		
Share of exports in total sales	-	-	-	-		
Purchase-related factors						
Share internal supply	.157* (.1)	-	-	.540*** (.443)	.541*** (.553)	.454*** (.378)
Share subcontracting	-	.276** (.303)	.235** (.242)	.335** (.347)	--	.411** (.435)
Single supplier per input	.269*** (17.9)	.305** (22.090)	.189* (13.301)		.311* (22.586)	
Model statistics						
R	.842	.409	.814	.793	.761	.775
R-squared	.709	.167	.663	.629	.580	.601
Adjusted *R*-squared	.685	.146	.640	.611	.560	.582
F-value model	29.649	7.738	28.355	35.628	28.945	31.645
Model significance	.000	.001	.000	.000	.000	.000

Notes: *** $p<0.001$, ** $p< 0.01$, * $p < 0.05$, + $p < 0.10$ all two-tailed. Values for independent variables are standardised beta-coefficients. Size (annual sales in US$) was also a possibility, but there are too many missing values for results to be reliable. NA: Not applicable

Through over-fitting, small sample sizes often result in high *R* and *R*-squared values, which might be misleading if authors do not report adjusted R-squared. Adjusted *R*-squared is the proportion of the variation of the dependent variable explained by the model. Adjusted *R*-squared attempts to correct the optimistic bias of the sample *R*-squared; it does not necessarily increase as additional variables are added to an equation and is the preferred measure of goodness of fit because it is not subject to the inflationary bias of unadjusted R-squared (Norusis, 1994: 346).

The multiple linear regression model is as follows:

$$\text{Degree of transfer} = \beta_0 +/- \beta_1\ \text{TIME} +/- \beta_2\ \text{PRODUCT} +/- \beta_3\ \text{MARKET} +/- \beta_4\ \text{SIZE} \\ +/- \beta_5\ \text{INPUT} +/- \beta_6\ \text{SUBCONTRACT} +/- \beta_7\ \text{SUPPLIER} +/- \beta_8\ \text{COUNTRY} + ei$$

where the ß terms are unknown parameters and the *ei* terms are independent random variables that are normally distributed with mean 0 and constant variance. Correlation levels among independent variables were shown to be low and insignificant.

Determinants of Degree of Transfer

The descriptive analysis of the numerical indicator 'transfer' was presented earlier in this chapter. Table 7.4 showed that there are disparities in the degree of transfer across firms. While the average for all firms of those that do no transfer at all is 33.7%, 37.8% of Japanese firms do not carry out any transfer, against 50% for firms from the NIEs, and a staggeringly low 5.6% for the category other countries. The average share of frequently transferred items in total items is low (Table 7.3), ranging from 11.41% for NIE firms to 43.3% for other countries, firms, with an average across all firms of 25.01%.

The series of regression analysis provides statistical evidence for the factors that do explain the degree of transfer between foreign affiliates and their local suppliers. Table 7.21 shows the results of the analysis for all three transfer indicators, that is 'transfer', 'transtec' and 'transtra'. All the regression models are highly significant, with a significance level of .001 or lower. There are differences between the models in terms of explained variance (adjusted *R*-squared), which run from a low 15% for 'transtec' (using all cases) to a high of 68% for 'transfer' (using all cases). The low *R*-squared for 'transtec' is not surprising, because the descriptive analysis has already shown that the degree of product and process technology transfer is very low across all cases. Still, 15% of explained variance is acceptable in the context of social science results (see Harzing, 1999: 305 for more details about appropriate levels of *R*-squared in social science).

Determinant factors of degree of overall transfer

In order to maximise the variables-to-cases ratio, only the variables appearing on the regression analysis tables were actually included in the regressions shown. There are two regressions related to overall transfer indicator; one carried out including all cases, and one with Japanese firms. When all the firms are included, regardless of their nationality of origin, the two most important factors in the explanation of the degree of transfer are the product manufactured and the single supplier factors. Two product dummy variables became significant factors in the regression analysis; the active components and the other products categories, both showing positive signs. The two product categories that emerge strongly are the categories including high-tech content in the products manufactured. Thus, the more technology-intensive the product manufactured is, the more likely the foreign affiliate is to transfer technology to its local suppliers. Products with low-tech content or mostly assembling plants included in the consumer electronics sub-category were found to be very important in the logistic regression analysis, which highlighted the factors explaining the existence of transfer, rather than the degree of transfer. Thus, products that are less technology-intensive are likely not to lead to transfer, and technology-intensive products will lead to a higher degree of transfer between foreign affiliates and their local suppliers.

The second most important explanatory factor is that of the single supplier per direct input purchased. The positive sign indicates that the purchasing strategy has an impact on the degree of transfer occurring among firms. The firms that favour a single supplier show a higher degree of transfer than other firms.

The size of the firm in terms of number of employees is another key factor in explaining the degree of transfer. Unlike the results of the logistic regression, the size of the firm is the only company characteristic factor to influence the degree of transfer. The export orientation explains the existence of transfer, but is not significant in the explanation of the degree of transfer.

One dummy variable related to the country of origin is significant. The other countries' category indicates that the higher levels of transfer occur for other countries, as opposed to the categories Japan and NIEs' countries. Thus, Asian firms do not tend to show the highest degrees of transfer. This clearly indicates that the country of origin of the foreign affiliates explains to some extent the degree of transfer, but it is not one of the most significant explanatory factors. Similarly, share of internal supply, although significant, is not one of the most important explanatory factors in the degree of transfer.

Running the same regression analysis with Japanese firms led to a different set of results. In the case of Japanese firms, the degree of overall transfer is only explained by purchase-related factors, with share of internal supplies and the share of subcontracting relations both being highly significant explanatory factors.

Determinant factors of degree of product and process technology transfer

Multiple regression analysis was run for the process and product technology transfer indicator, first with all cases, and then with Japanese cases only. The descriptive analysis showed that this type of transfer does not occur very often, and is less common than transfer through training. Table 7.21 highlights the distinction of this particular type of transfer in terms of the factors explaining the degree to which it occurs. In both regressions, the only explanatory factors are purchase-related. Thus, the degree of product and process technology transfer is solely explained by the purchasing strategy of the affiliates. The share of subcontracting and the single supplier strategy emerge as the two single explanatory factors when all the cases are considered. However, in the case of Japanese firms, the two explanatory factors are share of internal supply and the single supplier strategy. The degree of technology transfer is in this case clearly explained by the purchasing strategy adopted by the foreign affiliates, regardless of their size, their country of origin or the product they manufacture. These results contrast with the results of the logistic regression when the product manufactured by the affiliate became a significant explanatory factor in the existence of transfer. Clearly, the degree of transfer is then to be explained by a different set of factors, merely the purchase strategy carried out by the firm.

Determinant factors of degree of training between affiliates and suppliers

Transfer through training occurs more frequently than other types of transfers. The single most significant factor in the explanation of the degree of transfer through training is the size of the affiliate when all cases are taken into consideration. The larger the affiliate in terms of employees, the higher the degree of training provided by the affiliate to its local suppliers. The second most significant factor is the sub-category other products, indicating that the type of products manufactured by the affiliate explains the degree of training it will provide to its suppliers. The technology intensity of the product does not emerge as a strong explanatory factor, indicating that it is the type of product that matters, but not necessarily the capital intensity of the product. Purchase-related factors also explain the degree of training. Share of subcontracting becomes a strongly significant explanatory factor, while the single supplier strategy is somewhat less significant, but still to be accounted for. Finally, the origin of the affiliate is also significant, albeit less than other factors. The sub-category other countries shows a higher degree of transfer than Japanese or NIE firms. Thus, a greater variety of factors explain the degree of training than the degree of product and process technology transfer. Unlike the other type of transfers, the factors explaining the existence and the degree of training are similar.

The results described above are different for separate computation with Japanese firms in the multiple regression analysis for the degree of training. As highlighted in the case of the overall transfer effect and in the case of product and process technology transfer, the degree of transfer through training is explained solely by purchase-related factors. The most significant explanatory factor is share of internal supply, closely followed by share of subcontracting. This contrasts with the factors explaining the existence of transfer. The size of the company and the types of products manufactured were important explanatory factors in the existence of training, while purchase-related factors fully explain the degree of training of Japanese affiliates in the MEE sector.

Determinant Factors of Degree of Transfer: Discussion

The linear regression models described in Table 7.21 are summarised in Table 7.22 and help in giving a description of the key factors explaining the degree of transfer between foreign subsidiaries and their local direct suppliers. Table 7.22 is the basis for the discussion on the determinant role of individual independent variables in the degree of transfer.

Time spent in Malaysia

The time factor is insignificant in the explanation of the degree of knowledge transfer in all the linear regression results.

Share of export market

The 'export orientation' factor is insignificant in the explanation of the degree of knowledge transfer in all the linear regression results.

Size of subsidiary

The size of subsidiary is a significant factor explaining the degree of transfer between foreign subsidiaries and their local direct suppliers. This factor is significant in two of the six linear regressions. The size is highly significant when explaining the degree of training provided. A positive sign indicates that the larger the subsidiary, the higher the degree of transfer is. Incidentally, although the size was a significant factor in the explanation of the existence of transfer for Japanese firms, it is insignificant in the explanation of the degree of transfer by the Japanese subsidiaries.

Table 7.22 Summary of significant factors in explanation of degree of transfer between foreign subsidiaries and local suppliers

	All firms (83 firms)			Japanese firms (45 firms)		
Variable name	Transit	Tecdum.	Tradum.	Transit	Tecdum.	Tradum.
1. Time spent in Malaysia	-	-	-	-	-	-
2. Share of export market	-	-	-	-	-	-
3. Size: number of employees now	(+)**	-	(+)***	-	-	-
4. Share of internal inputs	(+)*	-	-	(+)***	(+)***	(+)***
5. Share of subcontracting	-	(+)**	(+)**	(+)**	-	(+)**
6. Product manufactured	-	-	-	-	-	-
(a) Active component	(+)***	-	-	-	-	-
(b) Passive component	-	-	-	-	-	-
(c) Electro-mechanical component	-	-	-	-	-	-
(d) Consumer electronics	-	-	-	-	-	-
(e) Other products	(+)**	-	(+)**	-	-	-
7. Single supplier	(+)***	(+)**	(+)*	-	(+)*	-
8. Country/region of origin	-	-	-	NA	NA	NA
(a) Japanese origin	-	-	-	NA	NA	NA
(b) Parent from NIEs	-	-	-	NA	NA	NA
(c) Parent from other countries	(+)*	-	(+)*	NA	NA	NA
R-squared	.685	.146	.640	.611	.560	.582

Notes:
NA: not applicable, country of origin cannot apply for Japanese sample only.
(+/-): sign of relationship.
Variables are significant at *** for p<0.001, ** for p<0.01, * for p<0.05 and + for p<0.1.

Share of internal inputs in total inputs

The share of internal input is highly significant in the explanation of the degree of transfer from Japanese subsidiaries. This factor has a high significance level for all three regressions related to Japanese firms. The higher the level of internal input, the higher the degree of transfer is with suppliers. In the case of Japanese firms, this may indicate that these firms tend to rely on internal supply within the company, and when they purchase from external suppliers they favour strong relationships with the latter. The share of internal inputs is also significant in the explanation of the overall transfer for all firms, with a positive sign.

Share of subcontracting agreements

The share of subcontracting agreements is a significant factor in the explanation of the degree of transfer. This factor is significant at the .01 level in four out of six linear regression models. A positive sign indicates that the higher the share of subcontracting relationships in the total contractual relations with local suppliers, the higher the degree of knowledge transfer is between foreign subsidiaries and their local suppliers.

Product manufactured

Two product sub-categories were significant in the regressions models, namely the active components' and the other products' sub-categories. For both sub-categories the sign of the relationship with the degree of transfer was positive, indicating that firms manufacturing these products are likely to provide a degree higher of transfer to their local suppliers. Interestingly, the type of product manufactured is not a significant factor for the degree of transfer from Japanese firms.

Single supplier

The single supplier purchasing strategy is a key factor in the explanation of the degree of transfer. It is highly significant in four of the six regression models. Contrary to the results highlighted in the logistic regressions, the single supplier strategy shows a positive sign. This indicates that transfer is more likely to occur when the subsidiary has a multiple purchasing strategy, but a higher degree of transfers occur for subsidiaries that favour a single supplier per item purchased.

Country/region of origin

One of the three sub-categories is significant in two of the three regression models, but at a low .01 significance level. This indicates that the country/region of origin of the firm is a significant factor in the explanation of the degree of transfer, but it is not highly significant. A positive sign in the case of other countries indicates that firms other than Japanese and from the NIEs show a higher degree of knowledge transfer.

OVERALL CONCLUSIONS ON DETERMINANTS OF EXISTENCE AND DEGREE OF KNOWLEDGE TRANSFER

Overview of the Results

In order to discuss all the regression models described in this chapter, we have classified independent variables in terms of their significance in the explanation of the existence and the degree of knowledge transfer. For each independent variable, we looked at the frequency of appearance in each six logistic and six linear regression models, and at the level of significance. Results obtained from the models vary widely according to a number of factors; thus stability in the significance of independent variables in part across a series of regression models ensures further the reliability of the results (Hair et al., 1998). From the results of Tables 7.20 and 7.21, the order of importance of each independent variable is shown in Table 7.23,[24] and schematised in Figure 7.1.

The upper part of Table 7.23 shows the factors influencing the existence of knowledge transfer between foreign subsidiaries and their local direct suppliers, ranked by order of significance. The lower part of the table shows the factors influencing the degree of transfer, again ranked by order of significance.

Looking at significance factors across all indicators and all companies, regardless of their origin, firm-related variables are the most important factors in the explanation of the existence of transfer (rank 1 order of significance), while purchase-related factors are most important in the explanation of the degree of transfer between partners. Knowledge transfer is likely to occur if the subsidiary is large (in terms of number of employees), the product manufactured is relatively high in technology content and if firms are from the other category (that is, neither from Japan, nor from the NIEs). Further, the degree of transfer will be higher if the subsidiary has a high share of internal inputs, a high share of subcontracting in total contractual relations with suppliers, if the single supplier strategy is favoured and if the company originates from the other country category.[25] Of course, this list of factors is not exhaustive, and ranks 2-4 factors also act as significant factors in the explanation of the existence and the degree of knowledge transfer.

Table 7.23a Determinant variables and order of significance

Significance level	Existence of transfer: results from logistic regression models							
	All indicators		Transit indicator		Tecdummy indicator		Tradummy indicator	
	Across all regressions	Japanese firms	Across both regressions	Japanese firms	Across both regressions	Japanese firms	Across both regressions	Japanese firms
Rank 1 Mostly firm-related variables (except for *Tecdummy*)	• Size • Product manuf. • Origin	• Size • Export share	• Size	• Size	• Product • Share of subcon-tracting	• Product • Share of subcon-tracting • Size • Single supplier • Export share	• Origin	• Size • Product manuf.
Rank 2 Firm and purchase-related	• Single supplier • Export share	• Product manuf. • Single supplier	• Origin • Single supplier • Export share	• Export share • Single supplier	• Size • Single supplier • Export share		• Size • Product	• Time • Export share
Rank 3 Mostly purchase-related and Time factor	• Share of subcon-tracting • Time	• Share of subcon-tracting • Time	• Product manuf.				• Time	
Rank 4 Mostly purchase-related	• Share of internal inputs						• Share of internal inputs • Export share • Single supplier	

Table 7.23b Determinant variables and order of significance

Significance level	Degree of transfer: results from linear regression models							
	All indicators		Transit indicator		Tecdummy indicator		Tradummy indicator	
	Across all regressions	Japanese firms	Across both regressions	Japanese firms	Across both regressions	Japanese firms	Across both regressions	Japanese firms
Rank 1 Purchase-related (and origin)	• Share of internal inputs • Share of subcon-tracting • Single supplier • Origin	• Share of internal inputs	Share of internal inputs	• Share of internal inputs	• Single supplier	• Share of internal inputs	• Share of subcon-tracting	• Share of internal inputs
Rank 2 Purchase and firm-related	• Size	• Share of subcon-tracting	• Single supplier • Product manuf. • Origin	• Share of subcon-tracting	• Share of internal inputs	• Single supplier	• Size • Share of internal inputs • Origin • Product manuf.	• Share of subcon-tracting
Rank 3 Purchase and firm-related	• Product manuf.	• Single supplier	• Share of subcon-tracting • Size		• Share of subcon-tracting			
Rank 4: Purchase-related							• Single supplier	

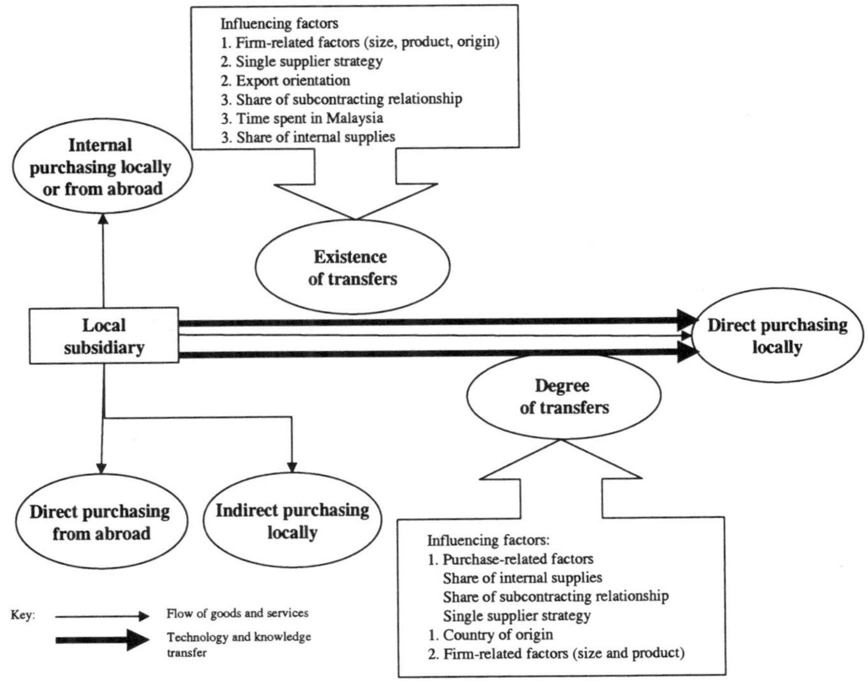

Figure 7.1 Schematised view of factors significantly influencing existence and degree of knowledge transfer
Note: Figure drawn from data presented in Table 7.23

If Japanese firms are analysed independently, a different set of explanatory factors is highlighted. Firm-related factors are still part of rank 1 for the explanation of the existence of transfer, but they mostly highlight the size of the company and the export share as key explanatory factors. Similar results occur for the explanation of the degree of transfer.

A similar classification was carried out for each transfer indicator independently. If the 'transit' indicator is looked at, then the size clearly becomes the key explanatory factor in the existence of transfer, regardless of the country of origin. The 'transfer' indicator is mainly explained by the share of internal inputs, that is the purchasing strategy and independence of the foreign subsidiary.

The transfer of process and product technology shows a distinctive set of results. In this type of transfer, purchase-related factors are key explanatory factors, whether it is for the existence of the transfer or its degree. Other key explanatory factors for the existence of product and process technology transfer include type of product manufactured and its technology content; size of subsidiary and export share are key factors in the case of a Japanese subsidiary.

Transfer through training proved to be more widely used by subsidiaries. Training by the subsidiaries of their suppliers can be explained by a wider set of factors than product and process technology transfer. The origin of the company is a key explanatory factor of the existence of training. The descriptive analysis clearly showed that the firms from the USA were providing substantial training to their suppliers, while such behaviour was less likely to occur among firms from the NIEs. The two other key factors in the explanation of existence of training are the type of product manufactured and the size of the company, as well as the time spent in Malaysia and the export orientation, as in the case of Japanese subsidiaries. The degree of training is mostly explained by purchase-related factors, namely the share of subcontracting across all firms, and the share of internal inputs for Japanese firms.

Discussion on the Results

While the statistical models were significant and introduced useful indicators of factors significantly influencing backward linkages' transfers, some results were unexpected and need further consideration. Overall validation of results was achieved through two means; an examination of the adjusted *R*-squared[26] and a comparison of results across sub-samples of the population.[27]

While some results were expected from the theoretical model designed in Chapter 2, and from the literature review in Chapter 3, others are unexpected. We can explain variation from expected results either by circumstances surrounding the firms, or by the specific construct chosen for the variable. In the case of time, results from the interviews highlighted that firms tend to keep suppliers over a long-time period, and having spent more time in the host-developing country will therefore not equate to creating more supplying relations. The linkages between buyer-supplier depend on each specific supply relation. We evaluated the time spent in Malaysia, and not the individual length of supply relationship. If we could calculate a time variable for each relationship, regressions' results may have shown different levels of significance.

In the case of nonmetric variables, these could not directly be included in the linear regression equation, and were therefore represented by a series of newly created variables (dummy variables[28]). We used dichotomous dummy variables (0 versus 1) in the linear regression models. For the logistic regression models, one dummy variable was dichotomous (single supplier variable), and two had more categories (*k*-1 category for the country of origin and product manufactured variables). First, the product manufactured category was found to be very significant in the logistic regression analysis, and significant for the linear regression analysis. For the former, *k*-1 categories helped comparing all types together, highlighting the categories less likely to lead to transfers with a negative relationship. For the latter, each dichotomous dummy variable was entered in the model, and only the significant ones were highlighted in the results (that is, other products and active components).

The independent variable single supplier is a dichotomous variable. It is a generalisation of each firm's main supply pattern. Firms that favour one single supplier for specific inputs were entered as 1, whereas those who favour several suppliers per item were entered as 0. A positive coefficient indicates that firms favouring single suppliers tend to transfer more than other firms. It can be argued that firms favouring one single supplier may also have several suppliers for some items, and vice versa. In this respect, results for this variable show a trend rather than an established fact. Models indicated a diverging sign of the relationship. The sign is negative for the logistic regression models, and positive for the linear regression models. This would indicate that transfer is more likely to occur when the subsidiary has a multiple purchasing strategy, but a higher degree of transfers occur for subsidiaries that favour a single supplier per item purchased. The signs were partly expected. They indicate the difference between transfer of technology and knowledge and the degree of transfer. Thus, if there is more than one supplier per item purchased, the company is more likely to create buyer-supplier relationships through transfers. On the other hand, companies that favour one supplier per item create more links with this supplier and the degree of transfer is higher.

Overall, our regression results all assist in addressing the basic research question: what affects buyer-supplier knowledge transfer? Our regression models have shown high levels of predictive accuracy,[29] and special care was taken to comply with the case-to-variable ratio. There were few dummy variables included in the models. Dummies entered in the models added to both the prediction and explanation of the research question.

CONCLUSION

In this chapter, we have performed a series of logistic and linear regression analyses in order to identify factors that significantly explain the existence and the level of backward linkages' transfer. We described in detail the methods chosen and the operationalisation of transfer indicators for overall transfers, product and process technology transfers and transfers through training. Each transfer indicator was proved reliable, and a descriptive analysis was provided. A series of logistic and linear regression models were computed, and led to a discussion on factors significantly influencing existence and degree of transfer.

All these results are further discussed in the next chapter, drawing on a comparison between previous surveys conducted by other researchers, and on the results obtained from the interviews with purchasing managers of 11 subsidiaries in the MEE in Chapter 6.

NOTES

1. The distinction between direct and indirect supply was made in Chapter 5. Direct supplies include parts and components that require some established knowledge from local suppliers. Indirect supplies would include boxes, pencils or items that would not directly enter into the production process. Supply of indirect material does not lead to buyer-supplier transfers, and for this reason, firms that only purchase these are not included for further analysis in this research.
2. The equation showing the odds ratio is included in this chapter. It allows differentiation between the probability that a firm will transfer technology and knowledge, as opposed to the probability that a firm will not.
3. Firms were given a list of technology and knowledge items, and were asked to rank them from being 'never', 'rarely' or 'frequently' transferred. Figures reported here are the percentages of firms that responded 'frequently' to the various transfers.
4. For a further analysis of these types and modes, and a comparison across firms from different nationalities of origin, see Giroud (1999b).
5. These initially included 'product specification', 'physical/technical

specifications', 'method of manufacture', 'input joint design' for product technology transfers; 'information on plant establishment', 'supply of machinery', 'supply of tools', 'information on input procurement', 'information on supply management and 'information on technical management' for process technology transfers; 'management training', 'operation staff training', 'training on the technology used' for transfer and the types of training; 'visits to the subsidiary', 'off-the-job training at the subsidiary', 'off-the-job training at the suppliers', 'off-the-job training in institutions', 'on-the-job training at the subsidiary', 'on-the-job training at the suppliers' for transfer through means of training.

6. The categories 'product specification' and 'physical/technical specifications' were found to occur in most cases, whereas the category 'management training', 'off-the-job training at suppliers' and 'off-the-job training in institutions' almost never occurred. These categories showed a small standard deviation, therefore not bringing additional information on technology transfer, since they either 'always' or 'never' took place. They were not used in the final computation.
7. The six variables included are 'operation staff training', 'training on technology used', 'visits to the subsidiary', 'off-the-job training at the subsidiary', 'on-the-job training at the subsidiary' and 'on-the-job training at suppliers'.
8. The 12 variables included are 'method of manufacture', 'supply of machinery', 'supply of tools', 'information on input procurement', 'information on supply management', 'information on technical management', 'operation staff training', 'training on technology used', 'visits to the subsidiary', 'off-the-job training at the subsidiary', 'on-the job training at the subsidiary' and 'on-the job training at the suppliers'.
9. The variable 'information on supply management' is removed.
10. These variables include 'method of manufacture', 'supply of machinery', 'supply of tools', 'Information on input procurement' and 'information on technical management'.
11. Equal weighting is given to all variables in the final combination.
12. Three independent variables are nominal variables. The type of product manufactured is divided into 5 sub-categories, namely active components, passive components, electro-mechanical components, consumer electronics and other products. The country/region of origin of the firms is divided into 3 sub-categories, namely Japan, NIEs and other countries. The purchasing strategy is a dummy variable indicating '1' if the affiliate favours a single supplier per item purchase, and '0' if not. Nominal variables were treated through the deviation method, in which each category of the predictor variable is compared to the overall effect.
13. Identical sizes almost never happen. The reliability analysis shows that results from the smallest group of non-transferrer are reliable. Another issue considered by the researcher is that of non-transferrer not filling in the questionnaire. This issue was considered unlikely because of the structure of the questionnaire. Transfer-related questions were included towards the end of the questionnaire, and it is likely that respondents would complete the questionnaire once they had started filling it in.
14. Hair et al. (1998) specify that the proportional chance criterion should be used when group sizes are unequal and the researcher wishes to identify correctly

members of the two groups. The formula for the proportional chance criterion is $Y = b^2 + (1\text{-}b)^2$, where '*b*' is the proportion of firms that do not transfer technology. In this model, $Y = (.337)^2 + (.663)^2 = .5531$. The classification accuracy of 84.21% is substantially higher than the proportional chance criterion of 55.31%. Because the total value analysis occurs 67.11% of the time, if we assigned all observations to this group, the maximum chance criterion is larger than the proportional test criterion, and our model should outperform 67.11%. The classification accuracy of 84.21% is substantially higher than the proportional chance criterion of 52.5%, and the maximum chance criterion of 67.11%. It also exceeds the suggested threshold of the value plus 25%, which in this case sets the threshold at (67.11*1.25) 83.89%. Thus, our classification accuracy exceeds the threshold, and it exceeds the maximum chance criterion by 16.7%. The calculation of the percentage improvement on the proportional chance criterion shows an improvement of (84.21/55.31) 52.25%.

15. The proportional chance criterion is $Y = (.337)^2 + (.663)^2 = .5531$. The classification accuracy of 85.90% is substantially higher than the proportional chance criterion of 55.31%. Because the total value analysis occurs 67.11% of the time, if we assigned all observations to this group, the maximum chance criterion is larger than the proportional test criterion, and our model should outperform 67.11%. The classification accuracy of 85.9% is substantially higher than the proportional chance criterion of 52.5% and the maximum chance criterion of 67.11%. It also exceeds the suggested threshold of the value plus 25%, which in this case sets the threshold at (67.11*1.25) 83.89%. Thus, our classification accuracy exceeds the threshold by 2%, and it exceeds the maximum chance criterion by 18.8%. The calculation of the percentage improvement on the proportional chance criterion shows an improvement of (85.90/55.31) 55.3%.
16. This category, as specified in Chapter 5, includes power sources, production and other miscellaneous equipment, computer and peripherals and other industrial equipment.
17. The proportional chance criterion is $Y = (.47)^2 + (.53)^2 = .5018$. The classification accuracy of 75% is substantially higher than the proportional chance criterion of 50.18%. Because the total value analysis occurs 55.26% of the time, if we assigned all observations to this group, the maximum chance criterion is larger than the proportional test criterion, and our model should outperform 55.26%. The classification accuracy of 75% is substantially higher than the proportional chance criterion of 50.18% and the maximum chance criterion of 55.26%. It also exceeds the suggested threshold of the value plus 25%, which in this case sets the threshold at (55.26*1.25) 69.07%. Thus, our classification accuracy exceeds the threshold by 6%, and it exceeds the maximum chance criterion by 24.82%. The calculation of the percentage improvement on the proportional chance criterion shows an improvement of (75/50.18) 49.46%.
18. The proportional chance criterion is $Y = (.47)^2 + (.53)^2 = .5018$. The classification accuracy of 74.07% is substantially higher than the proportional chance criterion of 50.18%. Because the total value analysis occurs 55.26% of the time, if we assigned all observations to this group, the maximum chance criterion is larger than the proportional test criterion, and our model should outperform 55.26%. The classification accuracy of 74.07% is substantially

higher than the proportional chance criterion of 50.18% and the maximum chance criterion of 55.26%. It also exceeds the suggested threshold of the value plus 25%, which in this case sets the threshold at (55.26*1.25) 69.07%. Thus, our classification accuracy exceeds the threshold by 5%, and it exceeds the maximum chance criterion by 23.89%. The calculation of the percentage improvement on the proportional chance criterion shows an improvement of (74.07/50.18) 47.6%.

19. The proportional chance criterion is $Y = (.458)^2 + (.542)^2 = .5035$. The classification accuracy of 80.26% is substantially higher than the proportional chance criterion of 50.35%. Because the total value analysis occurs 53.95 % of the time, if we assigned all observations to this group, the maximum chance criterion is larger than the proportional test criterion, and our model should outperform 53.95%. The classification accuracy of 80.26% is substantially higher than the proportional chance criterion of 50.35% and the maximum chance criterion of 53.95%. It also exceeds the suggested threshold of the value plus 25%, which in this case sets the threshold at (53.95*1.25) 67.44%. Thus our classification accuracy exceeds the threshold by 12.82%, and it exceeds the maximum chance criterion by 29.91%. The calculation of the percentage improvement on the proportional chance criterion shows an improvement of (80.26/50.35) 59.4%.
20. The proportional chance criterion is $Y = (.458)^2 + (.542)^2 = .5035$. The classification accuracy of 81.01% is substantially higher than the proportional chance criterion of 50.35%. Because the total value analysis occurs 53.95% of the time, if we assigned all observations to this group, the maximum chance criterion is larger than the proportional test criterion, and our model should outperform 53.95%. The classification accuracy of 81.01% is substantially higher than the proportional chance criterion of 50.35% and the maximum chance criterion of 53.95%. It also exceeds the suggested threshold of the value plus 25%, which in this case sets the threshold at (53.95*1.25) 67.44%. Thus, our classification accuracy exceeds the threshold by 13.58%, and it exceeds the maximum chance criterion by 30.66%. The calculation of the percentage improvement on the proportional chance criterion shows an improvement of (81.01/50.35) 60.89%.
21. The proportional chance criterion is $Y = (.378)^2 + (.622)^2 = .0.5298$. The maximum chance criterion is simply the percentage correctly classified if all observations were placed in the group with the greatest probability of occurrence. Because the total value analysis occurs 61.36% of the time, if we assigned all observations to this group, the maximum chance criterion is larger than the proportional test criterion, and our model should outperform 61.36%. The classification accuracy of 84.09% is substantially higher than the proportional chance criterion of 52.98% and the maximum chance criterion of 61.36%. It also exceeds the suggested threshold of the value plus 25%, which in this case sets the threshold at (61.36*1.25) 76.7%. Thus, our classification accuracy exceeds the threshold by 7.4%, and it exceeds the maximum chance criterion by 31.12%. The calculation of the percentage improvement on the proportional chance criterion shows an improvement of (84.09/52.98) 58.72%.
22. The proportional chance criterion is $Y = (.489)^2 + (.511)^2 = 0.5002$. The maximum chance criterion is simply the percentage correctly classified if all observations were placed in the group with the greatest probability of

occurrence. Because the total value analysis occurs 61.36% of the time, if we assigned all observations to this group, the maximum chance criterion is larger than the proportional test criterion, and our model should outperform 61.36%. The classification accuracy of 81.40% is substantially higher than the proportional chance criterion of 50.02% and the maximum chance criterion of 61.36%. It also exceeds the suggested threshold of the value plus 25%, which in this case sets the threshold at (61.36*1.25) 76.7%. Thus, our classification accuracy exceeds the threshold by 4.7%, and it exceeds the maximum chance criterion by 31.38%. The calculation of the percentage improvement on the proportional chance criterion shows an improvement of (81.40/50.02) 62.73%.

23. The proportional chance criterion is $Y = (.511)^2 + (.489)^2 = 0.5002$. The maximum chance criterion is simply the percentage correctly classified if all observations were placed in the group with the greatest probability of occurrence. Because the total value analysis occurs 61.36% of the time, if we assigned all observations to this group, the maximum chance criterion is larger than the proportional test criterion, and our model should outperform 61.36%. The classification accuracy of 86.67% is substantially higher than the proportional chance criterion of 50.02% and the maximum chance criterion of 61.36%. It also exceeds the suggested threshold of the value plus 25%, which in this case sets the threshold at (61.36*1.25) 76.7%. Thus our classification accuracy exceeds the threshold by 9.97%, and it exceeds the maximum chance criterion by 36.65%. The calculation of the percentage improvement on the proportional chance criterion shows an improvement of (86.67/50.02) 73.27%.
24. Table constructed from analysis of Tables 7.20 and 7.21.
25. As specified in Chapter 5, the other country category includes firms from Europe, the USA, Australia and New Zealand.
26. Adjusted *R*-squared indicates that the estimated model is not over-fitted to the sample and maintains an adequate ratio of observations to variables in the variate (Hair et al. 1998: 209).
27. Our results demonstrated a high level of similarity of the results in terms of *R*-squared, adjusted *R*-squared, and the standard error of the estimate.
28. Dummy variables represent the separate categories of the non metric variable. A dummy variable is an independent variables used to account for the effect that different levels of a non-metric variable have in predicting the dependent variable. The dummy variable is a dichotomous variable that represents one category of a non-metric independent variable. Any non-metric variable with *k* categories can be represented as *k*-1 dummy variables.
29. The amount of variance explained by linear regression models exceeds 58%, except for one model related to transfer of product and process technology, and it exceeds 40% for logistic regression models, except for one.

8. Conclusion

INTRODUCTION

As explained in Chapter 5, the objective of the triangulation method in this research is to obtain an in-depth understanding about the purchasing relationships of the foreign subsidiary, and to uncover hidden dimensions through the qualitative analysis of semi-structured interviews. Quantitative results are coupled with qualitative ones below, in order to describe all aspects of backward linkage creation. The multiple methods' strategy allowed us to gather information from various angles to reinforce reliability of the data. Both methodological approaches now converge into an integrated approach.

The results obtained from the literature review in Chapters 2, 3 and 4, semi-structured interviews and mail survey are integrated below. The qualitative analysis of the results obtained through 11 interviews appear in Chapter 6. In Chapter 7, the data compiled during the mail survey were computed for further statistical analysis. The analyses were based on 95 foreign firms in the Malaysian electrical and electronics sector (MEE), which represented 30% of all foreign firms operating in this sector at the time of the survey. Firms with no local purchase of direct supplies are not taken into account for the analysis of buyer-supplier transfers, but provide precious comparative information to evaluate the existence of backward linkages in the host country. A series of logistic and linear regression models[1] were evaluated, first to show which factors explain the existence of backward linkages' transfers, then to show factors important to explain the degree of such transfers. Both qualitative and quantitative paradigms are now used to reinforce the final results of this research.

In this concluding chapter, we first summarise our study's main findings on transnational corporations' (TNCs') backward linkages, and related knowledge transfer. These results are then discussed in the context of the integrative view of backward linkages formulated in Chapter 2. Additionally, we indicate the study's implications for TNCs, local suppliers and governments, and the study's contributions and limitations are identified. Finally, we offer some suggestions for further research.

INTEGRATION OF RESEARCH FINDINGS

This section is divided into three sub-sections. The first two provide a descriptive analysis of backward linkages and subsequent transfers, following the triangulation process. This leads to the third sub-section, in which we provide answers to the research questions posed in Chapters 1 and 5.

Evaluation of Transfers between Foreign Subsidiaries and Local Suppliers

The level of local purchasing is analysed, primarily drawing on information gathered during interviews. The types of knowledge transfer are then described, integrating both methodologies.

Level of local purchasing

Detailed assessment of local supplies by foreign subsidiaries was primarily gathered by means of interviews, to grasp the complexity of purchasing and its various facets. Tables 6.2-6.4 showed how little direct purchasing is carried out by foreign subsidiaries. Within the mail survey, seven firms (one-tenth of all firms in the sample) recorded that they had no local direct purchasing at all.[2] Table 5.5 provided details about firms that do not purchase any direct supplies locally. Slightly less than 10% of firms purchase no direct supply locally. Nearly one-fifth of firms from the newly industrialising economies (NIEs) do not purchase local supply in Malaysia, against 11% for European firms and 6.1% for Japanese firms. Results from the interviews showed some differences in purchasing patterns according to the origin of the firm. Japanese subsidiaries purchase few supplies locally, and when they do so, they primarily use other Japanese suppliers located in the MEE. European and North American affiliates rely more on local direct purchase, but for low-tech-intensive items only. Direct inputs primarily originate from overseas, either from suppliers external to the TNC, or from internal purchase. The share of internal supplies for respondents in the mail survey showed that on average, internal supplies account for 37.8% of total Japanese firms' supplies, 53.1% for firms from the NIEs, 21.6% for European firms and 24.4% for firms from the USA. The overall share of internal supplies across all firms accounts for 38.6% of total supplies. This seems to indicate that local purchasing increases with a better knowledge of the environment, as firms from the NIEs rely substantially on internal supplies, compared to firms from other origins that tend to have a greater experience in their internationalisation process.

The level of local direct supplies is low, and these are often purchased from other foreign affiliates, rather than from locally owned suppliers.

Furthermore, the key components sourced locally are often purchased from foreign-owned firms. Direct purchase from locally owned firms is mostly composed of standard/customised items, such as for instance, metal and plastic parts, standard electronic components, or subcontracting/assembly of parts and components. Problems faced by foreign firms with local suppliers were listed in Table 6.5. From a foreign firm's perspective, the main problems faced by locally owned firms are lack of flexibility, human resource management difficulties, lack of technical expertise, difficulty to meet quality requirements and a disadvantage of being a newcomer in a fast changing high-tech industry. The evaluation of the level of local purchasing is the initial step towards answering the first research question.

> *Research question 1*: Do TNCs have an impact on host-developing countries through backward linkages?

The discussion carried out above indicates 'yes', TNCs do impact on host-developing countries through backward linkages. However, as discussed further later in this chapter, the overall impact is small. Furthermore, nearly one-tenth of TNCs in the MEE will not have any impact on the host country through backward linkages. These firms may purchase indirect supply in Malaysia, but this is not associated with an impact on the production capabilities of local suppliers. Finally, when firms do have an impact through backward linkages, the degree of knowledge transfer that results from backward linkages varies. Such results show undeniably that the target of this research concerns a limited number of cases, and the subsequent impact from backward linkages is not expected to be substantial. Research question 1 will be further answered later in this chapter.

Types of knowledge transfer

Tables 6.6-6.8 showed a detailed account of the knowledge transfer carried out by the firms interviewed. Tables 7.1 and 7.2 reported transfers, as categorised by the researcher, operated by firms that answered the mail survey. These results help to answer our second research question:

> *Research question 2*: To what extent do backward linkages creation lead to knowledge transfer?

Knowledge transfer through backward linkages' is limited in the MEE. Some firms do not transfer any knowledge, while others transfer very little. In Chapter 6, company 3 clearly indicates that there is no transfer, mostly because the company only has local subcontractors, which do assembly work for the subsidiary. Table 7.4, shows that 33.7% of firms do not transfer technology frequently (and indeed transfer none at all). The split indicators show that 47% of firms do not transfer product and process technology, while 45.8% do not provide any training. There are more Asian firms without transfer than firms from other origin, and the share of firms from the NIEs that transfer technology is particularly low.

This discussion emphasises the fact that little transfer frequently occurs from subsidiaries to their local direct suppliers. Across all firms, the description of the indicators computed in Table 7.3 shows that, on average, 25% of all transfer categories are transferred frequently, against 28% for the product and process technology indicator, and 26% for the training indicator. This shows that little knowledge transfer frequently occurs. The benefit of this transfer to a local supplier will further depend on other factors, such as the capability of the supplier to assimilate the transfer and the actual usefulness of the transfer for the supplier.

The interviews highlighted the emphasis by foreign firms on product and process technology transfer, and the predominance of training as a means of transfer. Direct knowledge transfer is often limited to specifications provided by the foreign firm to its suppliers. Management transfers seldom take place, and often occur through providing training to suppliers on the relevant methods. Financial transfers[3] are virtually non-existent. These trends were subsequently confirmed by the mail survey results. The analysis of financial transfers was not taken further, because these seldom occur. Training is a common means of transfers, often linked to technology. As stated by firms interviewed, the most frequent knowledge transfer comes from exchange of information on the product to be purchased. Product specification and physical/technical specifications are provided frequently by 84.3% and 70.9%, respectively, of subsidiaries. More information or help in the manufacturing process is provided by nearly one-third of the firms, while closer cooperation with the suppliers rarely takes place; 29.4% of firms frequently provide training on technology. On-the-job training is the most common means of training, for all firms, either at the supplier or at the subsidiary, and visits to the subsidiary are frequent in all cases.

Factors Influencing Knowledge Transfer

A key lesson to be learnt from this research is that different factors may come into play in explaining first, whether or not transfers will take place, and second, the degree of transfers. Interviews have been insightful to understand how complex the situation behind the transfer process is. The quantitative analysis has allowed, for the first time, to highlight the specific factors behind the explanation of the existence of transfers. To date, no other research has differentiated transfers between their existence and their degree. Yet, results provided in our research show that a different set of factors explains both phenomena. Tables 6.6-6.8 and Table 7.23 summed up the major findings about knowledge transfer and the key explanatory factors behind the existence and degree of transfer.

It was emphasised in Chapter 3 that explanatory factors could be grouped within the predetermined categories, namely firm-related, purchase-related, investment-related and time factor categories. The investment-related factors included factors relative to the host country and government policies. Chapter 4 highlighted that government policies in Malaysia regarding spill-over effect and the set of soft policies were only recently established, and therefore the direct impact of these policies could not as yet be anticipated. Substantial impact of these policies, however, might occur in the future. Chapter 6 showed that government policy sometimes explains the creation of very strong and beneficial relationships between foreign subsidiaries and their local suppliers, but such relations occur in a very small number of cases, and positive results only emerge for these few suppliers involved. Thus, the three categories that are likely to make a substantial difference in the picture of transfers include the firm-related, purchase-related and time factor categories. We now answer research questions 3a and 3b.

> *Research question 3a*: What are the factors that may explain the *existence* of knowledge transfer through backward linkages?

Because many foreign firms do not transfer any technology and knowledge to their local supplier, the first step in looking at transfers is to analyse the conditions under which transfers do occur. This question was answered through the logistic regression models, and the results were summarised in Tables 7.20 and 7.23. For validation of results, the table lists factors that appear most often across logistic regression models. A series of explanatory factors are given according to whether firms are Japanese or not (mostly due to the number of responses available per category), and according to the type of transfer, whether it be product and process technology-related, or transfer through training. Figure 8.1 represents the factors that explain why knowledge transfer may take place or not. In the figure, three levels of importance for factors are provided. The most important set of explanatory factors is firm-related factors, that is size of the company (in terms of number of employees), type of product manufactured by the subsidiary (and subsequently the level of technology in the product) and origin of the firm.

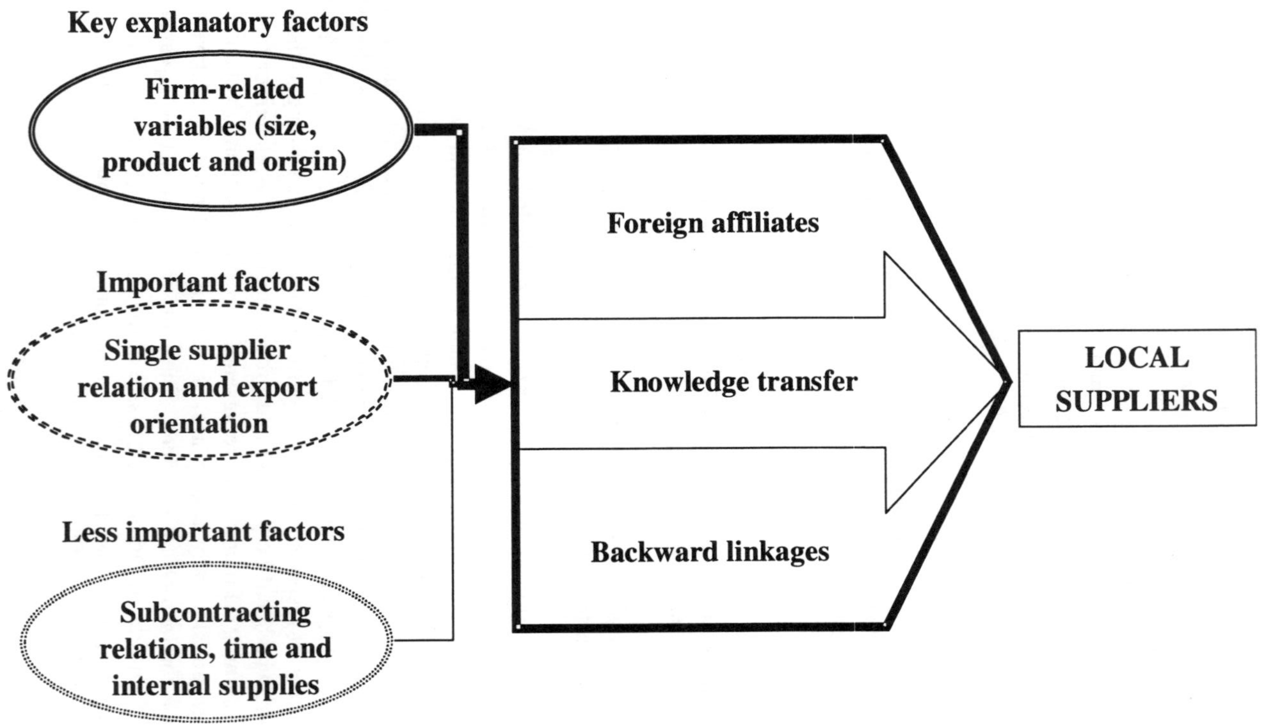

Figure 8.1 Factors explaining existence of knowledge transfer

Figure 8.1 presents a summary of our major findings. Some variation in the set of explanatory factors was identified in the quantitative analysis carried out in Chapter 7. Explanatory factors vary according to the origin of the firm. The ranking order of factors influencing transfers across all firms were: (1) firm-related factors (size, product manufactured, origin of firms); (2) single supplier strategy and export-orientation; and (3) share of subcontracting, time spent in Malaysia and share of internal supplies. However, if we consider the models that were computed independently with Japanese firms, a different set of explanatory factors is established. Overall transfer for Japanese firms, was explained by: (1) firm-related factors (size and export orientation); (2) product manufactured and single supplier strategy; and (3) share of subcontracting and time. Thus, conclusions are drawn across all models, but the specific nature of backward linkages' creation according to the country of origin of the firm is confirmed by additional computation that highlighted the specific behaviour of Japanese firms.

We have proved the specific role of origin of firms as a factor to explain backward linkages and knowledge transfer. Now, a similar analysis is possible for the various types of transfer. Figure 8.1 lists key factors for all transfer categories. Our analysis was more detailed and provided for the first time a differentiation according to the types of transfer. For all firms, the factors influencing the existence of product and process knowledge transfer included: (1) product manufactured and share of subcontracting; and (2) size, single supplier strategy and export orientation. However, the factors influencing the existence of transfer through training included: (1) origin of firms; (2) size and product manufactured; and (3) time spent in Malaysia; as well as a range of additional less significant factors.

For the first time, factors affecting the existence of knowledge transfer are studied. The answer to this question is summarised in Figure 8.1, and the key explanatory factors include firm-related factors, namely the size of the subsidiary, the type of product manufactured and the origin of the parent firm. Additionally, we proved that the situation clearly changes according to the country of origin of firms, and the type of transfers involved. Our results are therefore very important as a precedent for looking at transfers from different angles: as well as looking at the degree of transfer, it is important to look at the factors explaining the existence of such transfers.

Research question 3b: What are the factors that may explain the *degree* of knowledge transfer through backward linkages?

Chapter 7 generated a series of linear regression models in order to highlight the factors important in the explanation of the degree of knowledge transfer. The results of these linear models can be found in Tables 7.22 and 7.23. Again, the set of explanatory factors varies according to the origin of the firm, and the type of transfer. In the first column, however, results for the overall transfer indicator are gathered, through taking a look at the number and significance of occurrence of various factors. This table enables us to draw Figure 8.2, showing the key explanatory factors in the degree of the overall transfer between foreign subsidiaries and their local direct suppliers, namely purchase-related factors (share of internal supplies, share of subcontracting relationships and the single supplier strategy) and the origin of the parent firm. Since some firms do not transfer technology, the phenomenon under study is a small one, and therefore the significance levels obtained through the linear regression models are not as high and frequent as those obtained in the logistic analysis. There are less explanatory factors, but these are significant. Figure 8.2 summarises results for the overall indicator. The key explanatory factors are mostly purchase-related, they are share of internal supplies, share of subcontracting relationships and single supplier strategy. Origin of the firm is also a key explanatory factor of the degree of transfer. The second set of factors includes firm-related factors; these are size of the firm and product manufactured.

Figure 8.2 presents a summary of major findings related to the degree of transfer. However, we showed in Chapter 7 that explanatory factors vary according to the origin of the firm. The ranking order of factors influencing the degree of transfer across all firms were: (1) purchase-related factors (share of internal supplies, share of subcontracting and single supplier strategy); (2) origin of the firm; and (3) firm-related factors (size and product manufactured). If we now consider the models that were computed independently with Japanese firms, a different set of explanatory factors is established. Overall transfer for Japanese firms was explained by: (1) share of internal supplies; (2) share of subcontracting; and (3) single supplier strategy. Our conclusions are drawn across all models, but the degree of knowledge transfer varies according to the country of origin of the firm.

We have shown that the origin of firms is one of the factors explaining the degree of knowledge transfer. We can now analyse in a similar way other factors and the various types of transfer. Figure 8.2 lists key factors for all transfer categories. Our analysis was more detailed and provided for the first time a differentiation according to the types of transfer. For all firms, the factors influencing the degree of product and process technology transfer included: (1) single supplier strategy; (2) share of internal supplies; (3) share of subcontracting. However, the factors influencing the degree of transfer through training included: (1) share of subcontracting; (2) size, share of internal supplies and origin of the firm; (3) product manufactured; and (4) single supplier strategy.

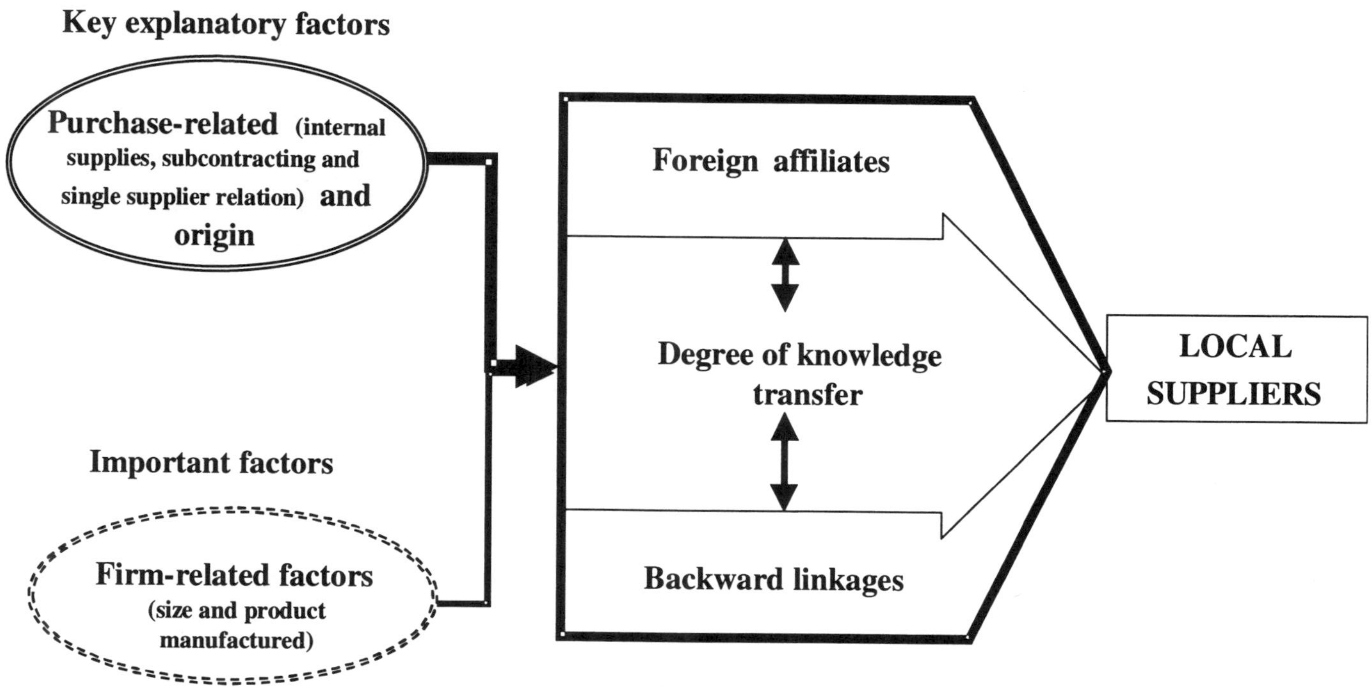

Figure 8.2 Factors explaining degree of knowledge transfer

Clearly, firm-related factors are key to the explanation of whether transfers take place or not, whereas purchase-related factors are the most significant factors to explain the degree of transfer, together with the origin of the firm. Thus, the firm's strategy has an impact on its relation with local suppliers, and the purchase-related factors explain the extent to which technology may be shared with suppliers.

The results obtained through interviews help to understand the process further. The country of origin and its impact on transfers is difficult to ascertain through the interview process. However, within the firm-related category, respondents identified some key factors. The type of product manufactured was clearly identified as a key factor in the explanation of transfers. Thus, high-tech companies manufacturing technology-intensive products were found to pay more attention to relationships with suppliers, so as to ensure the high quality of the products delivered. Firms manufacturing/assembling consumer electronics products may purchase more locally, but less emphasis on relationships with suppliers and less transfer may occur. These results were reinforced by findings in the quantitative analysis. Active components and other products categories showed a positive relationship with transfer indicators, while passive components, electro-mechanical components and consumer electronics clearly were either not significant, or showed a negative sign. The position of the product in the life cycle, the level of standardisation reached and the level of technology included in the product determine to a large extent the type of relations with external suppliers. The size of the firm is significant, inasmuch as the organisational structure of the firm has some impact on the way relationships with suppliers are organised. The interviews showed that within big firms, some specific departments are set up to deal with suppliers, which sometimes arrange for knowledge transfer to take place.

Interviews also identified the significance of purchase-related factors. The respondents mentioned the importance of regular meetings with suppliers, and their preference for long-term relationships. For 4 firms out of 11, knowledge transfer occurs mostly for subcontractors. Close suppliers, also called business partners, benefit from transfers in most companies of the sample. Although the majority of supply comes from foreign-owned firms, locally owned firms also benefit from transfers. The type of input purchased from locally owned firms is not usually technology-intensive, but foreign firms are keen to obtain good quality input and often are making regular contacts and evaluations to ensure this quality. Locally owned firms were often mentioned for the supply of plastic, metal and rubber parts, electrical components or support equipment.

The time factor was not clearly identified as a key factor in the explanation of transfer by respondents. However, long-term relationships were emphasised as important, and companies that have been established for a long time tend to keep the same suppliers. Therefore, although the time spent in the host country is not significant, the emphasis on long-term relations with suppliers is clear.

Findings on Research Sub-questions

The discussion above presents the major findings of this research project. These are now developed to provide answers to the sub-questions drawn from the literature review and specified in Chapter 5.

Firm-related factors

> *Question 3.1*: Does the type of product manufactured by the foreign firm have an impact on transfers generated to the local suppliers?

The type of product manufactured by the foreign subsidiaries has been found in previous studies to be of importance in the determination of backward linkages and subsequent knowledge transfer (Halbach, 1989; Rasiah, 1989; Dobson, 1993; O'Brien, 1993). Our study went into more details than most previous studies, looking at the technology-intensity of the product manufactured within the MEE. This emphasis on the product itself, which is helpful in showing how important the level of technology is in the product, also helps in explaining the existence and degree of knowledge transfer to local suppliers. Product manufactured was found to be highly significant in the logistic analysis, and in the linear regression models. Thus, type of product manufactured is a key explanatory factor in the existence of transfer between foreign subsidiaries and their local suppliers. Two sub-categories were significant in the linear regression models, namely the other products' and the active components' categories. Such results indicated that the subsidiaries that manufacture products with higher technology content tend to show a higher degree of transfer than other subsidiaries.

> *Question 3.2*: Are export-oriented foreign firms more or less likely to transfer technology to their local suppliers?

Export orientation is key to the analysis of buyer-supplier relations, because in the context of a developing country, subsidiaries that sell on the local market were found to pay somewhat less attention to quality. Second, foreign subsidiaries in the MEE are often export-oriented as a strategy from the parent firm. The level of local sourcing was found to be higher for firms with higher shares of domestic sale (Pangestu et al., 1992; Supapol, 1995;

Kelegama and Foley, 1999; Belderbos et al., 2001). To date, the literature has not provided any strong evidence of direct links between knowledge transfer to suppliers and export orientation. This research showed the significance of export orientation. This factor was significant in the logistic analysis, although the level of significance varied. The positive sign showed that the higher the share of the export market, the higher the chance that the foreign subsidiary will transfer some technology to local direct suppliers. Export orientation is not significant in the explanation of the degree of transfer between foreign subsidiaries and their local suppliers. Export orientated firms show a lower degree of local content, but are more likely to transfer knowledge to local suppliers.

Question 3.3: Does the size of the company have an impact on transfers to local suppliers?

The strategic orientation of the subsidiary, within the overall international strategy adopted by the parent firm, was found to be important in explaining the degree of local responsiveness by the foreign firm (Fukagwa, 1995). Thus, the emphasis on the size of the company was brought forward, inasmuch as the size explains the importance given by the parent firm to the local operations carried out by the subsidiary. In this research, the size of the company was found to be critical for organisational reasons, because often, larger-sized companies have specialised departments for the sourcing activities, and were found to be more likely to have specific policies to help their suppliers to reach higher quality standards. The interview results were reinforced by the quantitative analysis. The size of the subsidiary is a significant factor explaining the existence of transfer. Size was found less significant in explaining the degree of transfer, and mostly seemed significant in explaining the degree of transfer through training. The positive signs indicated that the larger the subsidiary in terms of the number of employees, the more likely it is to transfer knowledge, and the higher the degree of transfer through training. The sign of the relationship was found negative for one logistic regression, in the case of product and process technology transfer by other firms. Apart from US firms, firms in this category were found to be medium-sized. The fact that some European firms do proceed to some transfer, but are not of a relatively large size, might explain the negative sign in this case. Across all other regression models, the size of the company has a positive relationship with transfer. The larger the subsidiary, the more likely it is to transfer technology.

Question 3.4: Does the country/region of origin of the firm explain transfers to suppliers?

Studies looking at linkages and nationality of origin of the firm have not provided a clear picture of the impact of origin on linkages. Some studies found Japanese firms have few backward linkages in the host economy (Lim and Pang, 1982; Dobson, 1993; Dobson and Chia, 1997), and others found Japanese firms provide substantial links to local suppliers (IDE, 1994). Literature on backward linkages by firms from other origin is scarce, and again does not provide enough evidence. Our results have shown that the country of origin is important to explain the existence and degree of knowledge transfer. The origin is significant in the logistic analysis. Sub-categories representing Japan and NIEs had negative signs when computed against the other countries' category.[4] This indicates that Asian firms are less likely to generate transfers than firms from other countries. Such results must be taken with caution because of the insufficient number of firms from each origin. In the analysis on the degree of transfer, the sub-category other countries is a significant variable, with a positive sign. Across all regression models, the trend indicates that Asian firms are less likely to transfer knowledge, and the degree of transfers by Asian firms is likely to be smaller than that by other firms.

Purchase-related factors

> *Question 3.5*: Does the level of intra-firm supply have an impact on transfers to suppliers?

Chen (1990) found the level of dependence on parent firms and the share of intra-firm exchange of goods to be an important factor when looking at backward linkage creation. The level of intra-firm trade is high among companies interviewed and companies in the mail survey. The regression models showed that the share of internal supply is not often significant to explain the existence of transfer; however, it becomes significant when looking at the degree of transfer. The share of internal supplies is highly correlated to the degree of transfer in linear regression models. The sign of the relationship is negative, and it was discussed in Chapter 7 that this results from a careful external purchasing strategy on the part of foreign subsidiaries. When the share of internal supply is high, the subsidiary is less likely to resort to external suppliers; hence, if the subsidiary does use external suppliers, the relationship is stronger. There is no evidence in the literature on this factor, and the interview analysis has not provided any clear understanding of the process. Therefore, further study is needed to investigate the specific relationship between internal supply, firm's strategy and transfers to local suppliers.

Question 3.6: Does subcontracting as a contractual method explain transfers to suppliers?

There is no doubt that subcontracting relationships have attracted more attention from researchers (Halbach, 1989; Wong, 1991) than other types of supplying relations. The interviews have pointed out that firms are increasingly considering their suppliers as close business partners, often with specific systems to deal with such partners. Although subcontracting relations were found to lead to more transfers, links with close suppliers were also mentioned to be strong. The share of subcontracting relations is not highly significant in the logistic analysis. Thus, the share of subcontracting relations is not a key factor in the explanation of the existence of transfer. However, it is significant in linear regression models. Subcontracting relations are positively related to the degree of transfer, showing that subcontractors benefit from more transfers than other suppliers.

Question 3.7: Does the purchasing strategy, in terms of number of suppliers per item purchased, explain transfers to suppliers?

The purchasing strategy and the choice of suppliers is a significant factor in the logistic analysis, with a negative sign. The negative sign indicates that firms that do not favour a single supplier per item purchased are more likely to transfer technology. In contrast, in the linear regression models, this factor is positively significant. Thus, subsidiaries with several suppliers per input are more likely to operate transfer than others, but firms that favour a single supplier are likely to show higher levels of transfers. Because we used a dummy variable to indicate the single supplier strategy, the level of complexity in the purchasing strategy was not fully grasped, and the interpretation of the results needs to be taken with caution. However, this factor is clearly significant, and future research to clarify the specific nature of the single supplier relation with buyer-supplier knowledge transfer is needed.

Investment-related factors

Question 3.8: Do government policies help in explaining transfers to suppliers?

Government policies were highlighted in the interviews as being substantially beneficial, but only for a very limited number of suppliers. The development of soft policies is still new in Malaysia, and the government only recently tried to encourage foreign firms to link with local suppliers. As discussed in Chapter 4, it is therefore too early to perceive the full benefit of the government's effort to encourage greater inter-firm relationship in the economy. The companies interviewed mentioned that in each case the local supplier introduced to the subsidiary through a government initiative benefited greatly from a close relationship with the subsidiary and substantial knowledge transfer. These results were also confirmed by previous studies (Masterman, 1991; Meyanathan, 1994; Gultom-Siregar, 1995). Soft incentives bring benefits to the host country, but, on the contrary, it was found that local content regulations may lead to negative effect on FDI (Hackett and Srinivasan, 1998).

Time factor

> *Question 3.9*: Does the time spent by the foreign firm in the host country explain transfers to suppliers?

The time spent in Malaysia was considered as a separate factor from the others. Indeed, many aspects of firms' operations and the host economy come into play when looking at time. The host-developing economy may show increasing signs of development, and high-tech capabilities may be shown by the host indigenous firms. Moreover, the knowledge accumulated by the foreign subsidiary over the years spent in the host economy plays a role in the strategy it follows and in the knowledge it has of the local business environment. The time factor was not identified as important in the qualitative analysis, nor was it particularly significant in the quantitative analysis. Similarly, the number of years since operation started for Japanese subsidiaries was not found to have a clearly significant impact on the level of local content in previous studies (Belderbos et al., 2001). We found that time was significant in two of the six logistic regression models. The specific feature of this factor is that it was found significant only in the models related to the existence of transfer through training. Thus, although mostly not significant, time is key to explaining the existence of training between foreign subsidiaries and local suppliers. This factor was not found significant in the case of the degree of transfer.

BACKWARD LINKAGES WITHIN HOST COUNTRIES' STAGES OF GROWTH

We have now answered our research questions and sub-questions. To fully grasp the complexity of research question 1, however, we now incorporate our results into the theoretical framework designed in Chapter 2. We first evaluate backward linkages in the MEE, before placing our results within the stages of growth approach.

TNCs and backward linkages in the MEE

Figure 8.3 has been developed from Figure 2.4 to analyse the backward linkage creation in the MEE. The reliance on the parent firm and other subsidiaries appears through the average share of internal supply. The average share in the MEE is 38.6%, but this share is very high in the case of NIE firms (53%), and varies according to the origin of the firm. Similarly, the dependence of the subsidiary on its parent firm for decision-making varies according to the country of origin. In the mail questionnaire, 78% of firms mentioned that they do receive advice from their parent firms for their supply decisions, and half of the firms mention that their parent firm takes most decisions. Thus, the parent firms exercise substantial control over subsidiaries.

When analysing the level of local purchase, we mentioned that almost one-tenth of the firms have no local direct purchase. The interviews highlighted that firms with local direct supply often purchase from other foreign-owned firms, or supply specific products from locally owned firms, often low-tech-intensive products. Subsidiaries have on average more informal contractual relations, with 51% of all contracts being the straight exchange of goods with subcontractors and suppliers.[5] Twenty-three per cent of all contracts are subcontracting relationships, and in this case, the subsidiary does establish close links with the subcontractors. Similarly, key suppliers are often kept on a long-term basis by subsidiaries; these represent 26% of all contractual relationships. Transfers through backward linkages may concern subcontractors or key suppliers, but do not take place for informal contracts. This further limits the companies that may benefit from backward linkages created by TNCs in the MEE. Our empirical results showed that knowledge transfer takes place among all suppliers, whether subcontractors or key suppliers. However, subcontractors do benefit from higher degrees of transfer.

We can narrow down the potential impact of TNCs on local suppliers through backward linkage creation. One-tenth of firms have no local direct purchase. From those who do purchase locally, half of their contractual relations are informal and lead to no exchange of information between the two parties. Finally, key suppliers receive knowledge transfer, but less than subcontractors do. The impact on local suppliers' production capabilities is limited to a small number of suppliers, and it is likely to be substantial for an even smaller number of subcontractors.

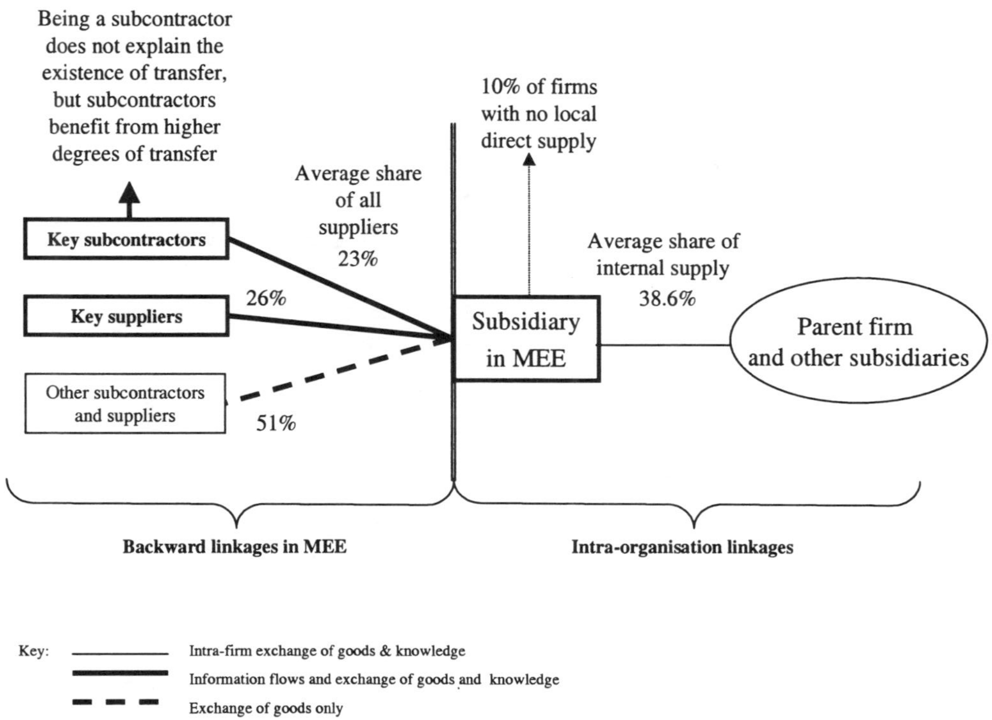

Figure 8.3 TNCs and backward linkages in MEE

Stages of Growth, TNCs and Backward Linkages in the MEE

Research question 1 relates to the potential impact of TNCs on host-developing countries through backward linkages. We have mentioned that TNCs will only impact substantially on a very few number of local suppliers. In Figure 8.4, we have placed Malaysia as the emerging host. The level of external direct supply is still very small (see '1' in Figure 8.4), and as Malaysia develops, the level will increase accordingly. Indeed, key factors that enhance backward linkages include the size of the host country's components' supply on industry and the quality of infrastructure. The level of external direct supply is closely related to the technological capabilities/entrepreneurial potential of the host country. In Malaysia, technological capabilities of local suppliers in the electrical and electronics sector are small, but the grey dotted arrow indicates that those increase as a result of TNCs' operations local sourcing agreement. Thirty-four per cent of foreign firms with local direct purchasing transfer no technology or knowledge. Across firms with some transfer, the average degree of transfer is 25%.[6] In Figure 8.4, TNCs' knowledge transfer (step '2') has an impact on the slope of the host country's technological capabilities (step '3'). However, this impact is very small and depends on a number of factors specific to the existence and the degree of backward linkages' transfers (see Figures 8.1 and 8.2).

This final analysis completes the answer to research question 1. TNCs have an impact on host-developing countries through backward linkage creation and related transfers. Overall, this impact is not substantial, and only concerns a small number of locally owned firms.

The Asian Crisis and its Impact on TNCs' Backward Linkages in MEE

Although we mentioned the Asian crisis in Chapter 4, it is important to mention that the data analysed for this research were gathered in 1996, before the outbreak of the crisis. This means that these results were not affected by the special circumstances faced by countries and companies during the crisis. However, we wish to discuss what this event means in the context of our results.

Malaysia is one of the countries hit by the financial crisis that spread across Asia in 1997 and 1998. A number of factors explain what happened in Asia; these include rapid financial liberalisation with inadequate attention to problems of corporate governance and of prudential regulation, poor macroeconomic management, and policy and institutional failures, the massive flight of capital from Malaysia and its neighbouring countries, as can be seen in Table 8.1.

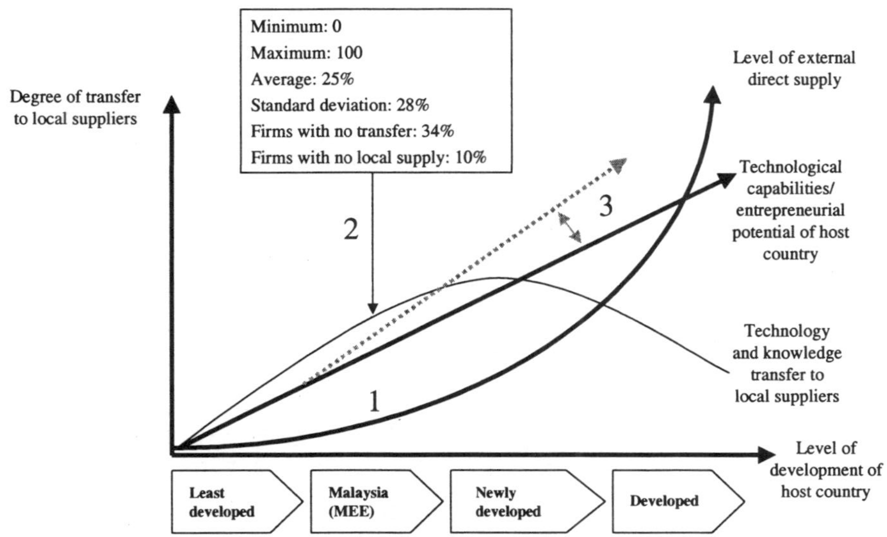

Figure 8.4 Stages of growth, TNCs and backward linkages in the MEE

Notes:

1. Figure 8.3 provides a record of the level of local direct supply.
2. Figures 8.1 and 8.2 showed the factors influencing the existence and degree of knowledge transfer
3. As well as Figures 8.1 and 8.2, the answers provided to the research questions and sub-questions indicate what factors might help increase the slope of the technological capability line.

Table 8.1 External financing of Indonesia, Malaysia, the Philippines, Korea and Thailand, 1994-98 (US$ billion)

	1994	1995	1996	1997	1998
Net external financing	47.4	80.9	92.8	15.2	15.2
Direct equity flows	4.7	4.9	7.0	7.2	9.8
Portfolio flows	7.6	10.6	12.1	-11.6	-1.9
Commercial bank lending	24.0	49.5	55.5	-21.3	-14.1
Non-bank private lending	4.2	12.4	18.4	13.7	-3.3
Net official flows	7.0	3.6	-0.2	27.2	24.6
Change in reserves	-5.4	-13.7	-18.3	22.7	-27.1

Source: Flynn (1999: 8).

Whereas most capital inflows decreased as a result of the crisis in Indonesia, Malaysia, the Philippines, Korea and Thailand, direct equity flows have kept a steady pace, and even increased. When studying TNCs in Malaysia, this has two consequences. First, the Malaysian government has had to find solutions for internal difficulties as a result of the Asian crisis, which it may not otherwise have addressed. This will improve the country's business environment. Second, the topic under study has gained importance as a result of the crisis. Direct equity flows still increased during the crisis, and the urge for local government to accelerate the upgrading of endogenous production capabilities is stronger. As a consequence, the role of TNCs in helping local firms' competitiveness is enhanced.

Additionally, the electrical and electronics sector is the largest contributor to the Malaysian's manufacturing output, employment and exports (see Table 4.7). Eighty-three per cent of firms in the MEE are foreign-owned. Thus, TNCs' operations and subsequent backward linkages are key to the long-term growth prospects of the country.

In Chapter 4, FDI inflow analysis showed a decrease in investment from Japan and the NIEs and an increase in investment from Europe and the USA. In our results, the country of origin factor is significant in explaining the existence and the degree of transfer. An increase in FDI from Europe and the USA in the manufacturing sector is a sign that increasing backward linkages' transfers will occur in the future. Thus, the impact of TNCs on local suppliers' capabilities should increase accordingly.

To conclude, the research on TNCs, backward linkages and impact on host countries is even more important currently, because of the changes within Asian economies, a reminder for governments to focus on capability development in endogenous firms, and the increased investment from Europe and the USA in the manufacturing sector.

IMPLICATIONS FOR TNCs, LOCAL SUPPLIERS AND GOVERNMENTS

Recommendations for further research were made while discussing the contributions and limitations of this book. Although our results do not appear to be practice-oriented, implications do arise for TNCs, local suppliers and governments. Knowledge building in any field is gradual and cumulative, and for the purpose of backward linkages' analysis, the results of one single study can hardly be used to offer firm recommendations. We have shown that TNCs' backward linkage creation depends on a multitude of factors, and factors affecting related knowledge transfers vary with each situation. Therefore, any isolated advice may be of limited use. In spite of these reservations, we endeavour to give some idea of the possible implications of our research for TNCs, local suppliers and governments.

TNCs

Implications for TNCs are considered first, because they were at the centre of this study, and it is their perspective that we have sought. Results showed that foreign firms rely very little on local suppliers for key parts and components in Malaysia. Yet, outsourcing is becoming increasingly strategic; firms world-wide tend to favour external purchasing, especially in the high-tech-intensive industries. We witnessed the importance of buyer-supplier relations when attending a supplier day event organised by a US TNC. First, the TNC was keen to provide incentives to its suppliers thus enabling them to improve their products and services. Second, the TNC was ready to share knowledge on technology with its suppliers in order to enhance their capabilities and help them to become better suppliers. This event proved the importance TNCs give to supplier development, even in a developing country such as Malaysia. The MEE is not attractive to TNCs because of the availability of locally owned suppliers, but if more local suppliers were competitive and able to provide adequate goods and services, the overall business environment would be greatly enhanced. This, in turn, would benefit TNCs.

Host governments in the developing world pay more and more attention to the development of created assets, and they have recently launched new policies to try and encourage backward linkages. By knowing how they contribute to the host country's development through their backward linkages, TNCs have an increased bargaining power in investment negotiations. During the interview process, we found that firms keep supplier records by products. The location of the supplier is often irrelevant for TNCs, they operate with specialised and reliable suppliers. Therefore, firms could not easily provide us with a breakdown of local purchasing by location. This is because the purchasing function is regarded as a strategic one within by firms. TNCs are often unaware of the positive impact they have on the host economy in which they operate. Our research raises the understanding of the role TNCs play in the development pattern of their developing host country.

Local Suppliers

This research focused on the TNCs' perspective, therefore locally owned firms can learn from the existing supply networks of TNCs, and the potential benefits they can gain by dealing with TNCs. In Chapter 6, we gave details on the types of products purchased in the MEE by TNCs, and on the problems of locally owned suppliers. This can help locally owned firms knowing, first, if they could become a supplier to TNCs, and second, what problems TNCs are most concerned with. Although locally owned firms do not need to have business relations with TNCs to be competitive on a national or international basis, there are some advantages to be gained for some firms in becoming suppliers to TNCs. In the MEE, TNCs have specific ownership advantages, among which are technological advantages. In Chapter 4, we showed that foreign firms dominate this industry, because local firms lack these advantages to compete. There is, therefore, a lot that locally owned firms could learn from dealing with TNCs.

This research highlighted the conditions under which backward linkages transfers are maximised. Our results can improve the awareness of locally owned suppliers in terms of what they can learn from TNCs and under which conditions. This allows suppliers to focus on the benefits they may gain from TNCs in host developing countries.

Governments

Host governments in developing Asia pay increasing attention to buyer-supplier development, and to the need to increase local firms' technological knowledge and capabilities. Our study provides a full account on how TNCs impact on host countries' capability development, and therefore provides precious information for governments eager to establish the most appropriate policies to encourage inter-firm linkages. Governments are keen to create a favourable environment for foreign firms to operate in. With a better understanding of backward linkage creation, and factors affecting transfers through backward linkages, governments are in a better position to improve the soft policy schemes mentioned in Chapter 4. In Chapter 7, the distinction between the existence and degree of transfer was made, and factors affecting those latter were identified. These results can help governments to focus on the right companies, in order to try and encourage buyer-supplier transfers.

Host governments have the dual position: they can create the most favourable environment for TNCs to operate in and for local entrepreneurial potential to increase; and they can encourage TNCs to enhance indigenous technological capabilities through backward linkages' transfers. For the Malaysian government, and all other governments attentive to buyer-supplier creation and TNCs activities, our models provide helpful information to evaluate the adequacy of their existing policies, and work on the most effective policies to be adopted in the future.

CONTRIBUTIONS, LIMITATIONS, AND SUGGESTIONS FOR FUTURE RESEARCH

In this section, we discuss the contributions and limitations specific to this research, before providing suggestions for future research.

Contributions

The main contribution of this book lies in its specific focus on backward linkages. This is an under-researched, but extremely important, area of work for both the international business subject area and for host government policy. The desire to capture knowledge transfer benefits from TNCs has been a long-standing objective of policy. This research provides important implications for the academic community, host governments, international organisations, TNCs and local suppliers.

Academic

The findings in this research add to existing literature and understanding of TNCs' activities and their impact on host countries, particularly in developing countries, and more specifically in the Association of Southeast Asian Nations.

Breaking myths. Literature on backward linkages has until now painted a positive picture of TNCs' impact on host countries' production capabilities. TNCs are said to have powerful (but possibly very uneven) effects on the development of local suppliers in developing host countries. It is often claimed that TNCs may be able to raise the capabilities and quality of domestic suppliers and subcontractors to international levels more effectively than domestic firms by transmitting technical information, skills, finance and other forms of assistance. The results presented in our study show that, overall, there is little knowledge transfer between TNCs and their local suppliers. Most studies on backward linkages focus on TNCs that do transfer technology, and fail to provide the overall supplying pattern of those firms. In our study, we show that many firms still do not purchase any direct materials in Malaysia, and although local purchasing increases, part of the increase is new supply from foreign-owned suppliers located in Malaysia. Of the firms that do purchase direct inputs locally, many do not transfer technology at all to their suppliers. Hence, while we acknowledge the existence of transfer, the degree of transfer remains limited, and the positive spillovers benefit few locally owned suppliers.

Bridging theoretical streams. This research was based on theoretical concepts developed by more than one theory. This approach brings in difficulties since it does not test one approach, but a construct of factors carefully designed from the analysis of theories linked to the analysis of the firm and the analysis of international trade. Key influential factors highlighted included industry, time and location. First, we consider the industry. The literature review shows that in the natural resources sector, TNCs tend to create fewer linkages with local suppliers. In our study, we target manufacturing companies in the electronics and electrical sector, where linkages are expected to be more significant. The industry concerned is studied, particularly in the context of the transaction cost theory, the competitive advantage of nations and the stages of growth approach. Rather than focusing attention on whether the industry is mature or not, we emphasise the importance of the level of technology input in the product manufactured. Thus, while we confirm that the industry is central to understanding backward linkages, the type of product manufactured, and the technology intensity are additional central factors to consider.

Second, the industrial net-work approach clearly identifies time as key to the understanding of the type and position of firms in a network, such as the type of linkages and the respective positions of TNCs and their local suppliers in the inter-firm network relations. Time is further associated with the learning curve and adjustments of TNCs to international activities and to the local business environment. The results of our study only partially conform to what is anticipated in the theory. Time shows low levels of significance when explaining the existence of transfer in the logistic regression analysis, and is insignificant when explaining the degree of transfer. Our study indicates that other factors explain close relations between TNCs and their local suppliers. In our quantitative study, the time variable represents the length of time spent in the host country, and not the length of each buyer-supplier relations. The results of the qualitative analysis show that TNCs favour long-term relations, and long-term suppliers benefit from stronger links with the foreign affiliates. We can therefore conclude that the learning curve and adjustment to the local environment is less important than the length of the supplying relationship in explaining close backward linkages between TNCs and their local suppliers. The industrial networks' analysis of definition of time is confirmed by our study, as it does lead to the existence of bonds between partners.

Finally, the location is reflective of the host country itself and the business potentials in the country. Our study confirms the need for appropriate local conditions to optimise the benefits from backward linkages. Companies that were interviewed pointed to the fact that suppliers are long-standing suppliers, contracted at an early stage, when the affiliate started operations in the host country. Although new suppliers develop, affiliates tend to secure and deepen existing supply relations. As a result, maximum benefit from backward linkages will tend to occur if the right factor conditions are in place at the time when the company invests.

Research tools. Backward linkages' studies focus on the degree of technology or knowledge transfer; they therefore concentrate on firms that have supplying relations in the host country. In our study, we include TNCs that do not have such linkages. The logistic regression analysis is used for the first time in the study of backward linkages and brings essential information about TNCs with no linkages as well as TNCs with linkages. The distinction between the two types of firms is essential to truly understand the concepts of buyer-supplier relationships in host countries. Our study sheds a new light on the research tools to be used, inasmuch as we highlight the crucial place of TNCs with no linkages.

Mixed research methodology. The use of the triangulation method has proved a powerful research methodology for the study of backward linkages and knowledge transfer. The qualitative method brings in-depth knowledge about the process through which TNCs interact with their local suppliers, while the two-way quantitative analysis (logistic regression and linear regression methods) adds substantially in terms of generalisation of the results, including a large number of firms in the study. Both methods complement each other to provide a complete picture of transfers through backward linkages, and through this process enable strong validation of the results.

Host country/Malaysian government

An understanding of how firms interact with economic development is fundamental in the effective formulation of policies and regulations related to FDI. In the current context of deregulation and liberalisation of economic policies, the study aims at determining whether such a trend is fully beneficial to the economic welfare of the host country. Also, as Malaysia is a major developing host of FDI, the findings have wider implications and provide examples for countries that try and attract FDI within their national boundaries.

From our results, we identified three essential steps for host government action. First, governments need to encourage all foreign affiliates to supply direct parts and components locally, by helping locally owned firms to become more competitive suppliers. Second, governments must try and encourage all foreign affiliates to transfer technology and knowledge to their local suppliers. Finally, governments ought to facilitate higher degrees of transfer between buyer and supplier.

This study showed that there are few backward linkages in the MEE. TNCs favour long-term supply relationships, and once a vendor is selected, TNCs will prefer to work closely with the chosen vendor than start a new relationship. This has important implications for the host government. It is necessary to provide the appropriate business environment with high-quality suppliers. If local suppliers are available, TNCs will start relationships with them, and contribute to the further development of the local firm. However, if there is a shortage of reliable vendors, TNCs will start relations with other foreign-owned suppliers, or will import direct materials. Hence, the primary role of the government is to set up systems to help develop local vendors. This refers to the location factor discussed in the theoretical framework. The local entrepreneurship potential must exist in the first place for TNCs to further contribute to its development. TNCs will then be in a position to contribute to further development of the existing endogenous production capabilities.

Different types of industry lead to various degrees of inter-firm linkages. Close links are expected in the manufacturing sector. Host governments ought to ensure that sufficient FDI in the manufacturing sector takes place. While the temptation is towards attractive high-tech-intensive firms, these firms may find it difficult to find suitable local suppliers, and fewer backward linkages are likely to occur than with low-tech-intensive firms. Although host governments are keen to develop technology-intensive sectors, they must also consider that low-tech-intensive firms might source more with locally owned firms. A greater number of local vendors would then benefit from close backward linkages and transfer of technology and knowledge.

In the case of Malaysia, new policy tools such as the Umbrella Marketing Scheme and the Vendor'Net Programme have proved beneficial, but for a small number of local vendors. The latter has now been replaced by the Global Supplier Programme, as indicated in Chapter 4. These soft policies generate very close relationships between TNCs and selected locally owned firms. However, in order to ensure that more local vendors benefit from linkages and related transfer, the host government ought to encourage linkages by taking part and acting as a facilitator. In Malaysia, foreign affiliates may obtain benefits when they invest in their employees through training programmes. Similar programmes could be facilitated and encouraged by the government between TNCs and local vendors.

International and inter-governmental bodies
For international organisations and inter-governmental bodies for example United Nations Centre on Transnational Corporations, United Nations Centre for Trade and Development, Organisation for Economic Cooperation and Development and Asian Development Bank that carry out research on the activities of TNCs and on FDI, the findings of this study complement current knowledge and lead to new directions for further research.

TNCs and international firms
Aware of the potential benefits from their activities for the country in which they operate, TNCs cannot only understand better how they can contribute to the indigenous development, but also how they can best benefit from such a development themselves.

Local suppliers
Similarly to TNCs, local suppliers are more aware of the process that determines inter-firm transfers, and what benefits they can gain from contractual relationships with foreign-owned customers. Because foreign firms often adopt different business practices and often try and influence local suppliers, the latter are not always aware of the long-term benefits to be gained from these practices, nor of the factors that influence linkage creation. The results of our study help local suppliers in understanding the nature, explaining factors and potential benefits of backward linkages' creation with TNCs established in their country.

Limitations

This section provides a discussion on the limitations of our study and, where applicable, offers some recommendations for further research. It focuses on linkages generated by TNCs in the electrical and electronics sector. Conclusions are therefore limited to this sector of activity. This sector is known to lead to a higher degree of inter-firm relations than other sectors. Thus, the results need to be taken with a degree of caution for generalisation.

Various sources for the description of FDI flows to Malaysia have been used. In the case of the International Monetary Fund (IMF) data on international financial statistics, they are too aggregated to allow detailed sectorial analysis. Other sources, such as the UN data on international investment, show similar problems for individual researchers. It is also virtually impossible to gather individual source country data; first, because of the scale of such a research and second, because of the difference in data collection and definition of concepts in each country. Additionally, data produced from the Malaysian government are published in terms of investment approval, and do not show the real number of investment projects implemented and completed. This leads to inaccuracies in the data analysis. The researcher is aware of this problem, and was careful in her conclusion building.

The collection of the microeconomic data should be carried out from leading and linked enterprises. The results of the study are incomplete since they are obtained from subsidiaries only. By reason of time and financial constraints, the sample size has been restricted. Further studies should include the suppliers' point of view on backward linkage creation by TNCs.

Time limitation: since the survey was carried out at a particular moment in time, it is a static research. Currently the need for a dynamic type of research is particularly acute with the recent downturn in the economic growth of many Asian countries, and the outburst of the Asian crisis in 1997, after the data were collected. Such a static approach remains the major drawback of a questionnaire survey, especially for researchers who work outside Asia and do not have the facility to renew their surveys at regular intervals.

This research narrows down to the analysis of only one type of linkages: backward linkages. The TNCs' linkage creation with consumers and competitors in the industry are therefore not analysed. Future research should include other types of TNC linkages in the host economy.

Suggestions for Future Research

Methodological consideration

The advantages of adopting a triangulation method were discussed; such a method can be adopted in researching the role of all actors in the linkage creation, that is, the host government, foreign firms and local suppliers. Future research is needed in other developing countries, as well as in other industries.

While doing research in a developing country, researchers need to be cautious about the reliability of data gathered from published materials. It is important to cross-check continuously the data gathered, notably in terms of statistical yearbooks, and company lists. There are inherent difficulties associated with doing research in a foreign developing environment, with which the researcher may not be familiar. For this reason, one must always ensure the reliability of sources. Advice on ways to carry out research in Malaysia can be found in Chapter 5. Each environment being different, such advice can only be indicative for researchers wishing to do some work in this country or in other developing countries.

Lack of data on the subject

Investigations on backward linkages' creation in developing countries and the subsequent knowledge transfer are scarce, and to date, little information has been gathered by academics on this subject area. This is an under-researched, yet crucial area of expertise. The first suggestion is to encourage researchers to consider this topic and add to the existing knowledge on buyer-supplier links in developing areas. The second suggestion addresses itself to government officials. There is a need for more data gathering on companies' activities within their countries, and a strengthening of data gathering on foreign investment flows.

Future research should pay particular attention to the way local suppliers perceive and manage their relationship with TNCs. The research in this book focused on the foreign firms' perspective, but, for a full two-way benefit to take place, a complete understanding of the local suppliers' role and position in the linkage is necessary. Research on the local suppliers' views about backward linkages would also enable a more complete development of an optimum government policy tool in this area.

Finally, a longitudinal study of the impact of new government policies aiming at increased inter-firm linkages would make it possible to highlight the policies that have a positive effect on creation of these linkages.

Comparative research

Comparative studies are essential for a better understanding of backward linkages' creation. Such comparison can take place across different host countries, or across industries.

First, there is a need for comparative studies across Asian developing countries. Such studies can be further complemented with comparative studies across various host-developing countries in different parts of the world, and across different levels of development achieved by the host country. Studies on backward linkages in host-developed countries show a different picture, notably because an existing base of already established competitive suppliers is often available. Hence, comparative analysis ought to be carried out, looking at backward linkages' creation in host-developed versus host-developing countries.

In our study, we focused on the electrical and electronics industry. This sector is known to give rise to substantial linkages with local suppliers and subcontractors. The mechanisms through which TNCs transfer knowledge in other industries need further research; the emphasis should be on a comparative analysis between firms' behaviour in different sectors of activity.

In terms of further work in Malaysia, the researcher wishes to highlight the necessity to deepen the knowledge about foreign firms' inter-firm relations and subsequent benefit for local firms. A large proportion of Malaysian exports originates from foreign-owned firms, and foreign investment accounts for a substantial share of gross domestic capital formation. While foreign investment contributes on a number of fronts to the Malaysian economic success, one must consider the need for locally owned firms to slowly increase their share in exports (notably in the electrical and electronics sector), and in capital formation. Inter-firm linkages between foreign affiliates and their local suppliers are one way through which local firms may become more competitive. More research in this field is therefore needed, as well as research in other sectors of the Malaysian economy.

NOTES

1. The logistic regression analysis allows differentiation between the probability that a firm will transfer technology and knowledge, as opposed to the probability that a firm will not. The linear regression analysis allows evaluation of factors that explain the degree of transfer through backward linkages.
2. These firms subsequently had to be removed from the analysis of local transfers.
3. Financial transfers were included in the early stage of this research, because they were mentioned in previous studies, and the researcher felt these ought to be investigated.
4. Other countries: this category includes firms from the USA and firms from the EU, one Australian firm and one firm from New Zealand.
5. Informal relations are the straight exchange of goods, leading to no further obligations on either part.
6. This is the share of technology and knowledge indicators that are frequently transferred to local suppliers.

References

Achrol, R.S. (1999), 'Forms of network organization', in Richter, F.-J. (Ed.) *Business Networks In Asia: Promises, Doubts And Perspectives*, Westport, Conn.: Quorum, pp. 3-38.

ADB (1996), *Asian Development Outlook 1996-1997*, Asian Development Bank, London: Oxford University Press.

Aggarwal, R. and T. Agmon (1990), 'The international success of developing country firms: the role of government-directed comparative advantage', *Management International Review*, **30** (2), 163-80.

Agmon, T. (1991), *Technology Transfer in International Business*, London: Oxford University Press.

Akamatsu, K. (1962), 'A historical pattern of economic growth in developing countries', *Developing Economies*, **1** 3-25.

Alavi, R. (1996), *Industrialisation in Malaysia: Import Substitution and Infant Industry Performance*, London, New York: Routledge.

Alburo, F.A., C.C. Bautista and M.S.H. Gochoco (1992), 'Pacific direct investment flows into ASEAN', *ASEAN Economic Bulletin*, **8** (3), 284-309.

Alibert, R. (1970), 'A theory of FDI', in Kindleberger, C.P. (Ed.) *The International Corporation*, Cambridge, MA: MIT Press.

Amsden, A.H. (1988), 'Private enterprise: the issues of business-government control', *Columbia Journal of World Business*, **21** (1), 37-51.

Amsden, A.H. (1989), *Asia's next giant: South Korea and late industrialization*, New York, Oxford: Oxford University Press.

Andersson, U. and M. Forsgten (1996), 'Subsidiary embeddedness and control in the multinational corporation', *International Business Review*, **5** (5), 487-508.

Anwar, T.S. (1997), 'Trends in international business thought and literature: reviewing East Asia's miracle, its economic dynamism, and future issues', *International Executive*, **39** (1), 83-109.

Aoki, M., B. Gustafsson and O.E. Williamson (Eds) (1990), *The Firm as a Nexus of Treaties*, London, Newbury Park, New Delhi: Sage.

APEC (1999), *Guide to the Investment Regime of APEC Member Economies*, APEC Committee for Trade and Investment, http://www.apecsec.org.sg.

Ariff, M. (1991), *The Malaysian Economy: Pacific Connections*, Oxford: Oxford University Press.

Ariff, M. and C.P. Lim (1987), 'Foreign investment in Malaysia', in Cable, V. and B. Persaud (Eds), *Developing with Foreign Direct Investment*, London: Croom Helm, pp. 101-22.

Arndt, S.W. (2001), 'Production networks in an economically integrated region', *ASEAN Economic Bulletin*, **18** (1), 24-34.

ASEAN (1999), *ASEAN Investment Report 1999: Trends and Developments in Foreign Direct Investment*, Jakarta: ASEAN Secretariat.

ASEAN (2000), *ASEAN Investment Report 2000: Challenges and Development*, Jakarta: ASEAN Secretariat.

ASEAN (2001), *ASEAN Investment Report 2001: Foreign Direct Investment and Regional Integration*, Jakarta: ASEAN Secretariat.

Athukorala, P. and J. Menon (1996), 'Foreign investment and industrialization in Malaysia: exports, employment and spillovers', *Asian Economic Journal*, **10** (1), 29-44.

Awuah, G.B. (1997), 'Promoting infant industries in LDCs: a network approach to analyse the impact of the exchange relationship between MNCs and their indigenous suppliers in LDCs' efforts to boost infant industries development', *International Business Review*, **6** (1), 71-87.

Balassa, B. (1989), *Comparative Advantage, Trade Policy and Economic Development*, New York, London: Harvester Wheatsheaf.

Bartlett, C.A. and S. Ghoshal (1987), 'Managing across borders: new strategic requirements', *Sloan Management Review*, **28** (7), 7-17.

Bartlett, C.A. and S. Ghoshal (1990), 'The multinational corporation as an inter-organisational network', *Academy of Management Review*, **15** (4), 603-25.

Bartlett, C.A. and S. Ghoshal (1998), *Managing Across Borders: The Transnational Solution*, Boston, MA: Harvard Business School Press.

Bartlett, C., Y. Doz and G. Hedlund (1990), *Managing the Global Firm*, London, New York: Routledge.

Battat, J., I. Frank and X. Shen (1996), Suppliers to Multinationals: Linkages Programs to Strengthen Local Companies in Developing Countries, Washington DC, Foreign Investment Advisory Service, World Bank, Occasional Paper N. 6.

Beamish, P.W. (1985), 'The characteristics of joint ventures in developed and developing countries', *Columbia Journal of World Business*, **20** 13-19.

Belderbos, R., G. Capannelli and K. Fukao (2001), 'Backward vertical linkages of foreign manufacturing affiliates: evidence from Japanese multinationals', *World Development*, **29** (1), 189-208.

Bende-Nabende, A. (1999), *FDI, Regionalism, Government Policy and Endogenous Growth: a Comparative Study of the ASEAN-5 Economies*

with Development Policy Implications for the Least Developed Countries, Aldershot UK, Brookfield USA: Ashgate.
Bernard, M. and J. Ravenhill (1995), 'Beyond product cycles and flying geese: regionalization, hierarchy and the industrialization of East Asia', *World Politics*, **47** (2), 171-209.
Blois, K.J. (1972), 'Vertical quasi-integration', *Journal of Industrial Economics*, **20** 253-72.
Blomström, M., A. Kokko and M. Zejan (2000), *Foreign Direct Investment: Firm and Host Country Strategies*, Basingstoke, New York: Macmillan.
Bloom, M. (1992), Technological Change in the Korean Electronics Industry, Paris, OECD Development Centre Studies.
Bloom, M. (1994), 'Globalisation and the Korean electronics industry: a Chandlerian perspective', in Schutte, H. (Ed.) *The Global Competitiveness of the Asian Firm*, Basingstoke: Macmillan, pp. 138-54.
Boissevain, J. (1974), *Friends of Friends: Networks, Manipulators and Coalitions*, Oxford: Blackwell.
Borensztein, E., J. De Gregorio and J.-W. Lee (1998), 'How does foreign direct investment affect economic growth?' *Journal of International Economics*, **45** (1), 115-35.
Borrus, M. (1993), 'The regional architecture of global electronics: trajectories, linkages and access to technology', *Business and the Contemporary World*, **5** (2), 151-74.
Brewer, T.L. and S. Young (1998), The World Trade Organisation: Global Rule-maker, Workshop on the Globalisation of Multinational Enterprise Activity and Economic Development: University of Strathclyde.
Brinberg, D. and J.E. McGrath (1985), *Validity and the Research Process*, Thousand Oaks, London, New Delhi: Sage.
Buckley, P.J. (1988), 'The limits of explanation: testing the internalisation theory of multinational enterprises', *Journal of International Business Studies*, **19** (2), 181-93.
Buckley, P.J. (1989), *The Multinational Enterprise: Theory and Applications*, Basingstoke: Macmillan.
Buckley, P.J. (1990), 'Problems and developments in the core theory of international business', *Journal of International Business Studies*, **21** (4), 657-66.
Buckley, P.J. (Ed.) (1992), *New Direction in International Business: Priorities in Research for the 1990's*, Aldershot, UK: Edward Elgar.
Buckley, P.J. (1994), 'International business versus international management: international strategic management from the viewpoint of internalisation theory', *Journal of Economics and Business*, **1** 95-104.
Buckley, P.J. and P. Artisien (1987), *North-South Direct Investment in the European Communities: The Employment Effect of Direct Investment by British, French and German Multinationals in Greece, Portugal and Spain*, Basingstoke: Macmillan.

Buckley, P.J. and M. Casson (1976), *The Future of the Multinational Enterprise*, Basingstoke: Macmillan.

Buckley, P.J. and M. Casson (1985), *The Economic Theory of the Multinational Enterprise*, Basingstoke: Macmillan.

Buckley, P.J. and M. Casson (1988), 'A theory of cooperation in international business', in Contractor, F.J. and P. Lorange (Eds), *Cooperative Strategies in International Business*, Lexington, MA: Lexington Books, pp. 31-54.

Buckley, P.J. and M. Casson (1991), 'Multinational enterprises in less developed countries: cultural and economic interactions', in Buckley, P.J. and J. Clegg (Eds), *Multinational Enterprise in Less Developed Countries*, Basingstoke: Macmillan, pp. 27-55.

Buckley, P.J. and M. Casson (1998a), 'Models of the multinational enterprise', *Journal of International Business Studies*, **29** (1), 21-44.

Buckley, P.J. and M. Casson (1998b), 'Analyzing foreign market entry strategies: extending the internalization approach', *Journal of International Business Studies*, **29** (3), 539-61.

Buckley, P.J. and M. Casson (2000), 'Foreign market entry: a formal extension of internalization theory', in Casson, M. (Ed.) *Economics of International Business*, Cheltenham: Edward Elgar, pp. 31-57.

Buckley, P.J. and J. Clegg (Eds) (1991), *Multinational Enterprise in Less Developed Countries*, Basingstoke: Macmillan.

Buckley, P.J. and P. Ghauri (Eds) (1993), *The Internationalisation of the Firm: A Reader*, London: Dryden Press.

Buckley, P.J., Z. Berkova and G. Newbould (1983), *Direct Investment in the UK by Smaller European Firms*, Basingstoke: Macmillan.

Buckley, P.J., G. Newbould and J.C. Thurwell (1988), *FDI by Smaller UK Firms: the Success and Failure of First-time Investors Abroad*, Basingstoke: Macmillan.

Buckley, P.J., J. Campos, H. Mirza and E. White (1997), *The International Transfer of Technology by Smaller Firms*, London: Macmillan.

Bürgenmeier, B. and J.-L. Mucchielli (Eds) (1991), *Multinationals and Europe 1992: Strategies for the Future*, London, New York: Routledge.

Cable, V. and B. Persaud (Eds) (1987), *Developing with Foreign Investment*, London, New York, Sydney: Croom Helm.

Cantwell, J. (1989), *Technological Innovation and the Multinational Corporation*, Oxford: Basil Blackwell.

Cantwell, J. and R. Mudambi (2000), 'The location of MNE R&D activity: the role of investment incentives', *Management International Review*, **40** (1), 127-49.

Cantwell, J. and P.E. Tolentino (1990), Technological accumulation and third world multinationals, Reading, University of Reading, Discussion Papers in International Investment and Business Studies, Series B, N.139.

Carrie, A.S. (2000), 'From integrated enterprises to regional clusters: the changing basis of competition', *Computers in Industry*, **42** (2-3), 289-98.

Carter, J.R. and R. Narasimhan (1996), 'Is purchasing really strategic?' *International Journal of Purchasing and Materials Management*, **32** (1), 20-8.
Casson, M. (1987), *The Firm and the Market: Studies on Multinational Enterprise and the Scope of the Firm*, Oxford: Basil Blackwell.
Casson, M. (1990), *Enterprise and Competitiveness: A System's View of International Business*, Oxford: Clarendon Press.
Casson, M. (Ed.) (1991), *Global Research Strategy and International Competitiveness*, Oxford: Basil Blackwell.
Casson, M. (1993), 'Internalisation theory and beyond', in Dunning, J.H. (Ed.) *The Theory of Transnational Corporations*, London: Routledge, United Nations Library on Transnational Corporations, pp. 361-86.
Casson, M. (1994), 'Why are firms hierarchical?', *Journal of the Economics of Business*, **1** (1), 47-76.
Casson, M. (2000), *Economics of International Business: A New Research Agenda*, Cheltenham: Edward Elgar.
Casson, M. and H. Cox (1992/93), 'Firms, networks and international business enterprise', Discussion Papers in International Investment and Business Studies, V, Series B, Number 23. University of Reading.
Casson, M. and R.D. Pearce (1987), 'The welfare effect of foreign enterprise: a diagrammatic analysis', *University of Reading: Discussion Papers in International Investment and Business Studies*, **98** (October).
Casson, M. and R.D. Pearce (1988), 'Intra-firm trade and developing countries', in Greenaway, D. (Ed.) *Economic Development and International Trade*, Basingstoke: Macmillan, pp. 99-126.
Caves, R.E. (1971), 'International corporations: the industrial economics of foreign investment', *Economica*, **38** 1-27.
Caves, R.E. (1982), *Multinational Enterprises and Economic Analysis*, Cambridge: Cambridge University Press.
Chalmers, I. (1991), 'International and regional integration: the political economy of the electronics industry in ASEAN', *ASEAN Economic Bulletin*, **8** (2), 194-209.
Chen, E.K.Y. (1983), 'Factor proportion of foreign and local firms in developing countries: a theoretical and empirical note', *Journal of Development Economics*, **12** 267-74.
Chen, E.K.Y. (1990), *Foreign Direct Investment in Asia*, Tokyo: Asian Productivity Organisation.
Chen, E.K.Y. (Ed.) (1994), *Transnational Corporations and Technology Transfer to Developing Countries*, UNCTC, London, New York: Routledge.
Chen, T.-J. (1992), 'Determinants of Taiwan's FDI: the case of a newly industrializing country', *Journal of Development Economics*, **39** (2), 397-407.

Chen, T.-J., Y.-P. Chen and Y.-H. Ku (1995), 'Taiwan's outward direct investment', in *The New Wave of Foreign Direct Investment in Asia*, Singapore: ISEAS/Nomura Research Institute, pp. 87-110.

Chia, Y.S. (1993), 'Foreign direct investment in ASEAN economies', *Asian Development Review*, (1), 60-94.

Chung, H.L. (1994), 'Korea's foreign direct investment in Southeast Asia', *ASEAN Economic Bulletin*, **10** (3), 280-97.

Coase, R.H. (1937), 'The nature of the firm', *Economica*, **4** 386-405.

Contractor, F.J. (1990), 'Contractual and cooperative forms of international business: towards a unified theory of modal choice', *Management International Review*, **30** (1), 31-54.

Contractor, F.J. (1995), 'Promoting FDI in developing countries', *International Trade Journal*, **9** (1), 107-42.

Contractor, F.J. and P. Lorange (1988), *Cooperative Strategies in International Business*, Lexington, MA: Lexington Books.

Contractor, F.J. and P. Lorange (1989), 'Inter-firm technology transfers and the theory of multinational enterprise', *International Trade Journal*, **4** (1), 21-48.

Creswell, J.W. (1994), *Research Design, Qualitative and Quantitative Approaches*, Thousand Oaks, London, New Delhi: Sage.

Davis, L. (1997), 'Cross-border buyer-supplier development collaborations', in Björkman, I. and M. Forsgren (Eds), *The Nature of the International Firm*, Copenhagen: Copenhagen Business School Press, pp. 391-410.

D'Cruz, J. and A. Rugman (1992), 'Business networks for international competitiveness', *Business Quarterly*, **56** (4), 101-07.

D'Cruz, J. and A. Rugman (2000), *Multinationals as Flagship Firms: Regional Business Networks*, Oxford: Oxford University Press.

De Mello, J.L.R. (1997), 'Foreign direct investment in developing countries and growth: A selective survey', *Journal of Development Studies*, **34** (1), 1-34.

De Mello, J.L.R. (1999), 'Foreign direct investment-led growth: evidence from time series and panel data', *Oxford Economic Papers*, **51** 133-51.

Dessler, G. (1994), *Human Resource Management*, Englewood Cliff, NJ: Prentice Hall International.

Dillman, D. (1978), *Mail and Telephone Surveys: The Total Quality Design Method*, New York, Chichester: Wiley Interscience Publications.

Dillman, D. (1991), 'The design and administration of mail surveys', *Annual Review of Sociology*, (17), 225-49.

Dobson, W. (1993), *Japan in East Asia: Trading and Investment Strategies*, Singapore: Institute of Southeast Asian Studies.

Dobson, W. and S.Y. Chia (Eds) (1997), *Multinationals and East Asian Integration*, Singapore: Institute of Southeast Asian Studies.

Dowling, M. and C.T. Cheang (2000), 'Shifting comparative advantage in Asia: new tests of the "flying geese" model', *Journal of Asian Economics*, **11** (4), 443-63.

Dunning, J.H. (1988a), *Explaining International Production*, London: Unwin Hyman.

Dunning, J.H. (1988b), *Multinationals, Technology and Competitiveness*, London: Unwin Hyman.

Dunning, J.H. (1993a), *The Globalization of Business*, London, New York: Routledge.

Dunning, J.H. (1993b), *Multinationals Enterprises and the Global Economy*, Wokingham: Addison Wesley.

Dunning, J.H. (1994), 'Re-evaluating the benefits of FDI', *Transnational Corporations*, **3** (1), 23-51.

Dunning, J.H. (1997), 'How should governments respond to globalization?' *International Executive*, **39** (1), 55-67.

Dunning, J.H. (1998a), 'Globalization and the new geography of foreign direct investment', *Oxford Development Studies*, **26** (1), 47-69.

Dunning, J.H. (1998b), 'Location and the multinational enterprise: a neglected factor', *Journal of International Business Studies*, **29** (1), 45-66.

Dunning, J.H. (2000), 'Globalization and the theory of MNE activity', in Hood, N. and S. Young (Eds), *The Globalization of Multinational Enterprise Activity and Economic Development*, Basingstoke: Macmillan Press, pp. 21-52.

Dunning, J.H. and R. Narula (Eds) (1996a), *Foreign Direct Investment and Governments: Catalysts for Economic Restructuring*, London, New York: Routledge.

Dunning, J.H. and R. Narula (1996b), 'The investment development path revisited: some emerging issues', in Dunning, J.H. and R. Narula (Eds), *Foreign Direct Investment and Governments: Catalysts for Economic Restructuring*, London, New York: Routledge, pp. 1-42.

Dwyer, R.F., P.H. Schurr and O. Sejo (1987), 'Developing buyer-seller relationships', *Journal of Marketing*, **51** 11-27.

Dyer, J.H. (1996), 'How Chrysler created an American keiretsu', *Harvard Business Review*, **74** 42-56.

Dyer, J.H. (1997), 'Effective interfirm collaboration: how firms minimize transaction costs and maximise transaction value', *Strategic Management Journal*, **18** (7), 535-56.

Dyer, J.H. and H. Singh (1998), 'The relational view: cooperative strategy and sources of interorganizational competitive advantage', *Academy of Management Review*, **23** 660-79.

Edgington, D.W. and R. Hayter (2001), 'Japanese direct foreign investment and the Asian financial crisis', *Geoforum*, **32** (1), 103-20.

England, W.B., M.R. Leenders and H. Lewis (1975), *Purchasing and Materials Management*, Homewood, IL: Irwin Press.

Enos, J.L. and W.-H. Park (1988), *The Adoption and Diffusion of Imported Technology: A Case Study of Korea*, London, New York, Sydney: Croom Helm.

Enright, M.J. (2000), 'Regional clusters and multinational enterprises: independence, dependence, or interdependence?' *International Studies of Management and Organization*, **30** (2), 114-38.

Ernst, D., T. Ganiatsos and L. Mytelka (1998), *Technological Capabilities and Export Success in Asia*, London, New York: Routledge.

European Round Table (1994), 'FDI as a tool for economic development and cooperation', *European Business Journal*, (l), 8-27.

Farmer, D. (Ed.) (1985), *Purchasing Management Handbook*, Aldershot: Gower.

Far Eastern Economic Review (1995), 'Malaysia: trade and investment', Far Eastern Economic Review, 158 (17), pp. 37-48.

Far Eastern Economic Review (1997), 'Special report: review of 200 Asia's leading companies', Far Eastern Economic Review, 160 (1), pp. 44-9.

Felix, D. (1994), Industrial development in East Asia: what are the lessons for Latin America?, UNCTAD/OSG/DP/84.

Fernandez-Arias, E. and R. Hausmann (2001), 'Is foreign direct investment a safer form of financing?' *Emerging Markets Review*, **2** (1), 34-49.

Finance and Development: A Quarterly Magazine of the IMF, http://www.imf.org/ (1999), 'Foreign direct investment in developing countries', **36** (1), pp. 34-7.

Financial Times (1996), 'Asia set to become top investment target', 6 February pp. 5.

Fink, A. (1995), *The Survey Handbook*, Thousand Oaks, London, New Delhi: Sage.

Fleury, A. (1999), 'The changing pattern of operations management in developing countries', *International Journal of Operations and Production Management*, **19** (5/6), 552-65.

Fleury, M.T., C. Ruy de Quadros and A. Fleury (1994), *The Role of Japanese Direct Investment in Developing Countries: The Case of Brazil*, Japan: Institute of Developing Economies.

Flynn, N. (1999), *Miracle to Meltdown in Asia: Business, Government, and Society*, Oxford: Oxford University Press.

Foddy, W. (1993), *Constructing Questions for Interviews and Questionnaires*, Cambridge: Cambridge University Press.

Fong, C.O. (1990), 'Industrialisation in Malaysia: role of small and medium scale industries', in Buang, A. (Ed.) *The Malaysian Economy in Transition*, Malaysia: Kuala Lumpur National Institute of Public Administration, pp. 71-134.

Forgsten, M. and J. Johanson (Eds) (1992), *Managing Networks in International Business*, Philadelphia: Gordon and Breach.

Fortanier, F. and M. Maher (2001), 'Foreign direct investment and sustainable development', *Financial Market Trends*, **79** 107-30.

Fowler, F.J. (1984), *Survey Research Methods*, Thousand Oaks, London, New Delhi: Sage.

Fujita, K. and R.C. Hill (1997), 'Auto Industrialization in Southeast Asia: National Strategies and Local Development', *ASEAN Economic Bulletin*, **13** (3), 312-32.

Fukagwa, Y. (1995), Japanese, American and European Transnationals in East Asia, Tokyo: Long Term Credit Bank of Japan.

Fukushima, K. and C.H. Kwan (1995), 'Foreign direct investment and regional industrial restructuring in Asia', in ISEAS/Nomura Research Institute (Ed.) *The New Wave of Foreign Direct Investment in Asia*, Singapore: Chapter 1, pp. 3-41.

Germidis, D. (Ed.) (1980), *International Subcontracting: A New Form of Investment*, Paris: OECD.

Giroud, A. (1997), 'Joint ventures versus wholly-owned subsidiaries: the Malaysian electrical and electronics sector', in LVMH Conference, 'Partnership and Joint Ventures in Asia: What have we Learned so far?' Euro-Asia Centre, Fontainebleau, France: INSEAD.

Giroud, A. (1999a), 'Manufacturing in Asia Pacific', in Warner, M.E. (Ed.) *Regional Encyclopedia of Business and Management: Management in Asia Pacific*, London: Business Press, Thomson Learning, pp. 156-67.

Giroud, A. (1999b), 'Upgrading endogenous production capabilities in less-developed countries: an evaluation of information flows between transnational corporations affiliates and their local suppliers', in Sang-Gon, L. and P.-B. Ruffini (Eds), *The Global Integration of Europe and East Asia: Studies of International Trade and Investment*, Cheltenham, UK, Northampton, USA: Edward Elgar, pp. 37-53.

Giroud, A. (2000), 'Japanese transnational corporations' knowledge transfer to Southeast Asia: the case of the electrical and electronics sector in Malaysia', *International Business Review*, **9** (5), 571-86.

Giroud, A. and H. Mirza (1998), 'Taiwanese firms' transfer of technology to ASEAN countries', in LVMH Conference, 'Asian Foreign Investment in Asia'. Euro-Asia Centre, Fontainebleau, France: INSEAD.

Glass, A.J. and S. Kamal (1998), 'International technology transfer and the technology gap', *Journal of Development Economics*, **55** (2), 369-98.

Grabner, G. (Ed.) (1993), *The Embedded Firm and the Socioeconomics of Industrial Networks*, London, New York: Routledge.

Graham, E.M. (1991), 'Strategic trade policy and the MNE in developing countries', in Buckley, P.J. and J. Clegg (Eds), *Multinational Enterprises in Less Developed Countries*, Basingstoke: Macmillan, pp. 79-92.

Graham, E.M. and P.R. Krugman (1993), 'The Surge of FDI in the 1980s', in Froot, K.A. (Ed.) *Foreign Direct Investment*, Chicago, London: University of Chicago Press.

Graham, T.S., P.J. Daughtery and W.N. Dudley (1994), 'The long term strategic impact of purchasing partnerships', *International Journal of Purchasing and Materials Management*, **30** (4), 12-26.

Granovetter, M. (1985), 'Economic action and social structure: the problem of embeddedness', *American Journal of Sociology*, **91** 481-510.

Gray, P.H. (Ed.) (1993), *Transnational Corporations and International Trade and Payments*, London, New York: UNLTC/Routledge.

Greene, J.C., V.J. Caracelli and W.F. Graham (1989), 'Toward a conceptual framework for mixed-method evaluation designs', *Educational Evaluation and Policy Analysis*, **11** (3), 255-74.

Gultom-Siregar, M. (1995), 'Indonesia', in Supapol, A.B. (Ed.) *Transnational Corporations and Backward Linkages in the Asian Electronics Industry*. UNESCAP, pp. 131-72.

Hackett, S.C. and K. Srinivasan (1998), 'Do supplier switching costs differ across Japanese and US multinational firms?', *Japan and the World Economy*, **10** 13-32.

Hair, J.F., R.E. Anderson, R.L. Tatham and W.C. Black (1998), *Multivariate Data Analysis*, London: Prentice-Hall International.

Håkanson, H. (Ed.) (1982), *International Marketing and Purchasing of Industrial Goods: An Interaction Approach*, Chichester: Wiley.

Håkanson, H. (Ed.) (1987), *Industrial Technological Development: A Network Approach*, London: Croom Helm.

Håkanson, H. and J. Johanson (1988), 'Formal and informal cooperative strategies in international industrial networks', in Contractor, F.J. and P. Lorange (Eds), *Cooperative Strategies in International Business*, Lexington, MA: Lexington Books, pp. 369-90.

Halbach, A.J. (1989), 'Multinational Enterprise and Subcontracting in the Third World: A Study of Inter-industrial Linkages', Geneva: ILO, International Labour Office Multinational Enterprises Programme.

Hallén, L. and F. Wiedersheim-Paul (1979), 'Psychic distance and buyer-seller interaction', *Organisation, Marknad och Samhälle*, **16** (5), 308-24. Reprinted in Buckley, P.J. and P. Ghauri (Eds) (1993), *The Internationalisation of the Firm*: A Reader, London: Dryden Press, pp. 291-303.

Hamel, J., S. Dufour and D. Fortin (1993), *Case Study Methods*, Thousand Oaks, London, New Delhi: Sage.

Han, S.-L., D.T. Wilson and S.P. Dant (1993), 'Buyer-supplier relationships today', *Industrial Marketing Management*, **22** (4), 331-38.

Harzing, A.-W.K. (1999), *Managing the Multinational*, Cheltenham: Edward Elgar.

Harzing, A.-W.K. (2000), 'Cross-national industrial mail surveys: why do response rates differ between countries?' *Industrial Marketing Management*, **29** (3), 243-54.

Hayes, R.H. and S.C. Wheelwright (1984), *Restoring our Competitive Edge: Competing through Manufacturing*, New York, Chichester, Brisbane, Toronto, Singapore: John Wiley and Sons.

Heide, J.B. and R.L. Stump (1995), 'Performance implications of buyer-supplier relationships in industrial markets: a transaction cost explanation', *Journal of Business Research*, **32** (1), 57-66.

Helleiner, G.K. (1981), *Intra-firm Trade and the Developing Countries*, Basingstoke: Macmillan.

Hennart, J.F. (1988), 'A transaction cost theory of equity joint venture', *Strategic Management Journal*, **9** (4), 361-75.

Hines, P. (1996), 'Purchasing for lean production: the new strategic agenda', *International Journal of Purchasing and Materials Management*, **32** (1), 2-10.

Hobday, M. (1994), 'Export-led technology development in the 4 dragons: the case of electronics', *Development and Change*, **25** (2), 333-61.

Hobday, M. (1995), *Innovation in East Asia: The Challenge to Japan*, Cheltenham: Edward Elgar.

Hobday, M. (1996), 'Understanding innovation in Southeast Asia: Malaysia's experience in Electronics', University of Sussex, Discussion Paper.

Holmlund, M. and S. Kock (1995), 'Buyer perceived service quality in industrial networks', *Industrial Marketing Management*, **24** (2), 109-21.

Hood, N. and S. Young (1979), *The Economics of Multinational Enterprise*, Singapore: Longman Publishers.

Hood, N. and S. Young (Eds) (2000), *The Globalization of Multinational Enterprise Activity and Economic Development*, Basingstoke: Macmillan.

Horst, L. (1972), 'Firms and industrial determinants of the decisions to invest abroad: an empirical study', *Review of Economics and Statistics*, **54** 258-66.

Howe, K. and M. Eisenhart (1990), 'Standards for qualitative (and quantitative) research: A prolegomenon', *Educational Researcher*, **19** (4), 2-9.

Humphrey, J., Y. Lecler and M.S. Salerno (Eds) (2000), *Global Strategies and Local Realities: The Auto Industry in Emerging Markets*, Basingstoke: Macmillan Press.

Hymer, S. (1976), *The International Operation of National Firms: A Study of Direct Foreign Investment*, Cambridge, MA: MIT Press.

Hyun, J.-H. (1994), 'Buyer-supplier relations in the European automobile component industry', *Long Range Planning*, **27** (2), 66-75.

IDE (1994), The Role of Japanese Direct Investment in Developing Countries: India, Indonesia, Taiwan, Tokyo, Institute of Developing Economies.

IFC (1997), Foreign Direct Investment, Washington, International Finance Corporation and Foreign Investment Advisory Service. Published by the World Bank.

ILO/UNCTC (1984), Technology Choice and Employment Generation by Multinational Entreprise in Developing Countries, Geneva, International Labour Office/United Nations Centre for Transnational Corporations.

ILO/UNCTC (1988), Economic and Social Effects of Multinational Enterprise in Export Processing Zone, Geneva, International Labour Office/United Nations Centre for Transnational Corporations.

IMF (1995-2001), *International Financial Statistics Yearbook*, International Monetary Fund.

ISEAS (1995), *The New Wave of Foreign Direct Investment in Asia*, Singapore: ISEAS/Nomura Research Institute.

Jarillo, J.C. (1988), 'On strategic networks', *Strategic Management Journal*, **9** (1), 31-41.

Jenkins, R. (1990), 'Comparing foreign subsidiaries and local firms in LDCs: theoretical issues and empirical analysis', *Journal of Development Studies*, **26** (2), 205-28.

Jick, T.D. (1979), 'Mixing qualitative and quantitative methods: triangulation in action', *Administrative Science Quarterly*, **24** 602-11.

Jobber, D. and D. O'Reiley (1998), 'Industrial mail surveys: a methodological update', *Industrial Marketing Management*, **27** 483-89.

Jobber, D., H. Mirza and K.H. Wee (1991), 'Incentives and response rates to cross-national business surveys: a logit model analysis', *Journal of International Business Studies*, **22** (4), 711-21.

Johanson, J. and L.G. Mattson (1987), 'Inter-organisational relations in industrial systems: a network approach compared with the transaction cost approach', *International Studies of Management and Organisation*, **18** (1), 34-48.

Johanson, J. and L.G. Mattson (1988), 'Internationalization in industrial systems-a network approach', in Hood, N. and J.E. Vahlne (Eds), *Strategies in Global Competition*, New York: Croom Helm. Reprinted in Buckley, P.J. and P. Ghauri (Eds) (1993), *The Internationalisation of the Firm: A Reader*, London: Dryden Press, pp. 303-22.

Johanson, J. and J.E. Vahlne (1977), 'The internationalisation process of the firm: a model of knowledge development and increasing foreign market commitment', *Journal of International Business Studies*, **8** (1), 23-32.

Jomo, K. (Ed.) (1993), *Industrializing Malaysia: Policy, Performance and Prospects*, London, New York: Routledge.

Jomo, K., F. Greg and R. Rasiah (Eds) (1999), *Industrial Technology Development in Malaysia: Industry and Firm Studies*, Routledge Studies in the Growth Economies of Asia, London, New York: Routledge.

Joshi, A.W. and R.L. Stump (1999), 'The contingent effect of specific asset investments on joint action in manufacturer-supplier relationships: an empirical test of the moderating role of reciprocal asset investments, uncertainty, and trust', *Academy of Marketing Science Journal*, **27** (3), 291-305.

Kamal, S., M.L. Young and R. Rasiah (1988), 'The changing face of the Electronics industry in the periphery: the case of Malaysia', *International Journal of Urban and Regional Research*, **12** (3), 375-404.

Kelegama, S. and F. Foley (1999), 'Impediments to promoting backward linkages from the garment industry in Sri Lanka', *World Development*, **27** (8), 1445-60.

Kindleberger, C.P. (Ed.) (1970a), *The International Corporation: A Symposium*, Cambridge, MA: MIT Press.

Kindleberger, C.P. (1970b), *Power and Money*, New York: Basic Books.

Kirk, J. and M.L. Miller (1986), *Reliability and Validity in Qualitative Research*, Thousand Oaks, London, New Delhi: Sage.

Kogut, B. (1983), 'FDI as a sequential process', in Kindleberger, C.P. and D. Audretsch (Eds), *The MNC in the 1980s*, Cambridge, MA: MIT Press,

Kojima, K. (1973), 'A macro-economic approach to foreign direct investment', *Hitotsubashi Journal of Economics*, **14** 1-21.

Kojima, K. (1996), *Trade, Investment and Pacific Integration*, Tokyo: Bunshindo.

Kojima, K. (2000), 'The "flying geese" model of Asian economic development: origin, theoretical extensions, and regional policy implications', *Journal of Asian Economics*, **11** (4), 375-401.

Kokko, A. (1996), 'Productivity spillovers from competition between local firms and foreign affiliates', *Journal of International Development*, **8** (4), 517-30.

Kosai, Y. and T.V. Tho (1994), 'Japan and industrialisation in Asia: an essay in memory of Dr. Saburo Okita', *Journal of Asian Economics*, **5** (2), 155-76.

Krickx, G.A. (2000), 'The relationship between uncertainty and vertical integration', *International Journal of Organizational Analysis*, **8** (3), 309-29.

Kumar, N. (1990), *Multinational Enterprises in India: Industrial Distribution, Characteristics and Performance*, London, New York: Routledge.

Kumar, N. (1998), *Globalization, Foreign Direct Investment and Technology Transfers: Impacts on and Prospects for Developing Countries*, London and New York: Routledge, in collaboration with the United Nations University, and INTECH: Institute for New Technologies.

Kwan, C.H. (1994), *Economic Interdependence in the Asia-Pacific Region: Toward a Yen Bloc*, London, New York: Routledge.

Lall, S. (1984), 'Export of technology by newly industrialising countries: an overview', *World Development*, **12** (5/6), 471-80.

Lall, S. (1993a), 'Promoting technological development: the role of technology transfer and indigenous effort', *Third World Quarterly*, **14** (1), 95-109.

Lall, S. (Ed.) (1993b), *Transnational Corporations and Economic Development*, London: Roultedge, United Nations Library on Transnational Corporations.

Lall, S. (1994), 'Industrial policy: the role of government in promoting industrial and technological development', *UNCTAD Review*, 65-89.

Lall, S. (1995), 'Malaysia: industrial success and the role of the government', *Journal of International Development*, **7** (5), 759-73.

Lall, S. (1996), 'The investment development path: some conclusions', in Dunning, J.H. and R. Narula (Eds), *Foreign Direct Investment and Governments: Catalysts for Economic Restructuring*, London, New York: Routledge, pp. 423-42.

Lall, S. (1998), 'Exports of manufactures by developing countries: emerging patterns of trade and location', *Oxford Review of Economic Policy*, **14** (2), 54-73.

Lall, S. and P. Streeten (Eds) (1977), *Foreign Investment, Transnational Corporations and Developing Countries*, Basingstoke: Macmillan.

Lecraw, D.J. (1991), 'Factors influencing FDI by TNC in host developing countries: a preliminary report', in Buckley, P.J. and J. Clegg (Eds), *Multinational Enterprises in Less Developed Countries*, Basingstoke: Macmillan, pp. 163-80.

Lecraw, D.J. (1992), 'MNE and developing countries', in Buckley, P.J. (Ed.) *Studies in International Business*, Basingstoke: Macmillan, pp. 28-51.

Lee, G.B. (1991), 'Transnational corporations in Malaysia: visible but unrevealing', *Regional Development Dialogue*, **12** (1), 21-44.

Lee, H. (1995), 'Globalization, foreign direct investment and competitive strategies of Korean electronics companies', in ISEAS (Ed.) *The New Wave of Foreign Direct Investment in Asia*, Singapore: ISEAS/Nomura Research Institute, pp. 64-87.

Legewie, J. and H. Meyer-Ohle (2000), *Corporate Strategies for Southeast Asia After the Crisis*, Basingstoke, New York: Palgrave.

Levy, B. (1988), 'Korean and Taiwanese firms as international competitors: the challenge ahead', *Columbia Journal of World Business*, **23** (1), 43-51.

Levy-Livermore, A. (Ed.) (1998), *Handbook on the Globalization of the World Economy*, Cheltenham: Edward Elgar.

Lewis, J.D. (1995), *The Connected Corporation: How Leading Companies Win Through Customer-Supplier Alliances*, New York: Free Press.

Lim, L.Y.C. and E.F. Pang (1977), The Electronics Industry in Singapore: Structure, Technology and Linkages, Singapore, Economic Research Centre, University of Singapore. Monograph Series No. 7.

Lim, L.Y.C. and E.F. Pang (1982), 'Vertical linkages and multinational enterprises in developing countries', *World Development*, **10** (7), 585-95.

Lim, L.Y.C. and E.F. Pang (1991), Foreign Direct Investment and Industrialization in Malaysia, Taiwan and Thailand, Paris, OECD Development Centre.

Linden, G. (2000), 'Japan and the United States in the Malaysian electronics sector', in Borrus, M., D. Ernst and S. Haggard (Eds), *International Production Networks in Asia: Rivalry or Riches?*, London, New York: Routledge, pp. 198-225.

Linder, S.B. (1961), *An Essay on Trade and Transformation*, New York: John Wiley.

Lindsey, C.W. (1986), 'Transfer of technology to the ASEAN region by U.S. corporations', *ASEAN Economic Bulletin*, **3** 225-47.

Lipsey, R.E. (1999), 'The role of foreign direct investment in international capital flows', NBER Working Paper Series, National Bureau of Economic Research, pp. 46.

Litwin, M.S. (1995), *How to Measure Survey Reliability and Validity*, Thousand Oaks, London, New Delhi: Sage.

Lok, I. and M. Abdullah (1992), 'Management dynamism in Malaysia', in APO (Ed.) *Management Dynamism: A Study of Selected Companies in Asia*, Tokyo: Asian Productivity Organization.

Loungani, P. and A. Razin (2001), 'How beneficial is foreign direct investment for developing countries?' *Finance and Development: A Quarterly Magazine of the IMF, http://www.imf.org/*, **38** (2), 6-9.

Lundgren, K. (1990), 'Vertical integration, transaction costs and "learning by using"', in Aoki, M., B. Gustafsson and O.E. Williamson (Eds), *The Firm as a Nexus of Treaties*, London, Newbury Park, New Delhi: Sage, pp. 112-32.

Lutz, J.M. (1987), 'Shifting comparative advantage, the NICs and the developing countries', *International Trade Journal*, **1** (4), 339-58.

Masterman, A.V. (1991), 'Japanese investment and management practices in Indonesia: towards a better partnership', *Journal of Southeast Asian Business*, **7** (1).

Mathison, S. (1988), 'Why triangulate?' *Educational Research*, **17** (2), 13-17.

McAleese, D. and D. McDonald (1978), 'Employment growth and the development of linkages in foreign-owned and domestic manufacturing enterprises', *Oxford Bulletin of Economic and Statistics*, **40** 321-40.

Meyanathan, S.D. (1994), Industrial Structures and the Development of Small and Medium Enterprise Linkages: Examples from East Asia, Economic Development Institute of the World Bank.

MIDA (1988 to 2001), Statistics on the Manufacturing Sector, Malaysian Industrial Development Authority, Malaysia.

MIDA (1996), Report on the Performance of the Manufacturing Sector in Malaysia, Malaysian Industrial Development Authority, Malaysia.

MIDA (1997a), Johor Technology Park Information Leaflet, Malaysian Industrial Development Authority, Malaysia.

MIDA (1997b), Perak Science Park: A Symbiosis of Minds and Hearts, Malaysian Industrial Development Authority, Malaysia.

MIDA (2001), Malaysia Industrial Digest, Kuala Lumpur, Malaysian Industrial Development Authority.
MIDA (2002), Report on the Performance of the Manufacturing Sector 2001, Kuala Lumpur, Malaysian Industrial Development Authority.
Miles, M.B. and A.M. Huberman (1994), *Qualitative Data Analysis*, Thousand Oaks, London, New Delhi: Sage.
Miles, R.E., C.C. Snow and H.J. Coleman (1992), 'Managing 21st century networks organisation', *Organisational Dynamics*, (Winter), 5-21.
Ministry of Finance (1996), Economic Report 1995/1996, Kuala Lumpur, Percetakan Nasional Malaysia.
Mirza, H. (1986), *Multinationals and the Growth of the Singapore Economy*, Beckenham: Croom Helm.
Mirza, H. (Ed.) (1998a), *Global Competitive Strategies in the World Economy*, Cheltenham, UK, Northampton, MA: Edward Elgar Publishing.
Mirza, H. (1998b), 'Transnational corporations as agents for the transmission of business culture to host countries', in Cook, P., C. Kirkpatrick and F. Nixson (Eds), *Privatisation, Enterprise Development and Economic Reform: Experiences of Developing and Transitional Economies*, Cheltenham: Edward Elgar, pp. 33-62.
Mirza, H. (2000), 'The globalization of business and East Asian developing-country multinationals', in Hood, N. and S. Young (Eds), *The Globalization of Multinational Enterprise Activity and Economic Development*, Basingstoke: Macmillan, pp. 202-25.
Mirza, H. and A. Giroud (1998), 'The investment strategies of Asian and non-Asian firms in ASEAN', in LVMH Conference, 'Asian Foreign Investment in Asia'. Euro-Asia Centre , Fontainebleau, France: INSEAD.
Mirza, H., F. Bartels, M. Hiley and A. Giroud (1997), The Promotion of Foreign Direct Investment into and within ASEAN: Towards the Establishment of an ASEAN Investment Area, Asia-Pacific Business and Development Research Unit, University of Bradford.
Mirza, H., A. Giroud, F. Bartels and K.H. Wee (1999), 'Foreign direct investment in ASEAN: a contemporary perspective', in Strange, R., J. Slater and C. Molteni (Eds), *The European Union and ASEAN: Trade and Investment Issues*, Basingstoke, New York: Macmillan, pp. 33-62.
MITI (1995), Malaysia: Guidelines for the Establishment of Representative Offices and Operational Headquarters, Kuala Lumpur, Ministry of International Trade and Industry and Malaysian International Chamber of Commerce and Industry.
MITI (1996), Research and Development Incentives in Malaysia, Ministry of International Trade and Industry, Malaysia.
Moosa, I.A. (2002), *Foreign Direct Investment: Theory, Evidence and Practice*, Basingstoke: Palgrave.

Morris, J. and R. Imrie (1992), *Transforming Buyer-Supplier Relations: Japanese Style Industrial Practices in a Western Context*, Basingstoke: Macmillan.

Morse, J.M. (1991), 'Approaches to qualitative-quantitative methodological triangulations', *Nursing Research*, **40** (1), 120-3.

Morsink, R.L.A. (1998), *Foreign Direct Investment and Corporate Networking: A Framework for Spatial Analysis of Investment Conditions*, Cheltenham, UK, Northampton, MA, USA: Edward Elgar.

Mucchielli, J.-L. (1985), *Les Firmes Multinationales: Mutations et Perspectives*, Paris: Economica.

Mucchielli, J.-L. (1997), Regionalisation and Globalisation in the New Strategies of Multinational Firms, Paris, French Commissariat au Plan.

Mucchielli, J.-L., P.J. Buckley and V. Cordell (Eds) (1998), *Globalization and Regionalization: Strategies, Policies and Economic Environment*, New York: International Business Press.

Muhlemann, A., J. Oakland and K. Lockyer (1992), *Production Operation Management*, London: Pitman.

Murray, R. (1981), *Multinationals Beyond the Market*, Brighton: Harvester Press Limited.

Narula, R. (1996), *Multinational Investment and Economic Structure: Globalization and Competitiveness*, London, New York: Routledge.

Natarajan, S. and T.J. Miang (1992), The Impact of MNC Investment in Malaysia, Singapore and Thailand, ISEAS-ASEAN Economic Research Unit.

Norusis, M.J. (1994), *SPSS Advanced Statistics 6.1*, Chicago: SPSS Inc.

O'Brien, L. (1993), 'Malaysian manufacturing sector linkages', in Jomo, K. (Ed.) *Industrialising Malaysia: Policy, Performance and Prospects*, London, New York: Routledge, pp. 147-62.

OECD (1985), The Semiconductor Industry: Trade Related Issues, Paris, Organization for Economic Co-operation and Development.

OECD (1992), Competing in the Electronics Industry: the Experience of Newly Industrialising Economies, Paris, Organization for Economic Co-operation and Development.

OECD (1995), Linkages: OECD and Major Developing Economies, Paris, Organization for Economic Co-operation and Development.

Okamoto, Y. (1994), 'Impact of trade and FDI liberalization policies on the Malaysian economy', *The Developing Economies*, **32** (4), 460-78.

Oman, C. (1989), New Forms of Investments in Developing Country Industries: Mining, Petrochemicals, Automobiles, Textiles and Food, Paris, OECD Development Center.

Oppenheim, A.N. (1992), *Questionnaire Design, Interviewing and Attitude Measurement*, London, New York: Pinter.

Ozawa, T. (1991), 'Japan in a new phase of multinationalism and industrial upgrading: functional integration of trade, growth and FDI', *Journal of World Trade*, **25** (1), 43-61.

Ozawa, T. (1992), 'Foreign direct investment and economic development', *Tansnational Corporations*, **1** (1), 27-54.

Ozawa, T. (1996), 'Japan: the macro-IDP, meso-IDPs and the technology development path (TDP)', in Dunning, J.H. and R. Narula (Eds), *Foreign Direct Investment and Governments: Catalysts for Economic Restructuring*, London, New York: Routledge, pp. 142-74.

Pangarkar, N. (1998), 'The Asian multinational corporation: strategic performance and key challenges', *Asia Pacific Journal of Management*, **15** (2), 109-18.

Pangestu, M. (1991), 'Foreign firms and structural change in the Indonesian manufacturing sector', in Ramstetter, E. (Ed.) *FDI in Asia's Developing Economies and Structural Change in the Asia Pacific Region*, Boulder, CO: Westview Press.

Pangestu, M., H. Soesastro and M. Ahmad (1992), 'A new look at intra-ASEAN economic co-operation', *ASEAN Economic Bulletin*, **8** (3), 333-52.

Pavitt, K. (1988), 'International patterns of technology accumulation', in Hood, N. and J.E. Vahlne (Eds), *Strategies in Global Competition*, Beckenham: Croom Helm.

Pearson, J.N., L.M. Ellram and C.R. Carter (1996), 'Status and recognition of the purchasing function in the electronics industry', *International Journal of Purchasing and Materials Management*, **32** (2), 30-7.

Penrose, E.T. (1956), 'Foreign investment and the growth of the firm', *Economic Journal*, **60** 220-235.

Penrose, E.T. (1959), *The Theory of the Growth of the Firm*, Oxford: Basil Blackwell.

Petersen, B. and P. Torben (1997), 'Twenty years after-support and critique of the Upssala internationalisation model', in Björkman, I. and M. Forsgren (Eds), *The Nature of the International Firm*, Copenhagen: Copenhagen Business School Press, pp. 117-34.

Plasschaert, S. (1996), 'Transfer pricing and taxation', in UNCTAD (Ed.) *Transnational Corporations and World Development*, London, Boston: International Thomson Business Press, pp. 394-417.

Porter, M. (1986), *Competition in Global Industries*, Boston, MA: Harvard Business School Press.

Porter, M. (1990), *The Competitive Advantage of Nations*, New York: Free Press, Macmillan.

Porter, M.E. (1998), 'Clusters and the new economics of competition', *Harvard Business Review*, **76** (6), 77-89.

Premachandra, A. and M. Jayart (1996), 'Foreign investment and industrialisation in Malaysia: export, employment and spillovers', *Asian Economic Journal*, **10** (1), 29-44.
Presutti, W.D. (1992), 'The single source issue: U.S. and Japanese sourcing strategies', *International Journal of Purchasing and Materials Management*, **28** (1), 2-9.
Ramstetter, E.D. (1994), 'Comparisons of Japanese multinational firms and other firms in Thailand's non-oil manufacturing industry', *ASEAN Economic Bulletin*, **11** (1), 36-58.
Rana, P.B. (1990), 'Shifting comparative advantage among Asian and Pacific countries', *International Trade Journal*, **4** (3), 243-58.
Rasiah, R. (1988a), 'The semiconductor industry in Penang: implications for new international division of labour theory', *Journal of Contemporary Asia*, **18** (1), 24-46.
Rasiah, R. (1988b), 'The changing face of the electronics industry in the periphery: the case of Malaysia', *International Journal of Urban and Regional Research*, **12** (3), 375-404.
Rasiah, R. (1989), 'Competition and restructuring in the semiconductor industry: implications for technology transfer and its absorption in Penang', *Southeast Asian Journal of Social Science*, **17** (2), 41-57.
Rasiah, R. (1992), 'Foreign manufacturing investment in Malaysia', *Economic Bulletin for Asia and the Pacific*, **63** (1), 63-77.
Rasiah, R. (1995), 'Malaysia', in Supapol, A.B. (Ed.) *Transnational Corporations and Backward Linkages in Asian Electronics Industries*. UN/ESCAP, pp. 175-206.
Rasiah, R. (1996), 'The changing organisation of work in Malaysia's electronics industry', *Asia Pacific Viewpoint*, **37** (1), 21-37.
Reinhardt, N. (2000), 'Back to basics in Malaysia and Thailand: the role of resource-based exports in their export-led growth', *World Development*, **28** (1), 57-77.
Reve, T. (1990), 'The firm as a nexus of internal and external contracts', in Aoki, M., B. Gustafsson and O.E. Williamson (Eds), *The Firm as a Nexus of Treaties*, London, Newbury Park, New Delhi: Sage, pp. 133-61.
Riedel, J. (1975), 'The nature and determinants of export-oriented direct foreign investment in a developing country: a case study of Taiwan', *Weltwirtschaftliches Archiv*, **111** (3), 505-29.
Riggins, F.J., C.H. Kriebel and T. Mukhopadhyay (1994), 'The growth of interorganizational systems in the presence of network externalities', *Management Science*, **40** (8), 984-98.
Riordan, M.H. (1990), 'What is vertical integration?' in Aoki, M., B. Gustafsson and O.E. Williamson (Eds), *The Firm as a Nexus of Treaties*, London, Newbury Park, New Delhi: Sage, pp. 94-112.

Robertson, M., J. Swan and S. Newell (1996), 'The role of networks in the diffusion of technological innovation', *Journal of Management Studies*, **33** (3), 333-59.

Robson, P. (1996), 'Transnational corporations and regional economic integration', in UNCTAD (Ed.) *Transnational Corporations and World Development*, London, Boston: International Thomson Business Press, pp. 448-72.

Rugman, A.M. (1986), 'New theories of the multinational enterprise: an assessment of internalisation theory', *Bulletin of Economics Research*, **38** (2), 101-18.

Rugman, A.R. (2001), 'The end of global strategy', *European Management Journal*, **19** (4), 333-43.

Salleh, I.M. (1995), 'Foreign direct investment and technology transfer in the Malaysian electronics industry', in ISEAS (Ed.) *The New Wave of Foreign Direct Investment in Asia*, Singapore: ISEAS/Nomura Research Institute, pp. 133-60.

Saunders, M., P. Lewis and A. Thornhill (1997), *Research Methods for Business Students*, Harlow: Pearson Education.

Schneiberg, M. and J.R. Hollingsworth (1990), 'Can transaction cost economics explain trade associations?' in Aoki, M., B. Gustafsson and O.E. Williamson (Eds), *The Firm as a Nexus of Treaties*, London, Newbury Park, New Delhi: Sage, pp. 320-46.

Schrader, S. (1991), 'Informal technology transfer between firms: cooperation through information trading', *Research Policy*, **20** 153-70.

Searle, P. (1999), *The Riddle of Malaysian Capitalism: Rent-seekers or Real Capitalists?*, Sydney, Singapore (New Asian Library): Allen and Unwin.

Segafi-Nejad, T. (1995), 'TNCs-host country relations and the changing FDI climate: towards 2000', *International Trade Journal*, **9** (1), 85-106.

Sikorski, D. and T. Menkhoff (2000), 'Internationalisation of Asian business', *Singapore Management Review*, **20** (11), 1-17.

Sivalingam, G. and S.P. Yong (1993), 'TNCs and industrialization in Malaysia: organisational behaviour and local and regional impacts', *Regional Development Dialogue*, **14** (4), 40-63.

Sjöholm, F. (1999), 'Technology gap, competition and spillovers from direct foreign investment: Evidence from establishment data', *Journal of Development Studies*, **36** (1), 53-73.

Slater, G. and D.A. Spencer (2000), 'The uncertain foundations of transaction costs economics', *Journal of Economic Issues*, **34** (1), 61-87.

South Centre (1997), Foreign Direct Investment, Development and the New Global Economic Order: a Policy Brief for the South, Geneva, South Centre.

Spekman, R.E., J.W. Kamauff and D.J. Salmond (1994), 'At last purchasing is becoming strategic', *Long Range Planning*, **27** (2), 76-84.

Sullivan, D. and A. Bauerschmidt (1990), 'Incremental internationalization: a test of Johanson and Vahlne's thesis', *Management International Review*, **30** (1), 19-30.

Supapol, A.B. (1995), Transnational Corporations and Backward Linkages in Asian Electronics Industries, New York, UNESCAP: United Nations Conference on Trade and Development, Economic and Social Commission for Asia and the Pacific. Monograph No. 5.

Swift, C.O. (1995), 'Preferences for single sourcing and supplier selection criteria', *Journal of Business Research*, **32** (1), 105-11.

Takachi, Y. (1993), 'Transition of theories and premises for development of developing countries: from import-substitution industrialization by government initiative capital formation through socialization of development objective to structural adjustment', *EXIM Review, Bank of Japan*, 1-67.

Takatoshi, I. and A.O. Krueger (2000), *The Role of Foreign Direct Investment in East Asian Economic Development*, Chicago: National Bureau of Economic Research (NBER).

Tallman, S.B. and O. Shengkar (1990), 'International cooperative venture strategies: outward investment and small firms from NIC's', *Management International Review*, **30** (4), 299-315.

Tan, B. and I. Vertinsky (1995), 'Strategic advantages of Japanese electronics firms and the scale of their subsidiaries in the US and Canada', *International Business Review*, **4** (3), 373-86.

Thorelli, H. (1986), 'Networks: between markets and hierarchies', *Strategic Management Journal*, **7** 37-51.

Tolentino, E. (1993), *Technological Innovation and Third World Multinationals*, London, New York: Routledge.

Tolentino, P.E. (2000), *Multinational Corporations: Emergence and Evolution*, London: Routledge.

Turnbull, P.W. (1987), 'A challenge to the stages theory of the internationalization process', in Rosson, P.J. and S.D. Reid (Eds), *Managing Export Entry and Expansion*, New York: Praeger, pp. 21-40.

Ulgado, F.M., C.J. Yu and A.R. Negandhi (1994), 'Multinational enterprises from Asian developing countries: management and organizational characteristics', *International Business Review*, **3** (2), 123-33.

UNCTAD (1992), World Investment Report 1992: Transnational Corporations as Engines of Growth, New York and Geneva, United Nations.

UNCTAD (1993), World Investment Report 1993: Transnational Corporations and Employment in the Workplace, New York and Geneva, United Nations.

UNCTAD (1995), World Investment Report 1995: Transnational Corporations and Competitiveness, New York and Geneva, United Nations.

UNCTAD (1996), World Investment Report 1996: Investment, Trade and International Policy Arrangements, New York and Geneva, United Nations.

UNCTAD (1997), World Investment Report 1997: Transnational Corporations, Market Structure and Competition Policy, New York and Geneva, United Nations.

UNCTAD (1998), World Investment Report 1998: Trends and Determinants, New York and Geneva, United Nations.

UNCTAD (1999), World Investment Report 1999: FDI and the Challenge of Development, New York and Geneva, United Nations.

UNCTAD (2001), *World Investment Report 2001: Promoting Linkages*, New York and Geneva: United Nations.

UNCTC (1981), Transnational Corporation Linkages in Developing Countries: The Case of Backward Linkages via Subcontracting, a Technical Paper on Automobile Sector in Peru, India and Morocco, New York, United Nations Centre for Transnational Corporations.

UNCTC (1986), TNCs and the International Semiconductor Industry, New York, United Nations Centre for Transnational Corporations, ST/CTC/39.

UNCTC (1988), Transnational Corporations in World Development: Trends and Prospects, New York, United Nations Centre for Transnational Corporations.

UNCTD (1996a), Incentives and Foreign Direct Investment, New York, Geneva, United Nations, UNCTAD/DTCI/28.

UNCTD (1996b), Sharing Asia's Dynamism: Asian Direct Investment in the European Union, New York, Geneva, United Nations.

UN-IDO (1991), *Malaysia, Sustaining the Industrial Investment Momentum*, Oxford, Cambridge, MA: Basil Blackwell. United Nations-Industrial Development Organization.

United Nations (1987), Transnational Corporations and the Electronics Industries of ASEAN Economies, New York, United Nations, ESCA/UNCTC Joint Unit on TNCs.

United Nations (1988), *TNCs in World Development: Trends and Prospects*, New York, Geneva: United Nations.

Vernon, R. (1966), 'International investment and international trade in the product cycle', *Quarterly Journal of Economics*, **2** (80), 190-207.

Vernon, R. (1971), *Sovereignty at Bay*, New York: Basic Books.

Wan, T.B. (1990), 'Organising purchasing in a competitive environment', *Asia Pacific Journal of Management*, **7** (1), 79-96.

Warr, P.G. (1987), 'Malaysia's industrial enclaves: benefits and costs', *Developing Economies*, **25** (1), 31-55.

Watanabe, S. (1980), 'International subcontracting and regional economic integration of the ASEAN countries: the role of multinationals', in Germidis, D. (Ed.) *International Subcontracting: A New Form of Investment*, Paris: OECD, pp. 210-27.

Watts, C.A., K.Y. Kim and C.K. Hahan (1995), 'Linking purchasing competitive strategy', *International Journal of Purchasing and Materials Management*, **31** (2), 2-24.

Weiermair, K. (1991), 'On the transferability of management systems: the case of Japan', in Buckley, P.J. and J. Clegg (Eds), *Multinational Enterprises in Less Developed Countries*, Basingstoke: Macmillan, pp. 56-76.

Wells, L.T.J. (1998), 'Multinationals and the developing countries', *Journal of International Business Studies*, **29** (1), 101-14.

Williamson, O.E. (1971), 'The vertical integration of production: market failure considerations', *American Economic Review*, **61** 112-13.

Williamson, O.E. (1975), *Markets and Hierarchies: Analysis and Antitrust Applications*, New York: Free Press.

Williamson, O.E. (1981), 'The economics of organization: the transaction cost approach', *American Journal of Sociology*, **87** 547-77.

Williamson, O.E. (1985), *The Economic Institutions of Capitalism: Firms, Markets, Relational Contracting*, New York: Free Press, London: Collier Macmillan.

Williamson, O.E. and S.G. Winter (Eds) (1991), *The Nature of the Firms: Origins, Evolution and Development*, New York, Oxford: Oxford University Press.

Wong, P.K. (1991), Technological Development through Subcontracting Linkages-A Case Study, Asian Productivity Organization, Faculty of Business Administration, National University of Singapore.

World Bank (1995), Industrial Structure and the Development of Small and Medium Enterprise Linkages: an Overview, New York, World Bank.

Yamashita, S. (Ed.) (1991), *Transfer of Japanese Technology and Management to the ASEAN Countries*, Tokyo: University of Tokyo Press.

Yean, T.S. (1997), 'Determinants of productivity growth in the Malaysian manufacturing sector', *ASEAN Economic Bulletin*, **13** (3), 333-43.

Yeung, H.W.-C. (1995), 'Qualitative personal interviews in international business research: some lessons from a study of Hong Kong transnational corporations', *International Business Review*, **4** (3), 313-39.

Yeung, H.W.-C. (1996), 'Sectoral specialization and competitive advantage: Hong Kong investments in the ASEAN region', *ASEAN Economic Bulletin*, **13** (1), 74-94.

Yeung, H.W.-C. (1998), *Transnational Corporations and Business Networks: Hong Kong Firms in the ASEAN Region*, London: Routledge.

Yoon, H.-D. (1994), 'An evaluation of the contribution of transnational corporations to the development of backward linkages of the Korean electrical and electronics industries', *Public Enterprise*, **14** (3/4), 379-403.

Young, S. and P. Lan (1997), 'Technology transfer to China through foreign direct investment', *Regional Studies*, **31** (7), 669-80.

Zhara, S.A., G. George and D.M. Garvis (1999), 'Networks and entrepreneurship in Southeast Asia: the role of social capital and membership commitment', in Richter, F.-J. (Ed.) *Business Networks In Asia: Promises, Doubts And Perspectives*, Westport, CT: Quorum, pp. 39-60.

Index